Studies in Computational Intelligence

Volume 1106

Series Editor

Janusz Kacprzyk, Polish Academy of Sciences, Warsaw, Poland

The series "Studies in Computational Intelligence" (SCI) publishes new developments and advances in the various areas of computational intelligence—quickly and with a high quality. The intent is to cover the theory, applications, and design methods of computational intelligence, as embedded in the fields of engineering, computer science, physics and life sciences, as well as the methodologies behind them. The series contains monographs, lecture notes and edited volumes in computational intelligence spanning the areas of neural networks, connectionist systems, genetic algorithms, evolutionary computation, artificial intelligence, cellular automata, self-organizing systems, soft computing, fuzzy systems, and hybrid intelligent systems. Of particular value to both the contributors and the readership are the short publication timeframe and the world-wide distribution, which enable both wide and rapid dissemination of research output.

Indexed by SCOPUS, DBLP, WTI Frankfurt eG, zbMATH, SCImago.

All books published in the series are submitted for consideration in Web of Science.

Arash Shaban-Nejad · Martin Michalowski ·
Simone Bianco
Editors

Artificial Intelligence for Personalized Medicine

Promoting Healthy Living and Longevity

Springer

Editors
Arash Shaban-Nejad
Center for Biomedical Informatics
Department of Pediatrics
College of Medicine
The University of Tennessee Health
Science Center–Oak-Ridge National Lab
(UTHSC-ORNL) Center for Biomedical
Informatics
Memphis, TN, USA

Martin Michalowski
School of Nursing
University of Minnesota
Minneapolis, MN, USA

Simone Bianco
Altos Labs—Bay Area Institute of Science
Redwood City, CA, USA

ISSN 1860-949X ISSN 1860-9503 (electronic)
Studies in Computational Intelligence
ISBN 978-3-031-36937-7 ISBN 978-3-031-36938-4 (eBook)
https://doi.org/10.1007/978-3-031-36938-4

This Springer imprint is published by the registered company Springer Nature Switzerland AG
The registered company address is: Gewerbestrasse 11, 6330 Cham, Switzerland

Preface

Personalized Medicine is transforming health care by using an individual's unique characteristics, environment, and genetic profile to guide the prevention, diagnosis, drug discovery, and treatment of disease more efficiently. Advances in Artificial Intelligence (AI) tools and techniques have provided a unique opportunity for researchers and physicians, and health organizations to collect, process, and exchange the vast amount of data necessary for complex decision-making in personalized medicine. Ageing research is one of the complex issues in personalized medicine and like many other complex challenges can benefit from advances in AI. AI and machine learning tools can detect and specify patterns and assist scientists in better understanding the ageing process. AI can also accelerate the development of effective interventions to improve the well-being of individuals and extend their lifespan and health span.

This book aims to highlight the latest achievements in the use of AI in personalized medicine and healthcare delivery. The edited volume contains selected papers presented at the 2023 Health Intelligence workshop, co-located with the Thirty-Seven Association for the Advancement of Artificial Intelligence (AAAI) conference, and presents an overview of the issues, challenges, and potentials in the field, along with new research results. This book provides information for researchers, students, industry professionals, clinicians, and public health agencies interested in the applications of AI in medicine and public health.

Memphis, USA — Arash Shaban-Nejad
Minneapolis, USA — Martin Michalowski
Redwood City, USA — Simone Bianco

Contents

Contributors

Hammam Abdelwahab Fraunhofer IAIS, Sankt Augustin, Germany

Maha M. Alwuthaynani University of Bristol, Bristol, UK;
College of Computer Science and Information Systems, Najran University, Najran, Saudi Arabia

José Luis Ambite Information Sciences Institute, University of Southern California, Los Angeles, CA, USA

Michael Ambler University of Bristol, Bristol, UK;
University Hospitals Bristol NHS Foundation Trust, Bristol, UK

C. Anantaram TCS Research, Chennai, India

Niklas Beck Fraunhofer IAIS, Sankt Augustin, Germany

Ranjeet S. Bhamber University of Bristol, Bristol, UK

Nan Bi Binghamton University, Binghamton, NY, USA

Simone Bianco Altos Labs - Bay Area, Institute of Science, Redwood City, CA, USA

Ioana Bica University of Oxford, Oxford, UK

Djallel Bouneffouf IBM Thomas J Watson Research Center, Yorktown Heights, NY, USA

Christopher P. Bourdeaux University Hospitals Bristol NHS Foundation Trust, Bristol, UK

Omar Boursalie Department of Electrical, Computer and Biomedical Engineering, Toronto Metropolitan University, Toronto, ON, Canada

Quoc Hung Cao Quy Nhon AI, FPT Software, Quy Nhon, Vietnam

Guillermo Cecchi IBM Thomas J Watson Research Center, Yorktown Heights, NY, USA

Kai-Wei Chang University of California, Los Angeles, USA

Jeffrey N. Clark University of Bristol, Bristol, UK

Hao Deng Massachusetts General Hospital, Boston, MA, USA

Sunipa Dev University of California, Los Angeles, USA

Thomas E. Doyle Department of Electrical and Computer Engineering, McMaster University, Hamilton, ON, Canada;
School of Biomedical Engineering, McMaster University, Hamilton, ON, Canada;
Vector Institute, Toronto, ON, Canada

Audrey Durand Université Laval, Quebec, QC, Canada

Ari Ercole University of Cambridge, Cambridge, UK

Marcos F. Vidal Melo Columbia University Irving Medical Center, New York City, NY, USA

Andrew Fisher Department of Mathematics and Computing Science, Saint Mary's University, Halifax, Canada

Ophir Frieder IR Lab, Georgetown University, Washington DC, USA

Nazli Goharian IR Lab, Georgetown University, Washington DC, USA

Aylin Gunal University of Michigan, Ann Arbor, MI, USA

Jesper Havsol Data Science and Advanced Analytics, BioPharmaceuticals R&D, AstraZeneca, Gothenburg, Sweden

Alexander Hepburn University of Bristol, Bristol, UK

Timothy Houle Massachusetts General Hospital, Boston, MA, USA

Emmette Hutchison Human-centered AI & ML, BioPharmaceuticals R&D, AstraZeneca, Gaithersburg, USA

Suparshva Jain TCS Research, Chennai, India

Sampath Jayarathna Old Dominion University, Norfolk, VA, USA

Dong H. Jeong University of the District of Columbia, Washington, DC, USA

Soo-Yeon Ji Bowie State University, Bowie, MD, USA

Katrina Kardiasmenos Bowie State University, Bowie, MD, USA

Richard Khoury Université Laval, Quebec, QC, Canada

Hrishikesh Kulkarni Georgetown University, Washington DC, USA

Alexandre Larouche Université Laval, Quebec, QC, Canada

Zhihua Li Binghamton University, Binghamton, NY, USA

Baihan Lin Columbia University, New York City, NY, USA

Dianbo Liu Mila AI Institute, Quebec, Canada

Pietro Liò University of Cambridge, Cambridge, UK

Sean MacAvaney University of Glasgow, Glasgow, UK

Vijay Mago Department of Computer Science, Lakehead University, Thunder Bay, Canada

Claudio Martens Fraunhofer IAIS, Sankt Augustin, Germany

Christopher J. McWilliams University of Bristol, Bristol, UK; University Hospitals Bristol NHS Foundation Trust, Bristol, UK

Martin Michalowski School of Nursing, University of Minnesota, Minneapolis, MN, USA

Rada Mihalcea University of Michigan, Ann Arbor, MI, USA

Alexander Nagrebetsky Massachusetts General Hospital, Boston, MA, USA

Quoc Khanh Nguyen Quy Nhon AI, FPT Software, Quy Nhon, Vietnam

Truong Thanh Hung Nguyen Friedrich-Alexander-Universität Erlangen-Nürnberg, Erlangen, Germany; Quy Nhon AI, FPT Software, Quy Nhon, Vietnam

Vo Thanh Khang Nguyen Quy Nhon AI, FPT Software, Quy Nhon, Vietnam

Anaelia Ovalle University of California, Los Angeles, USA

Anne M. Perrotti Old Dominion University, Norfolk, VA, USA

Aleksandra Petkova Human-centered AI & ML, BioPharmaceuticals R&D, AstraZeneca, Gaithersburg, USA

Verónica Pérez-Rosas University of Michigan, Ann Arbor, MI, USA

Sylvia Ranjeva Massachusetts General Hospital, Boston, MA, USA

Emma Rocheteau University of Cambridge, Cambridge, UK

Zahraa S. Abdallah University of Bristol, Bristol, UK

Reza Samavi Department of Electrical, Computer and Biomedical Engineering, Toronto Metropolitan University, Toronto, ON, Canada; Vector Institute, Toronto, ON, Canada

Amit Sangroya TCS Research, Chennai, India

Raul Santos-Rodriguez University of Bristol, Bristol, UK

Majid Sarrafzadeh University of California, Los Angeles, USA

Arash Shaban-Nejad Center for Biomedical Informatics, Department of Pediatrics, College of Medicine, The University of Tennessee Health Science Center – Oak-Ridge National Lab (UTHSC-ORNL), Memphis, TN, USA

Rajesh Sharma Institute of Computer Science, University of Tartu, Estonia and Department of Computer Science, Lakehead University, Thunder Bay, Canada

Caroline Sirois Centre d'excellence sur le vieillissement de Québec 1050 Chemin Ste-Foy, Quebec, QC, Canada

Ian Stewart Pacific Northwest National Laboratory, Richland, WA, USA

Dimitris Stripelis Information Sciences Institute, University of Southern California, Los Angeles, CA, USA

Xunzhu Tang University of Luxembourg, 2 Av. de l'Universite, Esch-sur-Alzette, Luxembourg

Ravi Tejwani MIT Media Lab, Cambridge, MA, USA

Mini Thomas Department of Computing and Software, McMaster University, Hamilton, ON, Canada

Van Binh Truong Quy Nhon AI, FPT Software, Quy Nhon, Vietnam

Lovekesh Vig TCS Research, Chennai, India

Charlotte von Numers Data Science and Advanced Analytics, BioPharmaceuticals R&D, AstraZeneca, Gothenburg, Sweden

Xinshi Wang Rensselaer Polytechnic Institute, Troy, NY, USA

Dennis Wegener Fraunhofer IAIS, Sankt Augustin, Germany

Enrico Werner University of Bristol, Bristol, UK

Lijun Yin Binghamton University, Binghamton, NY, USA

Yinan Yu Department of Computer Science and Engineering, Chalmers University of Technology, Gothenburg, Sweden

Jieyu Zhao University of California, Los Angeles, USA

Abbreviations

AD	Alzheimer's Diseases
Ada-SISE	Adaptive-Semantic Input Sampling for Explanation
ADHD	Attention-Deficit/Hyperactivity Disorder
ADNI	Alzheimer's Disease Neuroimaging Initiative
ADWIN	Adaptive Windowing
ANOVA	Analysis of Variance
ARDS	Acute Respiratory Distress Syndrome
AU	Action Unit
AUPRC	Area Under the Precision-Recall Curve
AUROC	Area Under the Receiver-Operator Curve
BART	Bidirectional and Auto-Regressive Transformer
BATM	Bidirectional Adversarial Training Model
BAUFER	Baseline-enabled Action Unit identification for Facial Expression Recognition
Bbox	Bounding box
BD	Bipolar Disorder
BERT	Bidirectional Encoder Representations from Transformers
BN	Bayesian Network
BoW	Bag of Words
CNN	Convolutional Neural Networks
CORD	COVID-19 Open Research Dataset
COVID-19	Coronavirus Disease 2019
CT	Computed Tomography
CUSUM	Cumulative Sum
CV	Cross Validation
DE	Differential Evolution
DFT	Discrete Fourier Transform
DISFA	Denver Intensity of Spontaneous Facial Action
DL	Deep Learning
DM	Density Map
D-RISE	Detector Randomized Input Sampling for Explanation

DT	Decision Tree
DWT	Discrete Wavelet Transform
EBPG	Energy-Based Pointing Game
ED	Emergency Department
EDDM	Early Drift Detection Method
EEG	Electroencephalogram
EHR	Electronic Health Records
EMR	Electronic Medical Records
ESAL	Expert System and Attention for Labeling
ETM	Embedded Topic Model
EWT	Empirical Wavelet Transform
FACS	Facial Action Coding System
FER	Facial Expression Recognition
FGM	Fast Gradient Method
FHE	Fully Homomorphic Encryption
FL	Federated Learning
FLINT	Federated Learning and Integration
FN	False Negative
FP	False Positive
FTL	Federated Transfer Learning
GLCM	Gray-Level Cooccurrence Matrix
Grad-CAM	Gradient-Weighted Class Activation Mapping
GSM	Gaussian Softmax Construction
HFL	Horizontal Federated Learning
ICD10	International Classification of Diseases, Tenth revision
ICU	Intensive Care Unit
IoU	Intersection over Union
IRB	Institutional Review Board
KDE	Kernel Density Estimation
KMMD	Kernel Maximum Mean Discrepancy
KNN	K-Nearest Neighbor
KS	Kolmogorov-Smirnov
LDA	Linear Discriminant Analysis
LIME	Local Interpretable Model-Agnostic Explanations
LOGAN	Local Group Bias Detection
LoS	Length of Stay
LR	Logistic Regression
LRP	Layer-wise Relevance Propagation
LSTM	Long Short-Term Memory
MAD	Mean Absolute Deviation
MATE	Mental Health Questions
MCI	Mild Cognitive Impairment
MDD	Major Depressive Disorder
MDIE	Medical Dialogue Information Extraction
MDS	Multi-dimensional Scaling

MGH	Massachusetts General Hospital
MIMIC	Medical Information Mart for Intensive Care
ML	Machine Learning
MLE	Maximum Likelihood Estimator
MMD	Maximum Mean Discrepancy
MMSE	Mini-mental State Examination
MoCA	Montreal Cognitive Assessment
MRI	Magnetic Resonance Imaging
MSE	Mean Squared Error
MSLE	Mean Squared Log Error
NASEM	National Academies of Sciences, Engineering, and Medicine
NIMH	National Institute of Mental Health
NLP	Natural Language Processing
OLV	One-Lung Ventilation
PCA	Principal Component Analysis
PD	Parkinson's Disease
PGD	Projected Gradient Descent
PIPs	Potentially Inappropriate Polypharmacies
PSD	Power Spectral Density
REM	Rapid Eye Movement
RISE	Randomized Input Sampling for Explanation
RNN	Recurrent Neural Networks
RR	Relative Risk
RSDD	Reddit Self-reported Depression Diagnosis
SAND	Semi-Supervised Adaptive Novel Class Detection
SAX	Symbolic Aggregate Approximation
SBERT	Sentence-BERT
SDOH	Social Determinants of Health
SFA	Symbolic Fourier Approximation
SLOGAN	patient Severity-based LOcal Group biAs detectioN
SMHD	Self-reported Mental Health Diagnoses
STBound	Soft Thresholding-based Boundary
SVM	Support Vector Machine
TimeLM	Time Language Models
TN	True Negative
TP	True Positive
TPC	Temporal Pointwise Convolution
UMAP	Uniform Manifold Approximation and Projection
VAE	Variational Autoencoders
VD	Ventilation Duration
VFL	Vertical Federated Learning
VT	Vision Transformer
WAE	Wasserstein Autoencoders
WAI	Working Alliance Inventory
WHO	World Health Organization

WM	Working Memory
WMD	Wearable Medical Devices
WT	Wavelet Transform
WTM	Wasserstein-based Topic Model
XAI	eXplainable Artificial Intelligence

Artificial Intelligence for Personalized Care, Wellness, and Longevity Research

Arash Shaban-Nejad, Martin Michalowski, and Simone Bianco

Abstract Artificial intelligence (AI) has the potential to transform personalized medicine by enabling healthcare professionals to deliver more precise, targeted treatments that are tailored to the individual needs of each patient. AI tools and techniques are also revolutionizing research and development of technologies that contribute to human longevity and healthy living in several ways, including, but not limited to, predictive analytics, disease diagnosis, treatment, and monitoring, and drug discovery and development. This chapter aims to explore the significance and applications of artificial intelligence tools and techniques to improve personal care and wellness and enhance human longevity research.

Keywords Artificial intelligence · Health AI · Personalized medicine · Ageing · Longevity · Healthy living

1 Introduction

Artificial intelligence (AI) and machine learning (ML) -based techniques are transforming healthcare and medicine by deriving novel insights from large sets of relevant data for generating reliable predictive models, discovering new and more effective treatments, reducing costs, and delivering a better quality of care. Nowadays AI

A. Shaban-Nejad (✉)
Center for Biomedical Informatics, Department of Pediatrics, College of Medicine, The University of Tennessee Health Science Center - Oak-Ridge National Lab (UTHSC-ORNL), Memphis, TN, USA
e-mail: ashabann@uthsc.edu

M. Michalowski
School of Nursing, University of Minnesota, Minneapolis, MN, USA
e-mail: martinm@umn.edu

S. Bianco
Altos Labs - Bay Area, Institute of Science, Redwood City, CA, USA
e-mail: sbianco@altoslabs.com

A. Shaban-Nejad et al. (eds.), *Artificial Intelligence for Personalized Medicine*, Studies in Computational Intelligence 1106,
https://doi.org/10.1007/978-3-031-36938-4_1

contributes significantly to improving different aspects of human life, well-being, healthy living, and longevity by enhancing medical diagnosis and treatment, chronic disease management, drug discovery, and lifestyle and preventive health. AI can further help healthcare professionals diagnose diseases earlier and more accurately and efficiently, leading to precision care and more effective treatment, and improved health outcomes. AI algorithms can also assist doctors and health providers in developing treatment plans that are personalized to each patient's unique medical condition, profile, and personal health history [1], taking into account the patient's behavior and lifestyle and various determinants of health and risk factors. Physicians use intelligent methods to track and analyze data from wearables, mobile, and other smart devices to monitor patients' health, detect potential health issues, and send alerts, notifications, and educational messages to improve patients' self-care, chronic disease management, and health outcomes [2]. Public health professionals and researchers can employ AI and ML methods to enhance disease surveillance [3] and improve public health messaging after detecting the source of mistrust and misinformation in online social media platforms [4, 5].

Moreover, AI and machine learning can accelerate the drug discovery process [6] by finding potential drugs and predicting their efficacy, safety, and dose–response characteristics. This in turn lowers the cost and time needed for drug discovery and manufacturing. In addition, interactive AI-driven intelligent apps and devices [7] can assist individuals in making better and more informed lifestyle choices through personalized recommendations and custom advice for exercise, diet, weight management, and other health-related actions.

Additionally, in recent years advances in the delivery of patient care [7] are being felt at the organizational and patient levels. Thanks to these advances, people typically live longer and have increased chances of suffering from multiple manageable health conditions. These patients suffer from multimorbidity, and AI methods are needed that address multimorbidity by identifying and mitigating adverse interactions occurring when standards of care for individual conditions are combined into a single treatment plan. [8–10] Given the complexity of developing a treatment plan for a multimorbid patient, explaining treatment decisions made by these methods establishes trust with the clinician and supports a shared decision-making process. As such, AI is becoming more and more explainable [11, 12], collaborative, multimodal [14], human-centered [15], equitable [16], ethical [17], and value-based [13]. In this chapter, we outline recent progress in the use and application of AI in personalized medicine and ageing research.

2 The Role of AI and Data Science in Ageing and Longevity Research

The ageing process is defined as "the progressive accumulation of changes with time that are associated with or responsible for the ever-increasing susceptibility to disease and death which accompanies advancing age" [18]. Ageing is a very complex process and like many other complex challenges can benefit from advances in AI. Nowadays artificial intelligence tools and techniques play an important role in ageing research and its relevant interventions. AI and machine learning can harvest, process, and analyze extremely large datasets to detect and specify patterns and assist scientists in better understanding the ageing process. Also, AI can accelerate the development of effective interventions to improve the well-being of individuals and extend their lifespan and health span. Artificial intelligence facilitates the discovery of new drugs, compounds, and biomarkers [19] that can be used to track the aging process and predict the risk of age-related health conditions and diseases, which is important in the targeted design and implementation of relevant interventions to delay or reverse the aging process.

Moreover, AI algorithms can analyze medical records, clinician's notes, lifestyle, social media use, environmental factors, social determinants of health risk factors, and other data sources to predict a person's risk of developing age-related diseases or issues such as Alzheimer's disease [20, 21], dementia [22], mild cognitive impairment [23], and neurological disorders [24]. This is especially important in assisting physicians and healthcare providers to implement personalized prevention, treatment, and care plans to reduce the risk and likelihood of developing such diseases. Another utility of AI is in image analysis to detect age-related physiological changes in the body and identify potential markers of age-related diseases and conditions. Artificial intelligence can also effectively identify digital biomarkers [25] from digital devices and integrate them with sociomarkers [26] and other biomarkers to improve our understanding and tracking of the ageing process and predict the risk of age-related conditions.

Especially important is the role of computational methods in characterizing biological age. A multitude of biomarkers have been recently established to correlate, often with very high accuracy, chronological and biological age, as well as aging in general [27]. More importantly, biological clocks that use multimodal -omics data are exceptional predictors of disease [28–30]. The use of AI in this field is, however, limited [31–33], and both a theoretical and computational foundation is currently missing. Another dimension of the work on aging is its therapeutic side, its latest research and technology trend exemplified by partial cellular reprogramming and rejuvenation [34–36]. To this date, there are no AI methods applied to the problem of predicting a rejuvenation or reprogramming strategy that is successful, either in the lab or therapeutically. This represents an exciting opportunity for research in this space to emerge and, hopefully, drive the next revolution in aging therapeutics.

3 Advances in AI Technologies and Data Analytics in Healthcare

AI techniques and applications in machine learning, natural language processing (NLP), knowledge representation, explainable AI, image processing, and pattern recognition along with several advanced methods in data science and engineering empowered researchers, healthcare providers, and public health organizations to detect human diseases in a timely manner and design and deliver preventive measures and therapeutic interventions in more efficient ways. Although new technologies bring new challenges, in general as a result of these AI innovations it is expected that more and more people live better, longer, and healthier.

Through several chapters of this book, we will explore studies on the significance of artificial intelligence tools and methods to tackle some of the pressing issues in public health and personalized medicine.

Nguyen et al. [37] applied some eXplainable artificial intelligence (XAI) methods to explain the prediction of the black-box AI models in the thyroid nodule diagnosis application. More specifically they proposed a statistic-based XAI method, namely Kernel Density Estimation and Density map, to explain the case of no nodule detected. Stripelis et al. [38] proposed an architectural vision for an end-to-end Federated Learning and Integration system, that enables geographically distributed data silos to collaboratively learn a joint machine learning model without sharing data, to spur further research on the intersection of health data management information systems and machine learning.

Abdelwahab et al. [39] conducted a case study on drift detection based on textual data from drug reviews created from the UCI ML Drug Review dataset. Moreover, they proposed the sub-sampling method to assess implementing drift detection with large datasets. Larouche et al. [40] proposed the OptimNeuralTS strategy to optimize the search for potentially inappropriate polypharmacies (PIPs), a prevalent phenomenon in older adults, defined as the simultaneous consumption of two or more drugs at once. This method mines claims datasets and builds a predictive model of the association between drug combinations and health outcomes.

Identifying patient subtypes with similar disease trajectories in a heterogeneous population is an important step in personalized medicine. Rocheteau et al. [41] presented an approach to clustering mechanical ventilation episodes, using a multi-task combination of supervised, self-supervised, and unsupervised learning techniques.

There is growing interest in quantifying stochastic and subjective concepts such as trust using Bayesian networks (BN). Thomas et al. [42] proposed a data-driven approach to estimate Bayesian parameters when trust needs to be quantified in the domain of wearable medical devices (WMD). By integrating wavelet transform and visual analysis on EEG signals, Ji et al. [43] introduced a method for understanding disorders, such as neurodevelopmental disorders, ADHD, autism spectrum disorder, depression, and other mental health diseases. Wavelet-based features are extracted to find informative information associated with any changes in the EEG signals to differentiate them from healthy subjects.

Ovalle et al. [44] proposed supplementing ML4H auditing frameworks with SLOGAN (patient Severity-based LOcal Group biAs detectioN), an automatic tool for capturing local biases in a clinical prediction task. SLOGAN adapts an existing tool, LOGAN (LOcal Group biAs detectioN), by contextualizing group bias detection in patient illness severity and past medical history. Werner et al. [45] presented a pipeline in which unsupervised machine learning techniques were used to automatically identify clinical subtypes of hospital in-patients in a large UK teaching hospital admitted between 2017 and 2021. With the use of explainability techniques, the identified subtypes were interpreted and assigned clinical meaning.

Machine learning can help radiologists to analyze CT scans faster and to detect lung cancer more accurately. However, it usually requires laboriously labeled training data. Wang and Tang [46] proposed the Expert System and Attention for Labelling (ESAL) model that uses a mix of experts and pre-trained BERT to retrieve the semantics of different categories, enabling the model to fuse the differences between them to improve medical information classification. Alwuthaynani et al. [47] proposed a transfer learning method using class decomposition to detect Alzheimer's disease from MRI images. They used the entropy-based technique to determine the most informative images for training the model. Kulkarni et al. [48] proposed an NLP-based method 'STBound' that intelligently determines the optimal region for knowledge augmentation and answers questions such as When to augment? for whom to augment? and how much to augment? This proposed selective knowledge augmentation method improves the early detection of depression. Lin et al. [49] proposed an analytical framework of directly inferring the therapeutic working alliance from the natural language within the psychotherapy sessions in a turn-level resolution with deep embeddings such as the Doc2Vec and SentenceBERT models.

Lin et al. [50] compared different neural topic modeling methods in learning the topical propensities of different psychiatric conditions from the psychotherapy session transcripts parsed from speech recordings. The authors also incorporated temporal modeling to improve interpretability by parsing out topic similarities as a time series in a turn-level resolution.

von Numers et al. [51] introduced an automated pipeline, named Baseline-enabled Action Unit identification for Facial Expression Recognition (BAUFER) consisting of (i) a personalized baseline component to calibrate for the neutral expression of a participant; (ii) predictions for anatomically-based facial muscle movement labels (Action Units) that enhance interpretability; and (iii) A multi-stage training approach with several types of annotations from different datasets. Jain et al. [52] looked at a popular deep-learning model for ECG classification and observed its performance on high-level perturbations. To improve model accuracy and adversarial robustness the authors performed adversarial training on these clinically perturbed ECG signals to enhance model robustness. Also, they used conventional adversarial training against low-level perturbations simultaneously to ensure robustness against adversarial attacks.

Nagrebetsky et al. [53] developed a deep-learning model to predict the occurrence and timing of one-lung ventilation (OLV) based on routinely collected intraoperative data. Their approach combines the variables' spatial and frequency domain features, using Transformer encoders to model the temporal evolution and convolutional neural network to abstract frequency-of-interest from wavelet spectrum images.

During the COVID-19 pandemic, news stations have used social media platforms such as Twitter to deliver information to the general public. To understand the trends as well as the impact of these posts, Fisher et al. [54] analyzed 500 k tweets and responses across 15 news outlets from the USA, Canada, and UK, and found that vaccine was the most popular topic discussed, audiences in the USA and UK have a considerable amount of differences in their responses and that the differences in political leanings strongly match with differences in audience response. Gunal et al. [55] introduced a new dataset of 1089 mental health-related post-response pairs from Reddit in which the responses contain questions and annotated these questions as rhetorical, information-seeking, or not applicable. Using linguistic features, the authors distinguished between rhetorical and information-seeking questions.

References

1. N. Ammar, J.E. Bailey, R.L. Davis, A. Shaban-Nejad, Using a personal health library-enabled mhealth recommender system for self-management of diabetes among underserved populations: use case for knowledge graphs and linked data. JMIR Form Res. **16;5**(3), e24738. (2021). https://doi.org/10.2196/24738
2. M. Barrett, J. Boyne, J. Brandts et al., Artificial intelligence supported patient self-care in chronic heart failure: a paradigm shift from reactive to predictive, preventive and personalised care. EPMA J. **10**, 445–464 (2019). https://doi.org/10.1007/s13167-019-00188-9
3. A. Shaban-Nejad, M. Michalowski, S. Bianco (eds.), *AI for Disease Surveillance and Pandemic Intelligence*, 1st edn. (Springer, Cham, 2022). https://doi.org/10.1007/978-3-030-93080-6
4. C.A. Melton, B.M. White, R.L. Davis, R.A. Bednarczyk, A. Shaban-Nejad, Fine-tuned sentiment analysis of COVID-19 vaccine-related social media data: comparative study. J. Med. Internet Res. **24**(10), e40408 (2022). https://doi.org/10.2196/40408
5. B.M. White, C. Melton, P. Zareie, R.L. Davis, R.A. Bednarczyk, A. Shaban-Nejad, Exploring celebrity influence on public attitude towards the COVID-19 pandemic: social media shared sentiment analysis. BMJ Health Care Inf. **30**(1), e100665 (2023). https://doi.org/10.1136/bmjhci-2022-100665
6. S. Dara, S. Dhamercherla, S.S. Jadav, C.M. Babu, M.J. Ahsan, Machine learning in drug discovery: a review. Artif. Intell. Rev. **55**(3), 1947–1999 (2022). https://doi.org/10.1007/s10462-021-10058-4. (Epub 2021 Aug 11)
7. V.R. Fuchs, Major trends in the U.S. health economy since 1950. New Engl. J. Med. **366**(11), 973–977 (2012)
8. A. Kogan, M. Peleg, S.W. Tu, R. Allon, N. Khaitov, I. Hochberg, Towards a goal-oriented methodology for clinical-guideline-based management recommendations for patients with multimorbidity: Gocom and its preliminary evaluation. J. Biomed. Inf. **112** (2020)
9. M. Michalowski, S. Wilk, W. Michalowski, M. Carrier, A planning approach to mitigating concurrently applied clinical practice guidelines. Artif. Intell. Med. **112** (2021)
10. D. Spruijt-Metz, C.K.F. Wen, G. O'Reilly et al., Innovations in the use of interactive technology to support weight management. Curr. Obes. Rep. **4**, 510–519 (2015). https://doi.org/10.1007/s13679-015-0183-6

11. A. Shaban-Nejad, M. Michalowski, J.S. Brownstein, D.L. Buckeridge, Guest editorial explainable AI: towards fairness, accountability, transparency and trust in healthcare. IEEE J. Biomed. Health Inf. **25**(7), 2374–2375 (2021)
12. A. Shaban-Nejad, M. Michalowski, D.L. Buckeridge, Explainability and interpretability: keys to deep medicine, in *Explainable AI in Healthcare and Medicine*, ed. by A. Shaban-Nejad, M. Michalowski, D.L. Buckeridge. Studies in Computational Intelligence, vol. 914 (Springer, Cham, 2021). https://doi.org/10.1007/978-3-030-53352-6_1
13. A. Shaban-Nejad, M. Michalowski, S. Bianco, J.S. Brownstein, D.L. Buckeridge, R.L. Davis, Applied artificial intelligence in healthcare: listening to the winds of change in a post-COVID-19 world. Exp. Biol. Med. (Maywood) **247**(22), 1969–1971 (2022). https://doi.org/10.1177/15353702221140406
14. A. Shaban-Nejad, M. Michalowski, S. Bianco, Multimodal artificial intelligence: next wave of innovation in healthcare and medicine, in *Multimodal AI in Healthcare*, ed. by A. Shaban-Nejad, M. Michalowski, S. Bianco. Studies in Computational Intelligence, vol. 1060 (Springer, Cham, 2023). https://doi.org/10.1007/978-3-031-14771-5_1
15. Shneiderman, B. Human-Centered AI. Oxford University Press, 1st edition, February 10, 2022. Neurocomputing. 2022 May 7 (2022);485:36–46. doi: https://doi.org/10.1016/j.neucom.2022.02.040
16. E. Gurevich, B. El Hassan, C. El Morr, Equity within AI systems: What can health leaders expect? Healthc. Manag. Forum. **36**(2), 119–124 (2023). https://doi.org/10.1177/08404704221125368
17. H. Mamiya, A. Shaban-Nejad, D.L. Buckeridge, Online public health intelligence: ethical considerations at the big data era, in *Public Health Intelligence and the Internet*, ed. by A. Shaban-Nejad, J. Brownstein, D. Buckeridge. Lecture Notes in Social Networks (Springer, Cham, 2017). https://doi.org/10.1007/978-3-319-68604-2_8
18. D. Harman, The aging process. Proc. Natl. Acad. Sci. USA **78**(11), 7124–7128 (1981). https://doi.org/10.1073/pnas.78.11.7124
19. A. Zhavoronkov, Artificial intelligence for drug discovery, biomarker development, and generation of novel chemistry. Mol. Pharm. **15**(10), 4311–4313 (2018). https://doi.org/10.1021/acs.molpharmaceut.8b00930
20. S. Qiu, M.I. Miller, P.S. Joshi et al., Multimodal deep learning for Alzheimer's disease dementia assessment. Nat. Commun. **13**, 3404 (2022). https://doi.org/10.1038/s41467-022-31037-5
21. S. El-Sappagh, J.M. Alonso, S.M.R. Islam et al., A multilayer multimodal detection and prediction model based on explainable artificial intelligence for Alzheimer's disease. Sci. Rep. **11**, 2660 (2021). https://doi.org/10.1038/s41598-021-82098-3
22. J.M. Ranson, M. Bucholc, D. Lyall, D. Newby et al., Harnessing the potential of machine learning and artificial intelligence for dementia research. Brain Inf. **10**(1), 6 (2023). https://doi.org/10.1186/s40708-022-00183-3
23. J. Harvey, R.A. Reijnders, R. Cavill et al., Machine learning-based prediction of cognitive outcomes in de novo Parkinson's disease. npj Parkinsons Dis. **8**, 150 (2022). https://doi.org/10.1038/s41531-022-00409-5
24. U.K. Patel, A. Anwar, S. Saleem et al., Artificial intelligence as an emerging technology in the current care of neurological disorders. J. Neurol. **268**(5), 1623–1642 (2021). https://doi.org/10.1007/s00415-019-09518-3
25. A. Coravos, S. Khozin, K.D. Mandl, Developing and adopting safe and effective digital biomarkers to improve patient outcomes. npj Digit. Med. **2**, 14 (2019). https://doi.org/10.1038/s41746-019-0090-4
26. E.K. Shin, R. Mahajan, O. Akbilgic et al., Sociomarkers and biomarkers: predictive modeling in identifying pediatric asthma patients at risk of hospital revisits. npj Digital Med **1**, 50 (2018). https://doi.org/10.1038/s41746-018-0056-y
27. L. Piovesan, P. Terenziani, G. Molino, Glare-sscpm: an intelligent system to support the treatment of comorbid patients. IEEE Intell. Syst. **33**(6), 37–46 (2018)
28. C. Lopez-Otin, M.A. Blasco, L. Partridge, M. Serrano, G. Kroemer, The hallmarks of aging. Cell **153**, 1194–1217 (2013). https://doi.org/10.1016/j.cell.2013.05.039

29. M.L. Levine, A. Higgins-Chen, K. Thrush, C. Minteer, P. Niimi, Clock work: deconstructing the epigenetic clock signals in aging, disease, and reprogramming. BioRxiv. https://doi.org/10.1101/2022.02.13.480245
30. S. Horvath, R. Kenneth, DNA methylation-based biomarkers and the epigenetic clock theory of ageing. Nat. Rev. Genet. **19**(6), 371–384 (2018)
31. R.E. Marioni, S.E. Harris, S. Shah, A.F. McRae, T. von Zglinicki, C. Martin-Ruiz, N.R. Wray, P.M. Visscher, I.J. Deary, The epigenetic clock and telomere length are independently associated with chronological age and mortality. Int. J. Epidemiol. **45**(2), 424–432 (2016). https://doi.org/10.1093/ije/dyw041
32. L.P. de Lima Camillo, L.R. Lapierre, R. Singh, A pan-tissue DNA-methylation epigenetic clock based on deep learning. NPJ Aging. **8**(1), 4 (2022). https://doi.org/10.1038/s41514-022-00085-y
33. F. Galkin, P. Mamoshina, K. Kochetov, D. Sidorenko, A. Zhavoronkov, DeepMAge: a methylation aging clock developed with deep learning. Aging Dis. **12**(5), 1252–1262 (2021). https://doi.org/10.14336/AD.2020.1202
34. F. Della Valle, P. Reddy, M. Yamamoto et al., LINE-1 RNA causes heterochromatin erosion and is a target for amelioration of senescent phenotypes in progeroid syndromes. Sci. Transl. Med. **14**(657), eabl6057 (2022). Aug 10. https://doi.org/10.1126/scitranslmed.abl6057
35. T. Hishida, M. Yamamoto, Y. Hishida-Nozaki et al., In vivo partial cellular reprogramming enhances liver plasticity and regeneration. Cell Rep. **39**(4), 110730 (2022). https://doi.org/10.1016/j.celrep.2022.110730
36. K.C. Browder, P. Reddy, M. Yamamoto et al., In vivo partial reprogramming alters age-associated molecular changes during physiological aging in mice. Nat Aging. **2**(3), 243–253 (2022). https://doi.org/10.1038/s43587-022-00183-2
37. T.T.H. Nguyen, V.B. Truong, V.T.K. Nguyen, Q.H. Cao, Q.K. Nguyen, Towards trust of explainable AI in thyroid nodule diagnosis, in *Artificial Intelligence for Personalized Medicine: Promoting Healthy Living and Longevity*. Studies in Computational Intelligence (Springer, 2023)
38. D. Stripelis, J.L. Ambite, Federated learning over harmonized data silos, in *Artificial Intelligence for Personalized Medicine: Promoting Healthy Living and Longevity*. Studies in Computational Intelligence (Springer, 2023)
39. H. Abdelwahab, C. Martens, N. Beck, D. Wegener, Investigation of drift detection for clinical text classification, in *Artificial Intelligence for Personalized Medicine: Promoting Healthy Living and Longevity*. Studies in Computational Intelligence (Springer, 2023)
40. A. Larouche, A. Durand, R. Khoury, C. Sirois, Neural bandits for data mining: searching for dangerous polypharmacy, in *Artificial Intelligence for Personalized Medicine: Promoting Healthy Living and Longevity*. Studies in Computational Intelligence (Springer, 2023)
41. E. Rocheteau, I. Bica, P. Liò, A. Ercole, Dynamic outcomes-based clustering of disease trajectory in mechanically ventilated patients, in *Artificial Intelligence for Personalized Medicine: Promoting Healthy Living and Longevity*. Studies in Computational Intelligence (Springer, 2023)
42. M. Thomas, O. Boursalie, R. Samavi, T.E. Doyle, Bayesian-based parameter estimation to quantify trust in medical devices, in *Artificial Intelligence for Personalized Medicine: Promoting Healthy Living and Longevity*. Studies in Computational Intelligence (Springer, 2023)
43. S.Y. Ji, S. Jayarathna, A.M. Perrotti, K. Kardiasmenos, D.H. Jeong, EEG analysis of neurodevelopmental disorders by integrating wavelet transform and visual analysis, in *Artificial Intelligence for Personalized Medicine: Promoting Healthy Living and Longevity*. Studies in Computational Intelligence (Springer, 2023)

44. A. Ovalle, S. Dev, J. Zhao, M. Sarrafzadeh, K.W. Chang, Auditing algorithmic fairness in machine learning for health with severity-based LOGAN, in *Artificial Intelligence for Personalized Medicine: Promoting Healthy Living and Longevity*. Studies in Computational Intelligence (Springer, 2023)
45. E. Werner, J.N. Clark, R.S. Bhamber, M. Ambler, C.P. Bourdeaux, A. Hepburn, C.J. McWilliams, R. Santos-Rodriguez, Identification, explanation and clinical evaluation of in-hospital patient subtypes, in *Artificial Intelligence for Personalized Medicine: Promoting Healthy Living and Longevity*. Studies in Computational Intelligence (Springer, 2023)
46. X. Wang, X. Tang, Automatically extracting information in medical dialogue: expert system and attention for labelling, in *Artificial Intelligence for Personalized Medicine: Promoting Healthy Living and Longevity*. Studies in Computational Intelligence (Springer, 2023)
47. M. Alwuthaynani, Z. Abdallah, R. Santos-Rodriguez, Transfer learning and class decomposition for detecting the cognitive decline of Alzheimer's Disease, in *Artificial Intelligence for Personalized Medicine: Promoting Healthy Living and Longevity*. Studies in Computational Intelligence (Springer, 2023)
48. H. Kulkarni, S. MacAvaney, N. Goharian, O. Frieder, Knowledge augmentation for early depression detection, in *Artificial Intelligence for Personalized Medicine: Promoting Healthy Living and Longevity*. Studies in Computational Intelligence (Springer, 2023)
49. B. Lin, G. Cecchi, D. Bouneffouf, Deep annotation of therapeutic working alliance in psychotherapy, in *Artificial Intelligence for Personalized Medicine: Promoting Healthy Living and Longevity*. Studies in Computational Intelligence (Springer, 2023)
50. B. Lin, D. Bouneffouf, G. Cecchi, R. Tejwani, Neural topic modeling of psychotherapy sessions, in *Artificial Intelligence for Personalized Medicine: Promoting Healthy Living and Longevity*. Studies in Computational Intelligence (Springer, 2023)
51. C. von Numers, Y. Yu, A. Petkova, E. Hutchison, J. Havsol, BAUFER: a baseline-enabled facial expression recognition pipeline trained with limited annotations, in *Artificial Intelligence for Personalized Medicine: Promoting Healthy Living and Longevity*. Studies in Computational Intelligence (Springer, 2023)
52. S. Jain, A. Sangroya, L. Vig, C. Anantaram, Robustness for ECG classification by adversarial training over clinical features, in *Artificial Intelligence for Personalized Medicine: Promoting Healthy Living and Longevity*. Studies in Computational Intelligence (Springer, 2023)
53. Z. Li, A. Nagrebetsky, S. Ranjeva, N. Bi, D. Liu, M.F.V. Melo, T. Houle, L. Yin, H. Deng, A transformer-based deep learning algorithm to auto-record undocumented clinical one-lung ventilation events, in *Artificial Intelligence for Personalized Medicine: Promoting Healthy Living and Longevity*. Studies in Computational Intelligence (Springer, 2023).
54. A. Fisher, R. Sharma, V. Mago, Analyzing the trends of responses to COVID-19 related tweets from news stations: an analysis of three countries, in *Artificial Intelligence for Personalized Medicine: Promoting Healthy Living and Longevity*. Studies in Computational Intelligence (Springer, 2023)
55. A. Gunal, I. Stewart, V. Pérez-Rosas, R. Mihalcea, Understanding the role of questions in mental health support-seeking forums, in *Artificial Intelligence for Personalized Medicine: Promoting Healthy Living and Longevity*. Studies in Computational Intelligence (Springer, 2023)

Towards Trust of Explainable AI in Thyroid Nodule Diagnosis

Truong Thanh Hung Nguyen, Van Binh Truong, Vo Thanh Khang Nguyen, Quoc Hung Cao, and Quoc Khanh Nguyen

Abstract The ability to explain the prediction of deep learning models to end-users is an important feature to leverage the power of artificial intelligence (AI) for the medical decision-making process, which is usually considered non-transparent and challenging to comprehend. In this paper, we apply state-of-the-art eXplainable artificial intelligence (XAI) methods to explain the prediction of the black-box AI models in the thyroid nodule diagnosis application. We propose new statistic-based XAI methods, namely Kernel Density Estimation and Density map, to explain the case of no nodule detected. XAI methods' performances are considered under a qualitative and quantitative comparison as feedback to improve the data quality and the model performance. Finally, we survey to assess doctors' and patients' trust in XAI explanations of the model's decisions on thyroid nodule images.

Keywords Explainable artificial intelligence · Object detection · Thyroid nodule · Medical imaging

---

T. T. H. Nguyen (✉)
Friedrich-Alexander-Universität Erlangen-Nürnberg, Erlangen, Germany
e-mail: hung.tt.nguyen@fau.de

T. T. H. Nguyen · V. B. Truong · V. T. K. Nguyen · Q. H. Cao · Q. K. Nguyen
Quy Nhon AI, FPT Software, Quy Nhon, Vietnam
e-mail: binhtv8@fsoft.com.vn

V. T. K. Nguyen
e-mail: khangnvt1@fsoft.com.vn

Q. H. Cao
e-mail: hungcq3@fsoft.com.vn

Q. K. Nguyen
e-mail: khanhnq33@fsoft.com.vn

A. Shaban-Nejad et al. (eds.), *Artificial Intelligence for Personalized Medicine*, Studies in Computational Intelligence 1106,
https://doi.org/10.1007/978-3-031-36938-4_2

1 Introduction

Thyroid cancer is one of the most common cancer types and is the leading cause of cancer death worldwide [5, 12], especially during the COVID-19 pandemic [7, 9]. Characterized by malignant cells formed in the thyroid gland tissues, the thyroid cancer prognosis depends on the type and the stage at which the disease is detected. Often, doctors rely on the medical images' interpretation, such as thyroid ultrasound images, to identify nodules' presence and provide a diagnosis. However, in routine cancer screening, the errors are mainly false negatives, in which a nodule is present but undetected [11]. Due to recent advances in AI, deep learning models can now serve as decision support means for medical experts. A medical diagnosis system must be accurate, transparent, and explainable to gain end-users trust. Considering the explainability capability, simple AI methods such as linear regression and decision trees are self-explanatory. Still, these methods lack the complexity required for tasks such as classifying two and three-dimensional medical images. Given the increasing ubiquity of advanced techniques such as deep neural networks (DNNs), a new challenge for medical AI is its so-called black-box nature, with decisions that seem opaque and inscrutable, even for experts to understand [8]. Thus, while their opacity is deeply intertwined with their success, it poses a challenge for applying DNNs to high-stakes problems such as medical imaging until we can develop methods that allow radiologists to develop understanding and appropriate trust. Furthermore, newer regulations like the European General Data Protection Regulation (GDPR) strictly require transparency in black-box models, especially in healthcare. Thus, there is a growing chorus of researchers calling for XAI methods. Therefore, in this paper, our main contributions are:

1. We applied several XAI methods, namely LIME [22], RISE [18], Grad-CAM [25], Grad-CAM++ [4], Ada-SISE [28], LRP [3], and D-RISE [19] to explain the two-stage model's classification and localization of nodules.
2. We proposed two statistic-based XAI methods, namely Kernel Density Estimation (KDE) and Density map (DM), to monitor the two-stage model's localization process from the first stage to the second stage and further explain the case of no nodule detected, especially false negative case.
3. We evaluated XAI methods' performance and suitability for the specific nodule detection cases with qualitative and quantitative results and a surveyed XAI's trust to end-users.

Our code is available for reproductivity on GitHub.[1]

[1] https://github.com/hungntt/xai_thyroid.

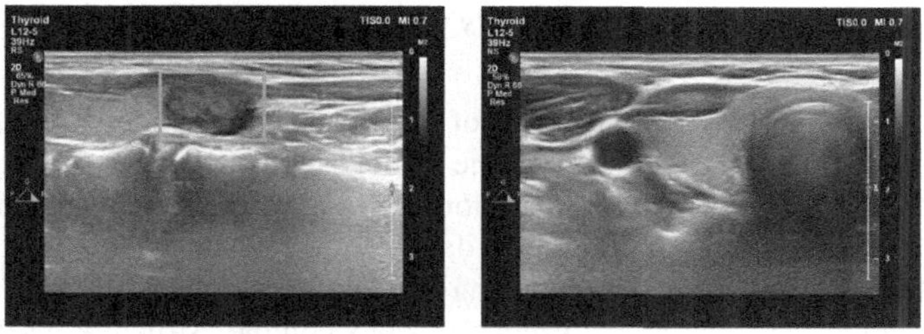

(a) With nodule (b) Without nodule

Fig. 1 Examples of the Vietnamese thyroid ultrasound dataset. The green box is the ground truth labeled by doctors

1.1 Dataset

We use the thyroid ultrasound dataset from [20] that contains 14171 thyroid ultrasound images of 970 Vietnamese patients. Samples are shown in Fig. 1, and medical experts label the nodule locations.

2 Related Work

2.1 Backpropagation-Based Methods

Backpropagation-based methods calculate the gradients of the model's output to the input features or hidden neurons. Hence, they utilize the backward pass of information flow to understand neuronal influence and relevance of input towards the output. The first gradients explanation technique proposed in [26] computes how much a change in each input dimension changes the predictions in a small neighborhood around the input. Some preceding backpropagation-based equations take relative importance given to gradient value during backpropagation to generate saliency maps [27, 33]. While LRP [3] modifies backpropagation rules to measure the relevance or irrelevance of the input features to the models' prediction.

2.2 CAM-Based Methods

Based on the CAM method [34], an extensive research effort on CAM-based methods has been put to blend high-level features extracted by CNNs into a unique explanation map. Grad-CAM [25] and Grad-CAM++ [4] are two improvements over CAM that utilize backpropagation to provide a better visual explanation for classifiers.

2.3 Perturbation-Based Methods

Perturbation-based methods are a class of techniques for explaining the decision-making process of DNNs by modifying the model's input and observing the output's changes. LIME [22] explains the prediction by learning an interpretable model that approximates the model locally around a data point using occlusions of superpixels. RISE [18] proposed a method for producing saliency maps using random perturbation techniques without having to analyze the model's complex structure. D-RISE [19] extended RISE to produce saliency maps for object detectors. SISE [23] improved upon RISE's fidelity and plausibility using attribution-based input sampling techniques. Still, it has high computational costs when there are many activation maps with positive slopes that are inefficient in the prediction process. Ada-SISE [28] was developed to solve this problem by removing unnecessary objects, which saves computational time and provides a better reasonable explanation.

2.4 Statistic-Based Methods

Kernel Density Estimation (KDE) is a non-parametric mathematical method for estimating the probability density function of a continuous variable [29, 32] which is one of the most common methods for estimating density level, set estimation, clustering, or unsupervised learning [17]. Recently, KDE has been made explainable with LRP for outlier and inlier detection in unsupervised learning models without the ground-truth labels [15]. Density map is commonly used in crowd counting literature, which is usually for estimating the distribution of objects, namely people in the scene [13]. However, the idea of counting the model's detected boxes to estimate the distribution of predicted boxes as an explanation has not been applied in previous works.

2.5 XAI in the Medical Diagnosis System

Several XAI applications for different cancer diagnoses are proposed to answer the black-box AIs. In recent years, there have been 37 publications on how XAI is applied in skin cancer detection [10]. More than half of the articles applied current XAI methods to their model, nearly 40% tried to solve specific problems such as bias detection and the effect of XAI on man-machine interactions, and the remaining 10% offered novel XAI methods or enhanced existing techniques. Recently, during the outbreak of COVID-19, LIME is also applied to explain the model's interpretability for screening patients with COVID-19 symptoms [2]. In the same context as our study, AIBx [30] employed the image similarity model and physicians to create an XAI model, increasing physicians' confidence in the predictions during the thyroid cancer diagnosis process.

However, the application of a wide-ranged number of XAI methods to nodule detection on thyroid ultrasound datasets has not been discovered yet. Urgently, very little is known about the influence of XAI on the predictive performance, confidence, and model trust of doctors and radiologists in an artificial setting, and nothing is known about its effects in a clinical setting.

3 Methodology

3.1 Object Detector and XAI Categorization

3.1.1 Analysis of Images with Nodule

Our object detector employs state-of-the-art object detection networks based on the Faster R-CNN [21] and RetinaNet [14] architectures to detect thyroid in ultrasound images. The model detection process comprises two stages. In the first stage, the Region Proposal Network generates object proposals from input images. Next, a bounding box is predicted for each object proposal, with a probability of whether the box contains a thyroid nodule. In the second stage, the Region-of-Interest pooling layer implements bounding box regression and bounding box classifier. We categorize XAI methods in terms of their applicability to three main blocks of an object detector (as shown in Fig. 2):

- Region Proposal Generation (Which proposals are generated by the model during the model's first stage?): Kernel Density Estimation (KDE), Density map (DM).
- Classification (Which features of an image make the model classify an image containing a nodule(s) at the model's second stage?): LRP, Grad-CAM, Grad-CAM++, LIME, RISE, Ada-SISE, D-RISE.
- Localization (Which features of an image does the model consider to detect a specific box containing a nodule at the model's second stage?): D-RISE.

Because XAI methods for the second stage (LRP, Grad-CAM, Grad-CAM++, LIME, RISE, Ada-SISE, D-RISE) require the model's output bounding boxes, they are only applicable to positive cases, where the model detects a nodule in the image.

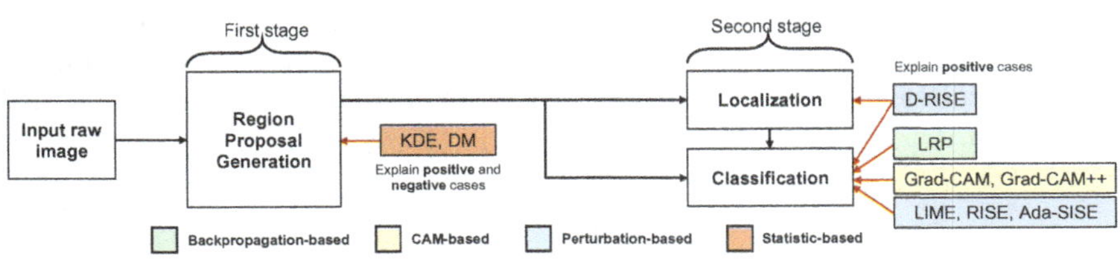

Fig. 2 Object detector's architecture and XAI methods' applicability to different tasks (red arrows)

While XAI methods for the first stage (KDE, DM) directly extract the model's attempts to find a nodule, they can be further applied to negative cases where the model does not detect any nodule.

3.1.2 Local Interpretable Model-Agnostic Explanation (LIME)

For any given instance and its corresponding prediction, LIME utilizes Simple Linear Iterative Clustering [1] as a segmentation algorithm to randomly sample data around the neighborhood of the input instance for which produced predictions. These generated data are used to train a local model. The local model's prediction generates an explanation by weighting each sample according to the instance. Then, LIME uses LASSO as a feature selection technique to choose the most important segments that contribute the most to the prediction for the explanation.

3.1.3 Random Input Sampling for Explanation (RISE)

In the masking generation, we firstly use binary-bit masking (0, 1) to generate $N = 500$ samples, where masks $M = \{M_i, 1 \leq i \leq N\}$ for each superpixel with a probability $p = 0.5$ [18]. We set each sample's size as 8×8. We upsample these masks using bilinear interpolation to ensure all values are in the range [0, 1]. Then, we feed samples into the model to get the bounding boxes and the corresponding score for each box S^b. Finally, RISE sums up all the masks using the box scores, which are predicted on each sample as the weight of each mask to explain the target box from the input image in the form of a saliency map.

3.1.4 Adaptive-Semantic Input Sampling for Explanation (Ada-SISE)

Ada-SISE selects multiple layers of the model and extracts feature maps by feeding the input image into the model. Then, it samples subsets of the feature maps that contain the most important features by partially backpropagating the signal to the layer and calculating the average gradient scores for each feature map. It then collects all feature maps with positive scores and applies an Otsu-based threshold [16] to remove those with lower scores. It produces attribution masks by bilinearly interpolating and normalizing the positive feature maps. For each selected layer, it obtains a layer visualization map by computing the weighted sum of the attribution masks. Finally, it combines obtained saliency maps in a fusion module to produce a final explanation.

3.1.5 Gradient-Weighted Class Activation Mapping (Grad-CAM) and Grad-CAM++

Grad-CAM [25] and Grad-CAM++ [4] are methods for producing a saliency map, which is a visual representation of the regions in an input image that is most important for a particular task or model.

Grad-CAM is a technique that uses the gradients of the target class with respect to the final convolutional layer of a CNN to produce a coarse localization map of the important regions in the input image. The map is then upsampled and weighted by the gradients to produce the final saliency map.

Grad-CAM++ is an extension of Grad-CAM that produces a more fine-grained and accurate localization map by using the gradients of the target class to the lower convolutional layers and the final convolutional layer. It also introduces a new way of weighting the upsampled map using a weighted combination of the gradients of the target class and the activations of the lower convolutional layers. The final saliency map is also produced by upsampling and weighting the localization map using the weighted combination of gradients and activations.

3.1.6 Layer-Wise Relevance Propagation (LRP)

LRP uses the weights and activations of a neural network to propagate relevance scores from the output layer back to the input layer according to the conservation rule, which states that a neuron that receives what is from the upper layer must be redistributed entirely to the lower layer in equal numbers. The relevance scores are propagated by calculating the quantity representing how much neuron j contributes to making neuron k, where j and k are neurons in two consecutive layers of the neural network. The propagation procedure terminates once the input features have been reached. LRP has several propagation rules with different non-linear rectifiers, but we use the epsilon rule, which adds a small positive term in the denominator of the equation used to propagate the relevance scores, because it helps to solve cases where the denominator is zero and typically leads to sparser explanations in terms of input features and less noise.

3.2 *XAI Methods for the Localization Task*

3.2.1 Kernel Density Estimation (KDE)

KDE creates a distribution density map by weighting distances of all data points for each specific location along the distribution. If more data points are grouped locally, the estimation is higher as the probability of seeing a point at that location increases.

In the object detector's first stage, we apply KDE to the center points of 300 independent and identically distributed $(b_0, b_1, \ldots, b_{300})$ boxes to obtain the distribution density map given by:

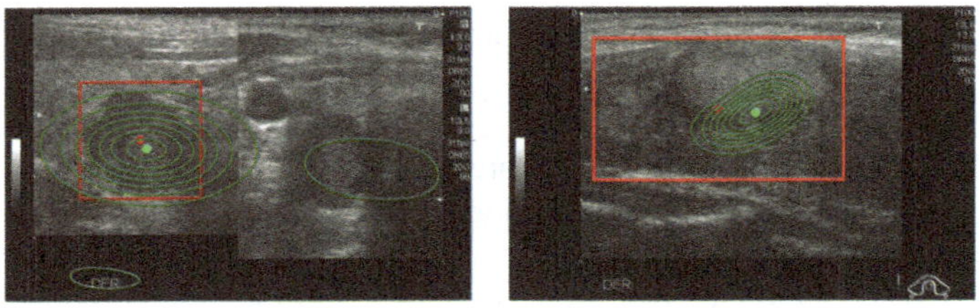

(a) Consistent (PCKDE=0.934)

(b) Non-consistent (PCKDE=0.297)

Fig. 3 **a** Consistent: the model detection's center point (red) is mostly at the same location as the point achieving the highest KDE score (green). **b** Non-consistent: the model detection's center point is far from the point having the highest KDE score

$$\widehat{p}_n(b) = \frac{1}{nh} \cdot \sum_{i=1}^{n} K\left(\frac{B_i - b}{h}\right) \tag{1}$$

where $K(b)$ is called the kernel function that is generally a smooth, symmetric function such as a Gaussian and $h > 0$ called the smoothing bandwidth controls the amount of smoothing.

We represent the distribution density map as multiple continuous probability density curves on the image. The KDE score of a point is computed as the log-likelihood of that point under the KDE model. The score reflects the likelihood that any given box has been drawn from the learned probability distribution. The higher the KDE score is, the more the given box matches the distribution.

The prediction's consistency KDE (PCKDE) is computed as the ratio between the KDE score of the final box's center point detected by the model and the KDE score of the point achieving the highest probability value on the distribution density map. Finally, we set the threshold as 0.5 to grade the PCKDE score. If the PCKDE score exceeds 0.5, the model's detection is considered consistent, as shown in Fig. 3a and non-consistent in Fig. 3b.

3.2.2 Density Map (DM)

DM is commonly used in crowd counting context to estimate objects' distribution in an image [13]. We propose DM as a new statistic-based method in the context of XAI by extracting the frequency value of each pixel from boxes predicted by the model in region proposal generation, as shown in Fig. 4a. The pixel's frequency value is calculated by the number of boxes containing that pixel. For an input image $X \in \mathbb{R}^{3 \times H \times W}$, the model detected n boxes after the first stage. For any box k, let $coor_k$ be the set of coordinates (i, j). Thus, DM's output D of X is the $H \times W$ matrix defined as $D_X = \sum_{i=1}^{n} B_k$, where B_k is computed:

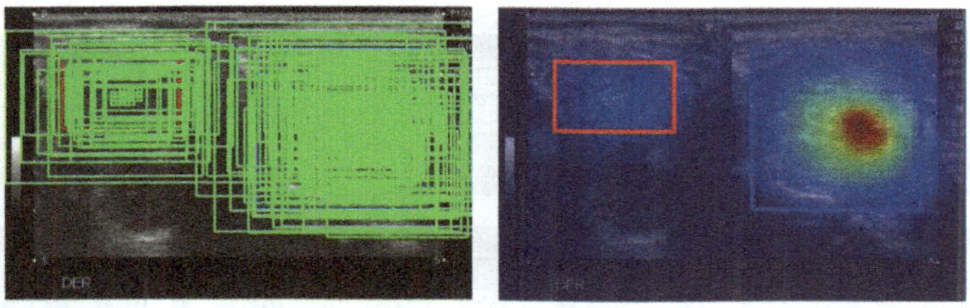

(a) (b)

Fig. 4 **a** DM extracts all model's regional proposals after the first stage. **b** DM's explanation as a saliency map

$$B_k = [a_{ij}]_{H \times W}, a_{ij} = 1_{coor_k}[(i, j)] \quad (2)$$

The more focused boxes a pixel has, the redder colors are indicated in the DM's explanation. In Fig. 4b, the model detects two boxes containing nodules where the blue box is correct with the ground-truth label, while the red box is false. The DM's saliency map can explain the blue box with redder colors, indicating that the model focuses on this region to detect the nodule.

3.2.3 Detector Randomized Input Sampling for Explanation (D-RISE)

D-RISE is a method for producing saliency maps that explain the regions of an input image that are important for a particular task or model. It is specifically designed to produce saliency maps for object detectors, making it the first method. It involves generating a set of binary-bit masks for each superpixel in the input image, upsampling the masks using bilinear interpolation, feeding the samples into a model to obtain bounding boxes and scores for each box, and summing up all the masks using the box scores as weights to produce the final saliency map. The regions of the input image that significantly impact the model's prediction appear as darker colors on the saliency map.

4 Results

4.1 Qualitative Evaluation

4.1.1 Analysis of Positive Cases (with Nodules)

In the positive case, we evaluate XAI explanations on two cases of true positive classification, namely correct localization, and mislocalization, as shown in Fig. 5. A correct localization is when the intersection over the union between the detected box for a nodule and the ground-truth box is larger than 0.5. All our implemented

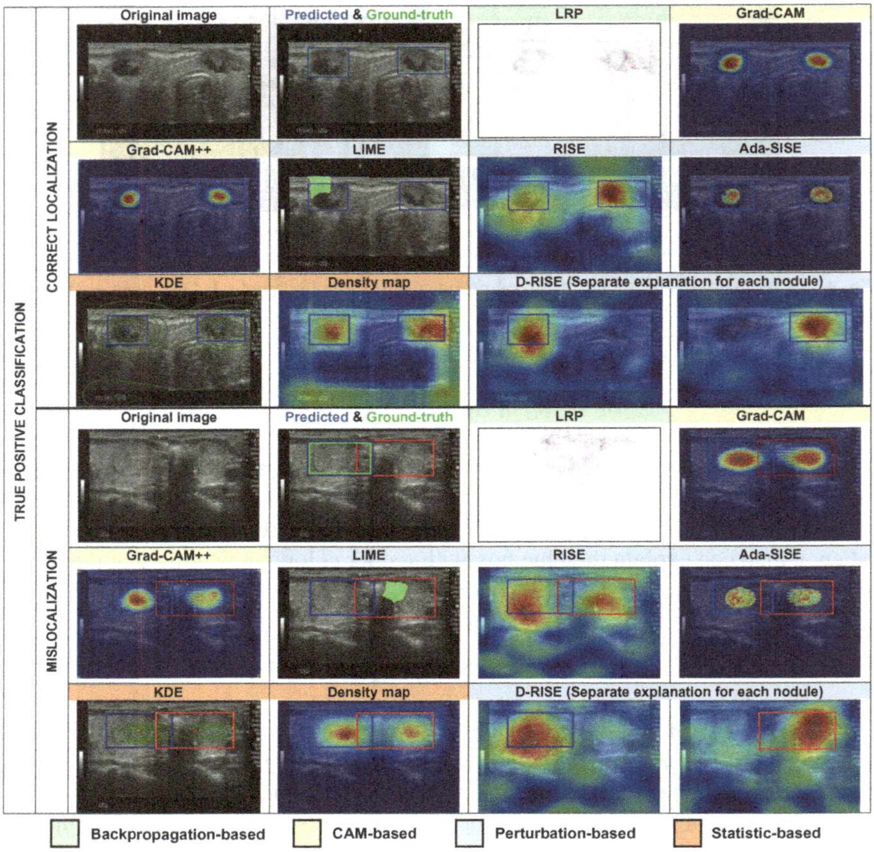

Fig. 5 Qualitative comparison between XAI explanations. In the input images, blue boxes are the correct model's predictions, red boxes are the wrong model's predictions, and green boxes are ground truths. The first row is the correct localization case where the predicted box overlaps the ground-truth label. The second row is the mislocalization case where the model predicted two boxes, but the nodule only exists in the left box

XAI methods' explanations are applicable in the correct localization case, and their explanations are consistent with the model's detected box in the correct localization case. Because D-RISE is the only method having access to the localization block, it can give end-users separate explanations for each detected object. Meanwhile, all other XAI methods show explanations for all nodules since they explain to end-users why the model classifies an image containing nodules.

To further observe D-RISE's advantages, we consider the mislocalization case where the model predicted two boxes, but the nodule only exists in the left box. In this case, D-RISE is the only method to show the explanation solely for the correct and incorrect boxes, while others explain both. Hence, D-RISE's results are more understandable to end-users when each nodule detection needs to be explained.

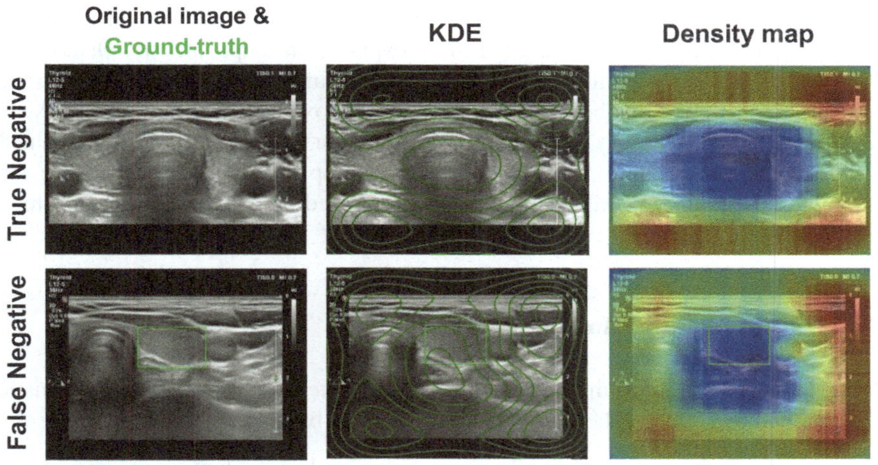

Fig. 6 Explanations of KDE and DM for true negative and false negative cases

4.1.2 Analysis of Negative Cases (Without Nodules)

One serious error of AI-assisted thyroid diagnosis is a false negative, where the model fails to detect any nodule. Hence, our proposed methods, KDE and DM, are the only applicable for explaining negative cases, namely true negative and false negative. As shown in Fig. 6, KDE estimates the distribution of the model's prediction placed around the image's corner. The same model's behavior is also reflected in DM, where the saliency map shows hot regions around the image's border. They both show that the model does not concentrate on any part inside the image, as it does not detect any nodules.

4.2 Quantitative Evaluation

Two critical aspects of XAI, plausibility, and faithfulness, are evaluated by quantitative metrics. These metrics are used to justify the model by assessing the extent to which the method satisfies the users by providing superior statistical explanations. All methods are evaluated on the whole dataset.

4.2.1 Plausibility Evaluation

- **Energy-Based Pointing Game (EBPG)** evaluates XAI methods' precision and denoising ability [31]. Extending the traditional Pointing Game, EBPG considers all pixels in the resultant saliency map S for evaluation by measuring the fraction of its energy captured in the corresponding ground truth G.

- **Intersection over Union (IoU)** analyses the localization ability and meaningfulness of the attributions captured in an explanation map. Initially, Otsu-based binarization [16] is applied to convert ultrasound images into binary images. We compute the mean IoU between the explanation box and the ground truth box.
- **Bounding box (Bbox)** is considered a size-adaptive variant of IoU [24], calculated by selecting the top N pixels in the saliency map significantly influencing the prediction results. It evaluates the regions captured by the bounding box which contains the object.

4.2.2 Faithfulness Evaluation

We evaluate XAI's faithfulness with the *Drop* and *Increase* [6]. The original image is perturbed by masking the important areas marked by the explanation.

- **Drop** checks the target audience's average predicted drop when using the explanatory as input.
- **Increase** measures the number of times the model's confidence increases when using the explanation as input.

4.2.3 Result

Table 1 shows the quantitative comparison of XAI methods concerning their plausibility and faithfulness. In detail, CAM-based methods achieved good results with plausibility metrics, especially EPBG and IoU, in a reasonably short time. While RISE and D-RISE perform better than other methods in terms of faithfulness metrics, such as Drop and Increase, as they faithfully reflect the model's behavior. Nevertheless, the computational time of D-RISE and RISE in specific and perturbation-based methods, in general, are the highest due to their perturbation process. LRP achieved the highest computation efficiency, which is its main advantage.

Table 1 Mean accuracy (%) of quantitative results and computational time of all XAI methods. For each metric, the arrow ↑ / ↓ indicates higher or lower scores are better. The best is shown in bold, and the second best is underlined

Metric	KDE	DM	LIME	Grad-CAM	Grad-CAM++	LRP	RISE	Ada-SISE	D-RISE
EPBG ↑	29.45	21.06	10.95	48.58	**52.11**	31.17	17.04	<u>50.31</u>	17.52
BB ↑	49.16	40.09	14.49	60.51	47.65	38.45	<u>62.41</u>	55.87	**63.42**
IoU ↑	18.94	18.73	10.09	<u>45.22</u>	**49.55**	44.98	12.06	14.99	12.07
Drop↓	34.39	27.31	16.21	26.88	45.76	65.81	<u>4.24</u>	12.43	**2.34**
Inc ↑	32.20	27.12	27.12	18.08	9.60	4.52	<u>46.33</u>	45.13	**53.67**
Time (s) ↓	66	28	380	<u>0.75</u>	0.8	**0.55**	245	295	319

Fig. 7 The number of participants from Vietnam (VN), European countries (EU), North American countries (NA), and other countries. Participants are medical students, radiologists, patients, and AI scientists

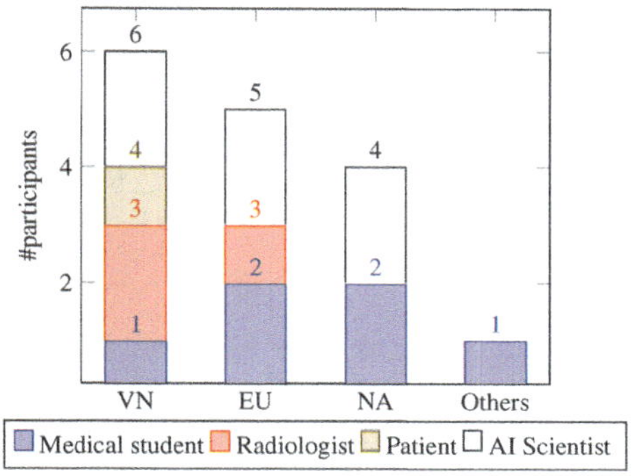

4.2.4 Evaluating User Trust

An essential question of XAI applications is whether explanations can build end-users trust in the AI system. We invited 16 participants with previous experience and familiarity with ultrasound images and AI models, namely three radiologists, one patient, six medical students, and 6 AI scientists from different countries, to take a survey. Six participants were self-reported as from Vietnam, five were from European countries (Germany, Romania, Ukraine), four were from North American countries (United States, Canada), and one was from Pakistan, as shown in Fig. 7. In this survey, the trust of humans towards XAI's explanation is evaluated in terms of understandability, where this explanation helped us understand why the model behaved as it did.

All participants answered seven questions categorized into four prediction cases, namely True Positive (TP), False Positive (FP), True Negative (TN), and False Negative (FN). In detail, TP cases have four questions for the following specific cases: one question for one nodule correctly detected, one question for two nodules correctly detected (correct localization), and two questions for one nodule correctly detected but another falsely detected (mislocalization). While there is one question for each FP, TN, and FN. Each question represents a specific case of the model's classification and localization where the model's prediction, ground truth, and explanations of applicable XAI methods are shown. Participants rated their trust of each XAI's explanation on a labeled, 5-point Likert scale, ranging from 1 (very unlikely) to 5 (very likely). The box plots in Fig. 8 statistically report the score of participants' ratings of XAI methods on four prediction cases. All box plots use the Altman whiskers to display the spread of participants' ratings, which can be particularly useful for showing outliers.

In general, XAI methods using a saliency map as the explanation have consistently higher understandability with higher medians towards humans than others in all cases. D-RISE gains the most trust from users in TP and FP cases, where its interquartile ranges are from 4 to 5, While in TN and FN cases, DM overall surpasses KDE. Still, despite not having a user-friendly explanation, KDE's maximum whisker reaches 5,

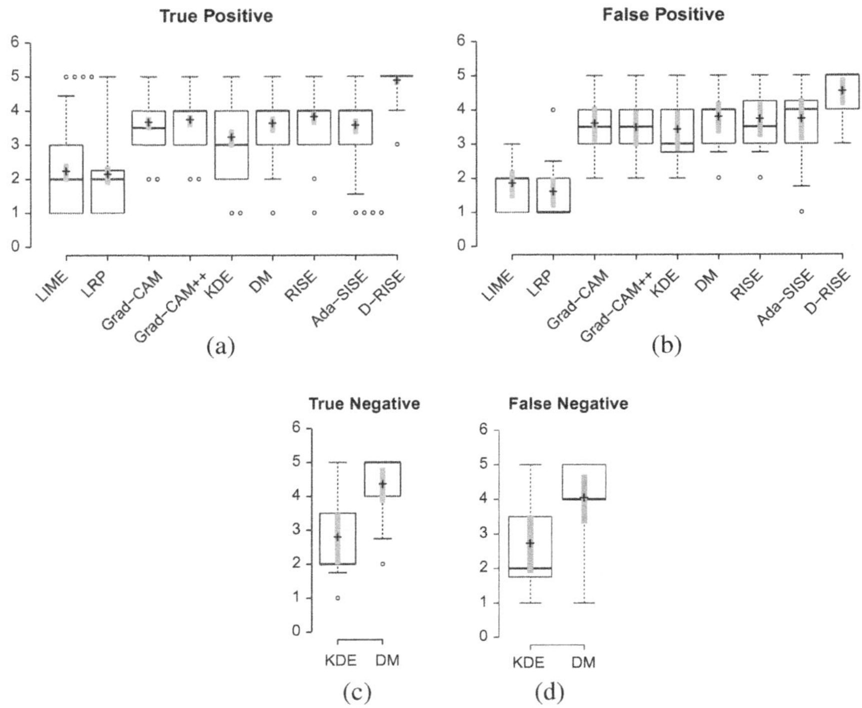

Fig. 8 Participants ratings of XAI methods in **a**True Positive, **b** False Positive, **c** True Negative, and **d** False Negative

which means that it still gains some high trust from users. Also, note that all methods contain low-rating outliers, which means that probably few users are still confused with explanations. Thus, the future of different explanation types is still wide-open.

5 Conclusion

We applied several XAI methods and proposed two new statistic-based methods, namely KDE and Density map, in the context of XAI to explain the model's predictions on the Vietnamese thyroid ultrasound dataset. Our implemented and proposed XAI methods can cover all prediction cases with high consistency with the object detector and doctors' knowledge. Consequently, according to our evaluations and surveys, we recommend end-users use Grad-CAM++ as the default method since it requires a very short time to explain plausibly per case. At the same time, D-RISE is suitable when we require explicitly separate explanations for each nodule due to its faithfulness but high computational time. In future works, we would like to integrate

XAI methods into the diagnosis process in real-time scenarios, so the transparency of AI decisions to doctors and patients can be improved. Also, we aim to conduct a more comprehensive survey that spans multiple countries and continents involving more various user subjects to increase the representativeness and reliability of XAI in the medical field.

References

1. R. Achanta, A. Shaji, K. Smith, A. Lucchi, P. Fua, S. Süsstrunk, Slic superpixels. Technical report (2010)
2. Md M. Ahsan, K.D. Gupta, M.M. Islam, S. Sen, Md Rahman, M.S. Hossain et al., Study of different deep learning approach with explainable ai for screening patients with covid-19 symptoms: using ct scan and chest x-ray image dataset (2020). arXiv:2007.12525
3. S. Bach, A. Binder, G. Montavon, F. Klauschen, K.-R. Müller, W. Samek, On pixel-wise explanations for non-linear classifier decisions by layer-wise relevance propagation. PloS one **10**(7), e0130140 (2015)
4. A. Chattopadhay, A. Sarkar, P. Howlader, V.N. Balasubramanian, Grad-cam++: generalized gradient-based visual explanations for deep convolutional networks, in *2018 IEEE Winter Conference on Applications of Computer Vision (WACV)* (IEEE, 2018), pp. 839–847
5. Y.J. Deng, H.T. Li, M. Wang, N. Li, T. Tian, W. Ying, X. Peng, S. Yang, Z. Zhai, L.H. Zhou et al., Global burden of thyroid cancer from 1990 to 2017. JAMA Netw. Open **3**(6), e208759–e208759 (2020)
6. R. Fu, Q. Hu, X. Dong, Y. Guo, Y. Gao, B. Li, Axiom-based grad-cam: towards accurate visualization and explanation of cnns (2020). arXiv:2008.02312
7. E. Giannoula, I. Iakovou, L. Giovanella, A. Vrachimis, Updated clinical management guidance during the covid-19 pandemic: thyroid nodules and cancer. Eur. J. Endocrinol. **186**(4), G1–G7 (2022)
8. R. Guidotti, A. Monreale, S. Ruggieri, F. Turini, F. Giannotti, D. Pedreschi, A survey of methods for explaining black box models. ACM Comput. Surv. (CSUR) **51**(5), 1–42 (2018)
9. M. Güven, H. Gültekin, The prognostic impact of thyroid disorders on the clinical severity of covid-19: Results of single-centre pandemic hospital. Int. J. Clin. Pract. **75**(6), e14129 (2021)
10. K. Hauser, A. Kurz, S. Haggenmüller, R.C. Maron, C. von Kalle, J.S. Utikal, F. Meier, S. Hobelsberger, F.F. Gellrich, M. Sergon et al., Explainable artificial intelligence in skin cancer recognition: a systematic review. Eur. J. Cancer **167**, 54–69 (2022)
11. C.R. Hebert, L.Z. Sha, R.W. Remington, Y.V. Jiang, Redundancy gain in visual search of simulated x-ray images. Atten. Percept. Psychophys. **82**(4), 1669–1681 (2020)
12. J. Kim, J.E. Gosnell, S.A. Roman, Geographic influences in the global rise of thyroid cancer. Nat. Rev. Endocrinol. **16**(1), 17–29 (2020)
13. C. Li, T. Yang, S. Zhu, C. Chen, S. Guan, Density map guided object detection in aerial images, in *proceedings of the IEEE/CVF Conference on Computer Vision and Pattern Recognition Workshops* (2020), pp. 190–191
14. T.-Y. Lin, P. Dollár, R. Girshick, K. He, B. Hariharan, S. Belongie, Feature pyramid networks for object detection, in *2017 IEEE Conference on Computer Vision and Pattern Recognition (CVPR)* (2017), pp. 936–944
15. G. Montavon, J. Kauffmann, W. Samek, K.-R. Müller, Explaining the predictions of unsupervised learning models, in *International Workshop on Extending Explainable AI Beyond Deep Models and Classifiers* (Springer, 2022), pp. 117–138
16. N. Otsu, A threshold selection method from gray-level histograms. IEEE Trans. Syst. Man Cybern. **9**(1), 62–66 (1979)

17. E. Parzen, On estimation of a probability density function and mode. Ann. Math. Stat. **33**(3), 1065–1076 (1962)
18. V. Petsiuk, A. Das, K. Saenko, Rise: randomized input sampling for explanation of black-box models (2018). arXiv:1806.07421
19. V. Petsiuk, R. Jain, V. Manjunatha, V.I. Morariu, A. Mehra, V. Ordonez, K. Saenko, Black-box explanation of object detectors via saliency maps, in *Proceedings of the IEEE/CVF Conference on Computer Vision and Pattern Recognition* (2021), pp. 11443–11452
20. T.-C. Pham, A. Doucet, V.-D. Hoang, Q.-H. Nguyen, T.-B. Phan, C.-T. Tran, T.-T. Bui, C.-M. Luong, V.-G. Bui, Evaluating the deep convolutional neural network for thyroid nodule detection on vietnamese ultrasound dataset, in *Advances in Intelligent Information Hiding and Multimedia Signal Processing* (Springer, 2021), pp. 358–366
21. S. Ren, K. He, R. Girshick, J. Sun, Faster r-cnn: towards real-time object detection with region proposal networks. IEEE Trans. Pattern Anal. Mach. Intell. **39**(6), 1137–1149 (2017)
22. M.T. Ribeiro, S. Singh, C. Guestrin, Why should i trust you? Explaining the predictions of any classifier, in *Proceedings of the 22nd ACM SIGKDD International Conference on Knowledge Discovery and Data Mining*, pp. 1135–1144 (2016)
23. S. Sattarzadeh, M. Sudhakar, A. Lem, S. Mehryar, K.N. Plataniotis, J. Jang, H. Kim, Y. Jeong, S. Lee, K. Bae, Explaining convolutional neural networks through attribution-based input sampling and block-wise feature aggregation, in *Proceedings of the AAAI Conference on Artificial Intelligence*, vol. 35 (2021), pp. 11639–11647
24. K. Schulz, L. Sixt, F. Tombari, T. Landgraf, Restricting the flow: information bottlenecks for attribution, in *International Conference on Learning Representations* (2019)
25. R.R. Selvaraju, M. Cogswell, A. Das, R. Vedantam, D. Parikh, D. Batra, Grad-cam: visual explanations from deep networks via gradient-based localization, in *Proceedings of the IEEE International Conference on Computer Vision* (2017), pp. 618–626
26. K. Simonyan, A. Vedaldi, A. Zisserman, Deep inside convolutional networks: visualising image classification models and saliency maps (2013). arXiv:1312.6034
27. D. Smilkov, N. Thorat, B. Kim, F. Viégas, M. Wattenberg, Smoothgrad: removing noise by adding noise (2017). arXiv:1706.03825
28. M. Sudhakar, S. Sattarzadeh, K.N. Plataniotis, J. Jang, Y. Jeong, H. Kim, Ada-sise: adaptive semantic input sampling for efficient explanation of convolutional neural networks, in *ICASSP 2021-2021 IEEE International Conference on Acoustics, Speech and Signal Processing (ICASSP)* (IEEE, 2021), pp. 1715–1719
29. G.R. Terrell, D.W. Scott, Variable kernel density estimation. Ann. Stat. 1236–1265 (1992)
30. J. Thomas, T. Haertling, Aibx, artificial intelligence model to risk stratify thyroid nodules. Thyroid **30**(6), 878–884 (2020)
31. H. Wang, Z. Wang, M. Du, F. Yang, Z. Zhang, S. Ding, P. Mardziel, X. Hu, Score-cam: score-weighted visual explanations for convolutional neural networks, in *Proceedings of the IEEE/CVF Conference on Computer Vision and Pattern Recognition Workshops* (2020), pp. 24–25
32. S. Węglarczyk, Kernel density estimation and its application, in *ITM Web of Conferences*, vol. 23 (EDP Sciences, 2018), p. 00037
33. M.D. Zeiler, R. Fergus, Visualizing and understanding convolutional networks, in *European Conference on Computer Vision* (Springer, 2014), pp. 818–833
34. B. Zhou, A. Khosla, A. Lapedriza, A. Oliva, A. Torralba, Learning deep features for discriminative localization, in *Proceedings of the IEEE Conference on Computer Vision and Pattern Recognition* (2016), pp. 2921–2929

Federated Learning over Harmonized Data Silos

Dimitris Stripelis and José Luis Ambite

Abstract Federated Learning is a distributed machine learning approach that enables geographically distributed data silos to collaboratively learn a joint machine learning model without sharing data. Most of the existing work operates on unstructured data, such as images or text, or on structured data assumed to be consistent across the different silos. However, silos often have different schemata, data formats, data values, and access patterns. The field of data integration has developed many methods to address these challenges, including techniques for data exchange and query rewriting using declarative schema mappings, and entity linkage. We propose an architectural vision for an end-to-end Federated Learning and Integration system, incorporating the critical steps of data harmonization and data imputation, to spur further research on the intersection of data management information systems and machine learning.

Keywords Federated learning · Data integration · Data imputation

1 Introduction

Federated Learning (FL) [35] and Analytics [37] is a distributed learning approach that allows to collaboratively train machine learning and other statistical models from decentralized data. Since different sources often cannot share their data due to competitiveness, legal, and privacy constraints, Federated Learning keeps the data at its original source, and pushes the learning process, usually training a neural network, down to each source. A central server coordinates the distributed training among the participating data sources and aggregates the locally learned neural models (or other statistics) to compute a global model. Training can be performed under strong privacy

D. Stripelis (✉) · J. L. Ambite
Information Sciences Institute, University of Southern California, Los Angeles, CA 90007, USA
e-mail: stripeli@isi.edu

J. L. Ambite
e-mail: ambite@isi.edu

A. Shaban-Nejad et al. (eds.), *Artificial Intelligence for Personalized Medicine*,
Studies in Computational Intelligence 1106,
https://doi.org/10.1007/978-3-031-36938-4_3

guarantees through homomorphic encryption [40]. Federated learning [23, 28, 55] has been very effective both for edge and mobile devices [31], and for federations of business organizations [32], or biomedical research consortia [24, 39].

The same privacy constraints that prevent data sharing across sources also inherently create isolated data environments, i.e., data silos [22]. Most work in Federated Learning focuses on solving challenges related to the distributed learning optimization problem [28, 49], but the core challenge of data harmonization across silos is overlooked. Existing systems [27] assume that the local data at the participating sources (which are the input into the learning model) follow the same schema, format, semantics, and storage and access capabilities. Such an assumption does not hold in realistic learning scenarios, where geographically distributed data sources have their own unique data specifications; a challenge that is commonly observed in Federated Database Management Systems [21, 43], Data Integration [12], and Data Exchange [14]. Therefore, we present an architecture for a Federated Learning and Integration with data harmonization and data imputation as its core components.

Figure 1 shows our high-level architecture. The *Data Harmonization* component maps each local schema (and values) to a common schema (and values) agreed by the federation, which we advocate should be done through declarative schema mappings [12–14, 17, 19, 52]. This common schema intends to support multiple learning scenarios over the domain of the data. Since not all sources may have values for all the attributes in the common schema, it is often necessary to impute missing values to improve the precision of statistical studies [26] and reduce prediction bias [3], specially in clinical studies. This has implications for the data integration methods used: instead of removing answers with skolems/labelled nulls, these can be preserved and

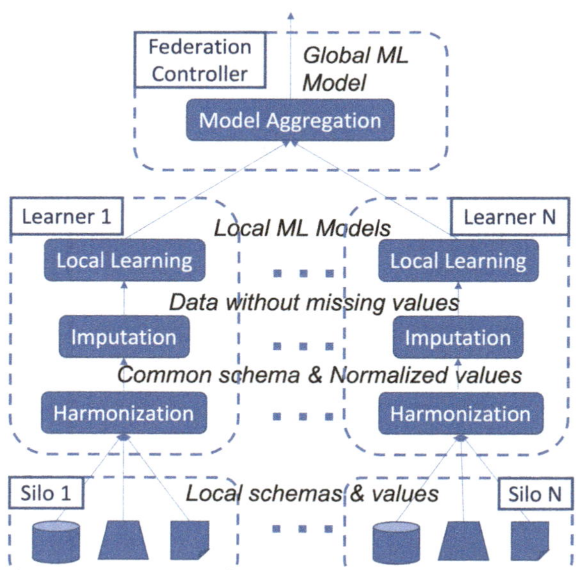

Fig. 1 A harmonized federated learning workflow

the missing values imputed. Altogether, we identify the following core challenges that need to be solved to facilitate the deployment of Federated Learning solutions in real-world settings:

- Private and secure data harmonization and normalization across federated learning silos.
- Enable federated training over missing values using imputation to improve learning efficiency and reduce bias.
- Improve data query access patterns for efficient ingestion of training data into siloed learning models.

Our architecture is general, but we focus on biomedical domains, since they require all the aspects of the architecture and have significant social impact. In the remainder, we provide the necessary background of Federated Learning optimization, discuss the need for federated data integration and imputation, and describe our proposed architecture.

2 Background

In a Federated Learning environment consisting of $\mathcal{N}$ participating learners, we want to minimize the objective function [35, 49]:

$$F(\boldsymbol{x}) = \mathbb{E}_{k \sim \mathcal{K}}[F_k(\boldsymbol{x})] \quad F_k(\boldsymbol{x}) = \mathbb{E}_{\xi \sim \mathcal{D}_k}[\ell_k(\boldsymbol{x}, \xi)] \tag{1}$$

where $\boldsymbol{x}$ represents the model parameters and $\mathcal{K}$ the learner distribution selection over the population of $\mathcal{N}$ learners. Every learner computes its local objective by minimizing the empirical risk F_k over its local data distribution $\mathcal{D}_k$, with $\mathcal{D}_i \neq \mathcal{D}_j, \forall i \neq j$ and $\ell_k(\boldsymbol{x}, \xi)$ the local loss function. The global objective function $F(\boldsymbol{x})$ can take the form of an empirical risk minimization objective $F(\boldsymbol{x}) = \sum_{k=1}^{P} \frac{p_k}{\mathcal{P}} F_k(w)$ with p_k denoting the contribution value of learner k in the federation and $\mathcal{P} = \sum p_k$ the normalization factor; hence $\sum_k^{\mathcal{K}} \frac{p_k}{\mathcal{P}} = 1$. In the original Federated Average (FedAvg) algorithm [35], the contribution value of every learner k equals its local training dataset size, $p_k = |\mathcal{D}_k|$ and a central server aggregates local models updates at given synchronization points known as federation rounds.

Federated **optimization** methods need to address several challenges that do not usually arise in centralized settings, namely learners' communication constraints, local data and computational heterogeneity, learning topology, and security and privacy. To provide convergence guarantees in the presence of these learning constraints, existing methods decouple the federated optimization problem into global (server-side) and local (client-side). Server-side optimization, e.g., [38], refers to the algorithm applied while merging learners' local models, and client-side optimization, e.g., [29], to the algorithm applied during learners' local training. Recent surveys [23, 49] provide further details.

Federated Learning systems can be structured with different **topologies** depending on the communication constraints [23, 39]. In a *centralized* (star) topology learners communicate through a central entity (controller). In a *decentralized* topology learners communicate directly with each other (peer-to-peer) without any central coordinator. Hierarchical and hybrid approaches [4, 39], where multiple sub-aggregators can co-exist [7], have also been explored.

Federated Learning has been applied in two **environments** with complementary properties: *cross-silo* FL consisting of tens or hundreds of reliable, stateful learners with ample computational resources (e.g., geo-distributed datacenters, hospital networks), and *cross-device* FL consisting of thousands or millions of unreliable, stateless learners with limited computation and communication capacity (e.g., IoT sensors, cell-phones) [23]. For simplicity of exposition, we focus on a cross-silo centralized topology where all learners train the same neural network.

Federated training can follow different **communication protocols**. In a synchronous protocol [35], each learner trains for a given number of epochs. The controller waits for all learners participating in the current round to finish their local training before computing a new global model, which may cause fast learners to remain idle while waiting for slow learners. In an asynchronous protocol [53], each learner trains for a given number of epochs and immediately sends its local model to the controller. There is no idle time, but the global model may be computed over stale local models. In a semi-synchronous protocol [47], all learners train for the same time period before sharing their local models with the controller. This approach avoids idle time, but learners may perform different amounts of work. Empirically, this approach often performs better.

To improve **privacy and security** during federated training, a range of privacy-preserving federated learning algorithms have been recently proposed [23] based on private-aware training and secure aggregation. To ensure protection against different attack scenarios, such as membership inference attacks [18] or collusion attacks [34], a Federated Learning solution needs to incorporate both approaches. In the case of private training, learners can train the global model on their local data using (differential) private-aware methods, such as DP-SGD [1]. In the case of secure aggregation, the aggregation of local models by the controller can be performed through Secure Multi-Party Computation [8] or Fully Homomorphic Encryption (FHE) [46, 60] schemes. In the FHE setting [46], the controller sends the encrypted global model weights (ciphertext) to each learner, a trusted entity generates the public and private key pair and shares the key pair with every other learner and only the public key with the controller. The learners decrypt the encrypted weights using a private key, train the decrypted model on their local dataset, encrypt the new local model (ciphertext) using the public key and send it back to the controller. Upon receiving the encrypted local models, the controller performs a private weighted-aggregation step over the ciphertexts to compute the global model.

Depending on the distribution of the training records used to train the federated model across the participating learners, different **data partitioning** schemes have been investigated [55, 57]. Let $\mathcal{I}$ denote the id (entity) space, $\mathcal{X}$ the feature space and $\mathcal{Y}$ the label space of the training records, these schemes are:

- *Horizontal Federated Learning (HFL).* Learners own data with the same feature and label space but different id space, e.g., participants in a research consortium with multiple sites.

$$\mathcal{X}_i = \mathcal{X}_j, \quad \mathcal{Y}_i = \mathcal{Y}_j, \quad I_i \neq I_j, \quad \forall \mathcal{D}_i, \mathcal{D}_j, i \neq j \tag{2}$$

- *Vertical Federated Learning (VFL).* Learners own data from the same id space but with different feature and/or label space, e.g., same patients across different hospitals.

$$\mathcal{X}_i \neq \mathcal{X}_j, \quad \mathcal{Y}_i \neq \mathcal{Y}_j, \quad I_i = I_j, \quad \forall \mathcal{D}_i, \mathcal{D}_j, i \neq j \tag{3}$$

- *Federated Transfer Learning (FTL).* Learners own completely disjoint datasets, with different id, feature and label space, e.g., different customers across different organizations.

$$\mathcal{X}_i \neq \mathcal{X}_j, \quad \mathcal{Y}_i \neq \mathcal{Y}_j, \quad I_i \neq I_j, \quad \forall \mathcal{D}_i, \mathcal{D}_j, i \neq j \tag{4}$$

Most federated learning algorithms focus on the HFL domain [23, 49] while VFL and FTL introduce additional federated optimization challenges that require more complex and expensive training protocols. Vertical federated learning requires an aggregation of the different features across learners and privacy-preserving computation of training loss and gradients across learners with [55] or without [51, 56] a third-party coordinator. The most critical step in this domain is record linkage at the start of federated training through privacy-preserving entity resolution techniques [20, 54]. Federated transfer learning needs to learn common representations from the diverse learners' feature spaces and obtain predictions through one-side features [55]. Depending on the minimization domain, FTL methods can be further categorized into instance-, feature- or parameter-based [57]. In this work, we focus on the data integration and imputation challenges that arise in the Horizontal Federated Learning domains, but we hope our proposed solutions to spur further research into the VFL and FTL domains as well.

3 Federated Learning and Integration

A scalable federated learning solution should adhere to the architectural principles of modularity, extensibility, and configurability [6]. *Modularity* refers to the development of functionally independent services (micro-services) that allow finer control of system components' interoperability. *Extensibility* refers to the functional interface expansion of each service. *Configurability* refers to the ease of deployment of new federated models and procedures. Following these principles, we propose a Federated Learning and Integration (FLINT) architecture (Fig. 2). We describe the core functions in service of learning, data harmonization, and data imputation.

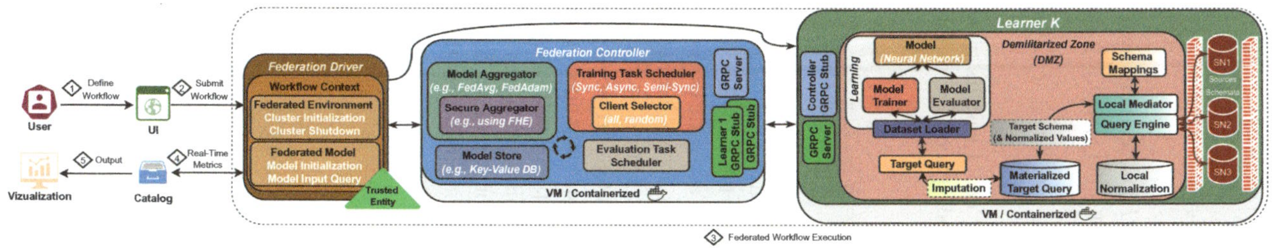

Fig. 2 Federated Learning and Integration architecture and internal components

3.1 Federated Learning Programming Model

Following the successful programming model of Apache Spark [59], the Federation Controller operates as the cluster manager of the federation, Learners as the computing nodes, and the Driver as the entry point of the federation launching various operations in parallel.

Federation Controller. The controller orchestrates the distributed training of the federated model across learners. It comprises four main components. First, the Model Aggregator mixes the local models of the learners to construct a new global model. Second, the Training Task Scheduler manages learners' participation and synchronization points and delegates local training tasks. The Model Store saves the local models and the contribution value of each learner in the federation to improve the efficiency of model aggregation over multiple training protocols [47]. The Model Store component can be materialized through an in-memory or on-disk key-value store, depending on the number of learners and the size of the models (key: learner id, value: model and its contribution p_k). The controller may also operate within an encrypted environment, in which case it needs to store encrypted local models and the global model aggregation function needs to be computed with homomorphic operations, such as the commonly-used weighted average methods [24, 46, 60]. Finally, the Evaluation Task Scheduler is responsible to dispatch the evaluation tasks to the learners and collect the associated metrics.

Learner. Every data silo acts as an independent learning entity that receives the global model and trains the model on its privately held local dataset through its Model Trainer component. A learner can also support a Model Evaluator component to evaluate incoming models on its local training, validation and/or test datasets. Such evaluation can provide a score to weigh the models on actual learning performance [44]. The Dataset Loader feeds *harmonized* data to the training and evaluation components with the appropriate format. Section 3.2 discuss the harmonization and imputation process in detail.

Driver. The Driver defines the high-level control flow of the federated application. Its main tasks are to initialize the Federation Controller and Learner services, and define and initialize the neural network architecture (with a random or a pretrained model). The driver also collects real-time metadata associated with the federated training process and stores them inside the Catalog for further bookkeeping. In our design, we consider the driver to be an independent trusted entity that can generate the key pairs of the encryption scheme.

Evaluation. An evaluation of the presented Federated Learning Programming Model is shown in Fig. 3. The figure shows the convergence of federated learning training policies for different communication protocols and model aggregation functions (FedAvg, FedRec, FedAsync) on standard benchmarks and a neuroimaging domain using the MetisFL system [47]. For this evaluation, we consider a horizontal data partitioning scheme with all datasets conforming to the same schema, and training performed over complete data records with no missing values. CIFAR is a benchmark for object detection in images (with 10 or 100 object classes). Extend-

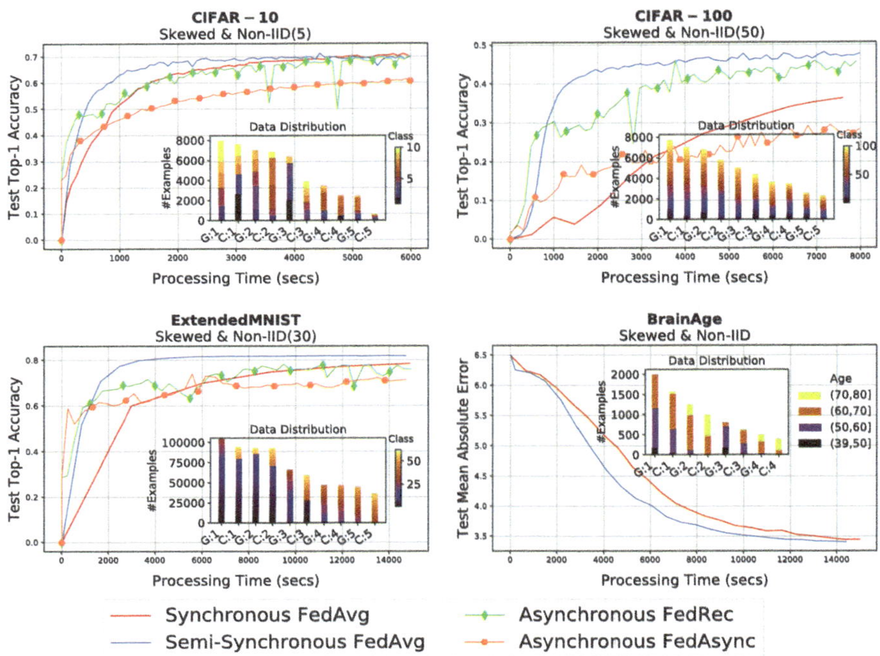

Fig. 3 Federated models convergence in MetisFL

edMNIST [9] is a benchmark for recognition of handwritten letters and digits (we use ExtendedMNST ByClass with 62 unbalanced classes). BrainAge [45] is a neuroimaing task for predicting the age of a human brain from a structural MRI scan. The difference between the predicted and chronological age value is a biomarker of brain pathologies. For CIFAR and ExtednedMNIST, we consider a computationally heterogeneous federation (5 CPUs and 5 GPUs) and for BrainAge a computationally homogeneous federation (8 GPUs). In all environments, the training data is non-IID across learners and learners hold different amounts of training samples (rightly skewed assignment; see plots' insets). The SemiSync policy has faster convergence, particularly in heterogeneous data and computational environments [47].

3.2 *Data Harmonization and Imputation*

All recently proposed Federated Learning systems [27] assume that the local training dataset of every silo conforms to the same data specifications. Such a scenario is not always true in real-world settings. Each silo is an independent entity and it is therefore natural to have its own unique data specifications. For example, in an international federation of hospitals, each institution may adhere to data specifications unique to the geographical region it operates on [11, 33]. Creating a consensus data model that can harmonize the nuances of such regional data specifications is a not an easy task, but it is critical for meaningful data analysis. Therefore, we propose to incorporate a data harmonization and integration component as a core feature of our architecture.

Source Modeling/Schema Mapping. The federated machine learning model needs a harmonized input across all participating sites (sources). Although this could be accomplished by ad-hoc ETL pipelines at each site, such pipelines introduce maintenance and extensibility challenges. To mitigate this, we advocate for a more principled declarative approach based on formal schema mappings following the vast work in data integration [12–14, 17, 19, 52]. First, we define a common schema (aka global, domain, mediated, or target schema) that represents an agreed-upon view of the application domain for the purposes of the federation. Such a common model may follow established standards (e.g., the OMOP Common Data Model and Vocabularies [42], or be defined pragmatically by the members of the federation. The target schema is a degree of freedom of the formalism. It does not necessarily need to provide the "perfect" model for the domain, but it needs to provide sufficient details to support the expected queries and analysis, and a reasonable expectation of extensibility as new sources are added. Second, we define declarative schema mappings to translate the data from the sources into the common schema. These mappings are existential formulas of the form: $\forall \mathbf{x}, \mathbf{y}\ \phi_S(\mathbf{x}, \mathbf{y}) \rightarrow \exists \mathbf{z}\ \psi_G(\mathbf{x}, \mathbf{z})$, where ϕ_S and ψ_G are conjunctions of predicates from the source and global (common) schemas, respectively, and $\mathbf{x}$, $\mathbf{y}$, $\mathbf{z}$ are tuples of variables. These mappings can be used for virtual data integration using query rewriting [12] or for data warehousing/data exchange [14]. Complex constraints can be enforced on the target schema (i.e., being an ontology) and corresponding query answering methods exist [17, 52]. Declarative mappings have the advantage of being easier to generate, maintain, and provide opportunities for automatic learning and optimization (e.g., [25]). Figure 4 shows an example of a global schema, schema mapping rules, and how queries on the domain schema support multiple learning tasks.

Entity Linkage/Data Normalization. It is also important to recognize when objects from different sources correspond to the same entity in the real world. For example, a patient may have interacted with several doctors, hospitals, testing facilities, pharmacies, etc., each of which may have created different records of these interactions. The data integration system must recognize that all theses records refer to the same patient, and link them into a complete medical history for the patient. When we deal with complex structured objects, such as patients, this problem is called entity or record linkage [13, 15, 36]. A simpler version of the problem also occurs with atomic values; different sources may use different strings to refer to the same value. For example, in a radiation oncology domain, one source may code an anatomical structure with a value "LTemp lobe," while another uses the value "LTemporal." To provide clear semantics for analysis, we need to map these two values to a normalized value such as "Left Temporal Lobe" (UBERON:0002808).

Figure 2 shows the detailed data harmonization components in our architecture. Each learner has an instance of a local mediator [12, 50] with access to the schema mappings from its local source/s to the global schema. We envision that the federation will tackle different learning problems, at different times, over the same common view of the data that the mediator produces. Each learning problem would require a different neural network with different input. Thus, we obtain the required input data for each problem through a *query* over the common schema, as opposed to ad-hoc ETL processes. The local data is never changed; however, our system can answer such queries using the schema mappings (and target schema constraints, if any) through query rewriting and data exchange techniques [12–14, 17, 19, 52]. For data normalization, in simple cases we can use a local database with mappings between the values used in each source and normalized values, or use functional predicates that compute a similarity function between the source and the global values in more complex cases. These normalization relations can easily be added to the schema mappings (Fig. 4) and the query answering procedure as interpreted predicates, e.g., [2]. The target query, which computes the input to the neural network, is materialized, so that the model trainer can efficiently operate over it.

Entity linkage across learners is more complex. So far we have discussed horizontal federated learning, where sites have similar id, feature and label space that is needed as input to the learning algorithm, possibly with imputed values. However, the required input data (i.e., data of a single learning example) may be distributed across several sites, so called vertical federated learning, where true private cross-source record linkage is needed [20, 51, 54–56].

Data Imputation. After data harmonization, the machine learning model input is uniform and meaningful, since it is the output of a query over the common schema and values haven been normalized. However, real sources often have missing values, either missing at random or systematically. One option is to remove rows or columns with missing values, but that diminishes the utility of the data and the quality of the learned models. It is generally preferable to *impute* the missing values, that is, to find the most likely value for that given attribute and example [48, 58]. Models learned with imputed values can lead to better performance [5]. In the context of federated learning, participating sources may have limited information, and statistically diverse data distributions, and therefore their local records may not be used to impute missing values/attributes. In these learning settings, an imputation function can be learned at the federation level. By training such a federated imputation function we can leverage the information from all sources, improve data quality and provide better data distribution coverage.

Data imputation interacts with formal query rewriting methods in an interesting way that opens new avenues for research. Since formal schema mappings have existential variables in the consequent, the query rewriting process may generate null values (skolems) in the answers to a query. Tuples with such null values are discarded, since they are not *certain answers* (i.e., true in all possible worlds) [12, 14]. However, for the purpose of learning, such null values can be imputed probabilistically. Therefore, we advocate to modify query answering algorithms to preserve null values and incorporate imputation procedures. Interestingly, the target query may need to retrieve attributes beyond those required by the input of the machine learning algorithm in order to improve the quality of the imputation.

```
Global schema
subject(id, sex, re)                    # demographics, re = race/etchnicity
clinical(id, visit, age, moca, dx)      # clinical data, visit = date of the assessment/dx, dx should be icd10 codes
imaging(id, visit, type, image)         # medical imaging of different types

Schema mappings
s1(id, dob, sex, re, visit, mmse, dx, mri) ∧            # s1 only has MRIs, dx has missing values
minus(dob, visit, age) ∧                                 # compute age at assessment as date of birth minus visit date
impute_f1(sex, age, re, mmse, moca_imp, dx_imp)          # imputation of MoCA (full column) and missing values of dx
→ subject(id, sex, re) ∧ clinical(id, visit, age, moca_imp, dx_imp) ∧ imaging(id, visit, "MRI", mri)

s2_dem(id, sex, re) ∧
s2_image(id, visit_image, age_image, image_type, scan) ∧ image_type = "MRI" ∧    # only interesred in MRIs
s2_dx(id, visit_dx, age_dx, dx) ∧ dx in ["CT", "MCI", "AD"] ∧                    # and on Alzhemer's diagnoses
normalize(dx, icd10) ∧                                                           # normalize diagnostic codes
impute_f2(sex, age_dx, re, icd10, moca_imp)                                      # imputation of MoCA values,
→ subject(id, sex, re) ∧ clinical(id, visit_dx, age_dx, moca_imp, icd10) ∧ imaging(id, visit_image, "MRI", image)

Alzheimer's prediction query
q(sex, re, age, mri, dx) ← subject(id, sex, re) ∧ imaging(id, visit1, "MRI", mri) ∧
   clinical(id, visit2, age1, moca, dx) ∧ ( |visit1 − visit2| < 60 )

Cognitive decline query
q(sex, re, mri1, diff_age, diff_moca) ← subject(id, sex, re) ∧ imaging(id, visit1, "MRI", mri1) ∧
   clinical(id, visit1, age1, moca1, dx1) ∧ clinical(id, visit2, age2, moca2, dx2) ∧ visit2 > visit1
   minus(age1, age2, diff_age) ∧ minus(moca1, moca2, diff_moca)
```

Fig. 4 Global schema and schema mapping rules

Example. Figure 4 shows a notional example of horizontal federated learning and integration (FLINT) on sources with medical data. The federation designers define a (harmonized) global schema with 3 relations: subject, which models subject demographics; clinical, which models clinical assessments and diagnoses, and imaging, which models different types of medical imaging. The federation expects normalized icd10 codes for diagnoses, and standardizes on the Montreal Cognitive Assessment (MoCA) as a measure for dementia. There are two sources: s1, which represents a clinic specializing in the treatment of Alzheimer's Disease that captures magnetic resonance imaging (MRI) and administers a Mini-Mental State Examination (MMSE) for each patient, both on a single visit; and s2, which represents a hospital that treats a wider variety of diseases. The two sources are mapped to the global schema using formal schema mappings that include both data transformation and imputation functional predicates. The first mapping uses a simple functional predicate (minus) to compute the age at assessment from the difference of the patient date of birth and the visit date, as well as an imputation procedure (impute_f1) that imputes both the MoCA score and possibly missing diagnosis codes from the MMSE score, age, sex, race/ethnicity, and existing diagnosis. Since s1 contains only MRIs, this is the recorded type of the resulting imaging in the harmonized schema. The second mapping joins 3 tables from source s2 comprising demographics, imaging, and diagnoses. Assume the federation is only interested in neuroimaging of Alzheimer's Disease, so it chooses to populate the global schema only with MRI scans and relevant diagnoses (AD: Alzheimer's Disease, MCI: mild cognitive impairment, and controls). An interpreted predicate maps the source's diagnoses to appropriate ICD10 codes. Finally, a second imputation function (impute_f2) imputes the MoCA values from sex, age, race/ethnicity and diagnosis. Note that since the MoCA score is not produced by the source, but it is required by the global schema predicate clinical, it would have been represented by an skolem function in a traditional data integration system, and query tuples with such skolem would have been removed. Here, we impute the MoCA score, so no tuples are lost.

The global schema, through the schema mappings, enables a variety of queries that support different learning tasks within the federation. Figure 4 shows two such queries. The first computes the training data for a *classification* learning task that predicts an AD status (AD/MCI/CT) diagnosis based on the MRI, sex, age, and race/ethnicity of a subject. The second query computes the training data for a *regression* learning task to predict cognitive decline based on an MRI at an initial time point and the ages and cognitive assessment values at two timepoints.

Privacy. Federated learning assumes that data from one site cannot be leaked to a different site. Therefore, in our system, federated training encrypts the neural network parameters (weights, gradients) and the aggregation of neural models from different sites is done under homomorphic encryption. We refer the reader to [46] for details. To enforce data privacy, query rewriting and data normalization need to be performed locally at each site and therefore the schema mappings and the normalization tables need to be kept local at each site and only the global model schema is shared across sites. Similarly, record linkage can be done in a privacy-preserving manner [16, 30, 41] but federated training becomes significantly more complex, which is an active area of research [10, 20, 54, 55].

4 Discussion

We presented an architecture for a Federated Learning and Integration (FLINT) platform for distributed training across a federation of data silos, including data integration and imputation components, which are critical for meaningful analysis. We advocated using principled data harmonization methods, leveraging the vast literature on data integration [12–14, 17, 19, 52]. Specifically, we proposed to model the application domain through a target schema and formal schema mappings, and to execute target queries to provide the input data for the federated learning model. Since the purpose of data integration is analysis, we propose new research directions for query answering techniques to incorporate statistical imputation (instead of discarding answers with labeled nulls). We plan to release MetisFL as an open-source prototype FLINT to stimulate further research on the interaction of databases and machine learning.

Acknowledgements This research was supported in part by the Defense Advanced Research Projects Agency (DARPA) under contract HR00112090104, and in part by the National Institutes of Health (NIH) under grant R01DA053028.

References

1. M. Abadi, A. Chu, I. Goodfellow, H.B. McMahan, I. Mironov, K. Talwar, L. Zhang, Deep learning with differential privacy, in *Proceedings of the 2016 ACM SIGSAC Conference on Computer and Communications Security* (2016), pp. 308–318
2. J.L. Ambite, M. Tallis, K.I. Alpert, D.B. Keator, M.D. King, D. Landis, G. Konstantinidis, V.D. Calhoun, S.G. Potkin, J.A. Turner, L. Wang, Schizconnect: virtual data integration in neuroimaging, in *Proceedings of the 11th International Conference on Data Integration in the Life Sciences (DILS 2015)*, Los Angeles, CA (2015), pp. 37–51
3. O.F. Ayilara, L. Zhang, T.T. Sajobi, R. Sawatzky, E. Bohm, L.M. Lix, Impact of missing data on bias and precision when estimating change in patient-reported outcomes from a clinical registry. Health Qual. Life Outcomes **17**(1), 1–9 (2019)
4. P. Bellavista, L. Foschini, A. Mora, Decentralised learning in federated deployment environments: a system-level survey. ACM Comput. Surv. (CSUR) **54**(1), 1–38 (2021)
5. D. Bertsimas, C. Pawlowski, Y.D. Zhuo, From predictive methods to missing data imputation: an optimization approach. J. Mach. Learn. Res. **18**(1), 7133–7171 (2017)
6. D.J. Beutel, T. Topal, A. Mathur, X. Qiu, T. Parcollet, P.P. de Gusmão, N.D. Lane, Flower: a friendly federated learning research framework (2020). arXiv:2007.14390
7. K. Bonawitz, H. Eichner, W. Grieskamp, D. Huba, A. Ingerman, V. Ivanov, C. Kiddon, J. Konecny, S. Mazzocchi, H.B. McMahan et al., Towards federated learning at scale: system design (2019). arXiv:1902.01046
8. K. Bonawitz, V. Ivanov, B. Kreuter, A. Marcedone, H.B. McMahan, S. Patel, D. Ramage, A. Segal, K. Seth, Practical secure aggregation for federated learning on user-held data (2016). arXiv:1611.04482
9. S. Caldas, S.M.K. Duddu, P. Wu, T. Li, J. Konečnỳ, H.B. McMahan, V. Smith, A. Talwalkar, Leaf: a benchmark for federated settings (2018). arXiv:1812.01097
10. D. Cha, M. Sung, Y.R. Park, Implementing vertical federated learning using autoencoders: practical application, generalizability, and utility study. JMIR Med. Inf. **9**(6) (2021). DOI:https://doi.org/10.2196/26598
11. R.J. Cruz-Correia, P.M. Vieira-Marques, A.M. Ferreira, F.C. Almeida, J.C. Wyatt, A.M. Costa-Pereira, Reviewing the integration of patient data: how systems are evolving in practice to meet patient needs. BMC Med. Inf. Decis. Making **7**(1), 1–11 (2007)
12. A. Doan, A. Halevy, Z. Ives, *Principles of Data Integration* (Morgan Kauffman, 2012)
13. X.L. Dong, D. Srivastava, *Big Data Integration*. Synthesis Lectures on Data Management (Morgan & Claypool Publishers, 2015). https://doi.org/10.2200/S00578ED1V01Y201404DTM040
14. R. Fagin, P.G. Kolaitis, R.J. Miller, L. Popa, Data exchange: semantics and query answering. Theor. Comput. Sci. **336**(1), 89–124 (2005). https://doi.org/10.1016/j.tcs.2004.10.033
15. I.P. Felligi, A.B. Sunter, A theory for record linkage. J. Amer. Stat. Assoc. **64**(328), 1183–1210 (1969)
16. T. Ghai, Y. Yao, S. Ravi, P. Szekely, Evaluating the feasibility of a provably secure privacy-preserving entity resolution adaptation of ppjoin using homomorphic encryption (2022). https://doi.org/10.48550/ARXIV.2208.07999. https://arxiv.org/abs/2208.07999
17. G. Gottlob, T. Lukasiewicz, A. Pieris, Datalog+/-: questions and answers, in *14th International Conference on Principles of Knowledge Representation and Reasoning KR* (2014)
18. U. Gupta, D. Stripelis, P.K. Lam, P. Thompson, J.L. Ambite, G. Ver Steeg, Membership inference attacks on deep regression models for neuroimaging, in *Medical Imaging with Deep Learning* (PMLR, 2021), pp. 228–251
19. A.Y. Halevy, Answering queries using views: a survey. VLDB J. **10**(4), 270–294 (2001)
20. S. Hardy, W. Henecka, H. Ivey-Law, R. Nock, G. Patrini, G. Smith, B. Thorne, Private federated learning on vertically partitioned data via entity resolution and additively homomorphic encryption (2017)
21. D. Heimbigner, D. McLeod, A federated architecture for information management. ACM Trans. Inf. Syst. (TOIS) **3**(3), 253–278 (1985)

22. R. Jain, Out-of-the-box data engineering events in heterogeneous data environments, in *Proceedings 19th International Conference on Data Engineering (Cat. No. 03CH37405)* (IEEE, 2003), pp. 8–21
23. P. Kairouz, H.B. McMahan, B. Avent, A. Bellet, M. Bennis, A.N. Bhagoji, K. Bonawitz, Z. Charles, G. Cormode, R. Cummings et al., Advances and open problems in federated learning (2019). arXiv:1912.04977
24. G.A. Kaissis, M.R. Makowski, D. Rückert, R.F. Braren, Secure, privacy-preserving and federated machine learning in medical imaging. Nat. Mach. Intell. **2**(6), 305–311 (2020)
25. C.A. Knoblock, P. Szekely, J.L. Ambite, A. Goel, S. Gupta, K. Lerman, M. Muslea, M. Taheriyan, P. Mallick, Semi-automatically mapping structured sources into the semantic web, in *Proceedings of the Extended Semantic Web Conference*, Crete, Greece (2012)
26. T. Köse, S. Özgür, E. Coşgun, A. Keskinoğlu, P. Keskinoğlu, Effect of missing data imputation on deep learning prediction performance for vesicoureteral reflux and recurrent urinary tract infection clinical study. BioMed Res. Int. (2020)
27. Q. Li, Z. Wen, Z. Wu, S. Hu, N. Wang, Y. Li, X. Liu, B. He, A survey on federated learning systems: vision, hype and reality for data privacy and protection. IEEE Trans. Knowl. Data Eng. (2021)
28. T. Li, A.K. Sahu, A. Talwalkar, V. Smith, Federated learning: challenges, methods, and future directions. IEEE Signal Process. Mag. **37**(3), 50–60 (2020)
29. T. Li, A.K. Sahu, M. Zaheer, M. Sanjabi, A. Talwalkar, V. Smith, Federated optimization in heterogeneous networks (2018). arXiv:1812.06127
30. G. Liang, S.S. Chawathe, Privacy-preserving inter-database operations, in *Intelligence and Security Informatics*, ed. by H. Chen, R. Moore, D.D. Zeng, J. Leavitt (Springer, Berlin, Heidelberg, 2004), pp.66–82
31. W.Y.B. Lim, N.C. Luong, D.T. Hoang, Y. Jiao, Y.C. Liang, Q. Yang, D. Niyato, C. Miao, Federated learning in mobile edge networks: a comprehensive survey. IEEE Commun. Surv. & Tutor. **22**(3), 2031–2063 (2020)
32. Y. Liu, A. Huang, Y. Luo, H. Huang, Y. Liu, Y. Chen, L. Feng, T. Chen, H. Yu, Q. Yang, Fedvision: an online visual object detection platform powered by federated learning, in *Proceedings of the AAAI Conference on Artificial Intelligence*, vol. 34 (2020), pp. 13172–13179
33. B. Louie, P. Mork, F. Martin-Sanchez, A. Halevy, P. Tarczy-Hornoch, Data integration and genomic medicine. J. Biomed. Inf. **40**(1), 5–16 (2007)
34. J. Ma, S.A. Naas, S. Sigg, X. Lyu, Privacy-preserving federated learning based on multi-key homomorphic encryption (2021). arXiv:2104.06824
35. B. McMahan, E. Moore, D. Ramage, S. Hampson, B.A. y Arcas, Communication-efficient learning of deep networks from decentralized data, in *Artificial Intelligence and Statistics* (PMLR, 2017), pp. 1273–1282
36. F. Naumann, M. Herschel, *An Introduction to Duplicate Detection*. Synthesis Lectures on Data Management. (Morgan & Claypool Publishers, 2010)
37. D. Ramage, S. Mazzocchi, Federated analytics: collaborative data science without data collection (2020). https://ai.googleblog.com/2020/05/federated-analytics-collaborative-data.html
38. S.J. Reddi, Z. Charles, M. Zaheer, Z. Garrett, K. Rush, J. Konečný, S. Kumar, H.B. McMahan, Adaptive federated optimization, in *International Conference on Learning Representations* (2020)
39. N. Rieke, J. Hancox, W. Li, F. Milletari, H. Roth, S. Albarqouni, S. Bakas, M.N. Galtier, B. Landman, K. Maier-Hein et al., The future of digital health with federated learning. npj Digital Med. **3**(119) (2020)
40. R.L. Rivest, L. Adleman, M.L. Dertouzos et al., On data banks and privacy homomorphisms. Found. Secure Comput. **4**(11), 169–180 (1978)
41. M. Scannapieco, I. Figotin, E. Bertino, A.K. Elmagarmid, Privacy preserving schema and data matching, in *Proceedings of the 2007 ACM SIGMOD International Conference on Management of Data, SIGMOD '07* (Association for Computing Machinery, New York, NY, USA, 2007), pp. 653–664. https://doi.org/10.1145/1247480.1247553

42. O.H.D. Sciences, Informatics: the Book of OHDSI. OHDSI (2019). https://ohdsi.github.io/TheBookOfOhdsi/
43. A.P. Sheth, J.A. Larson, Federated database systems for managing distributed, heterogeneous, and autonomous databases. ACM Comput. Surv. (CSUR) **22**(3), 183–236 (1990)
44. D. Stripelis, J.L. Ambite, Accelerating federated learning in heterogeneous data and computational environments (2020). arXiv:2008.11281
45. D. Stripelis, J.L. Ambite, P. Lam, P. Thompson, Scaling neuroscience research using federated learning, in *IEEE International Symposium on Biomedical Imaging*, Nice, France (2021)
46. D. Stripelis, H. Saleem, T. Ghai, N. Dhinagar, U. Gupta, C. Anastasiou, G. Ver Steeg, S. Ravi, M. Naveed, P.M. Thompson et al., Secure neuroimaging analysis using federated learning with homomorphic encryption, in *17th International Symposium on Medical Information Processing and Analysis*, vol. 12088 (SPIE, 2021), pp. 351–359
47. D. Stripelis, P.M. Thompson, J.L. Ambite, Semi-synchronous federated learning for energy-efficient training and accelerated convergence in cross-silo settings. ACM Trans. Intell. Syst. Technol. (TIST) (2022)
48. S. Van Buuren, K. Groothuis-Oudshoorn, mice: multivariate imputation by chained equations in r. J. Stat. Softw. **45**(1), 1–67 (2011)
49. J. Wang, Z. Charles, Z. Xu, G. Joshi, H.B. McMahan, M. Al-Shedivat, G. Andrew, S. Avestimehr, K. Daly, D. Data et al., A field guide to federated optimization (2021). arXiv:2107.06917
50. G. Wiederhold, Mediators in the architecture of future information systems. IEEE Comput. **25**(3), 38–49 (1992)
51. Y. Wu, S. Cai, X. Xiao, G. Chen, B.C. Ooi, Privacy preserving vertical federated learning for tree-based models (2020). arXiv:2008.06170
52. G. Xiao, D. Calvanese, R. Kontchakov, D. Lembo, A. Poggi, R. Rosati, M. Zakharyaschev, Ontology-based data access: a survey, in *27th International Joint Conference on Artificial Intelligence* (IJCAI, 2018), pp. 5511–5519
53. C. Xie, S. Koyejo, I. Gupta, Asynchronous federated optimization (2019). arXiv:1903.03934
54. R. Xu, N. Baracaldo, Y. Zhou, A. Anwar, J. Joshi, H. Ludwig, Fedv: privacy-preserving federated learning over vertically partitioned data (2021)
55. Q. Yang, Y. Liu, T. Chen, Y. Tong, Federated machine learning: concept and applications. ACM Trans. Intell. Syst. Technol. (TIST) **10**(2), 1–19 (2019)
56. S. Yang, B. Ren, X. Zhou, L. Liu, Parallel distributed logistic regression for vertical federated learning without third-party coordinator (2019). arXiv:1911.09824
57. X. Yin, Y. Zhu, J. Hu, A comprehensive survey of privacy-preserving federated learning: a taxonomy, review, and future directions. ACM Comput. Surv. (CSUR) **54**(6), 1–36 (2021)
58. J. Yoon, J. Jordon, M. Schaar, Gain: missing data imputation using generative adversarial nets, in *International Conference on Machine Learning* (2018), pp. 5689–5698
59. M. Zaharia, M. Chowdhury, M.J. Franklin, S. Shenker, I. Stoica et al., Spark: cluster computing with working sets. HotCloud **10**(10–10), 95 (2010)
60. C. Zhang, S. Li, J. Xia, W. Wang, F. Yan, Y. Liu, Batchcrypt: efficient homomorphic encryption for cross-silo federated learning, in *2020* {USENIX} *Annual Technical Conference* ({USENIX} {ATC} 20) (2020), pp. 493–506

Investigation of Drift Detection for Clinical Text Classification

Hammam Abdelwahab, Claudio Martens, Niklas Beck, and Dennis Wegener

Abstract Today, machine learning models are applied in various healthcare applications in productive use. The availability of extensive patient information in electronic formats makes it possible to utilize them and develop machine learning-based models for data analysis. However, the performance of an operational model is continuously subject to degradation due to unforeseen changes in the input data flow. Therefore, monitoring data drift becomes essential to maintain the desired performance of the trained models. In the context of monitoring and drift detection, statistical hypothesis testing enables us to examine whether incoming data deviate from training data. Recent studies show that Kernel Maximum Mean Discrepancy (KMMD) and Kolmogorov–Smirnov (KS) can reliably measure the distance between multivariate distributions, hence drift detection. In this work, we conduct a case study on drift detection based on textual data from drug reviews and propose the sub-sampling method to stabilize drift detection. The results of our experiments show that both KMMD and KS detect changes in the text reviews with a limited number of these reviews in both the reference and test data.

H. Abdelwahab (✉) · C. Martens · N. Beck · D. Wegener
Fraunhofer IAIS, Sankt Augustin, Germany
e-mail: hammam.abdelwahab@iais.fraunhofer.de

C. Martens
e-mail: claudio.martens@iais.fraunhofer.de

N. Beck
e-mail: niklas.beck@iais.fraunhofer.de

D. Wegener
e-mail: dennis.wegener@iais.fraunhofer.de

A. Shaban-Nejad et al. (eds.), *Artificial Intelligence for Personalized Medicine*,
Studies in Computational Intelligence 1106,
https://doi.org/10.1007/978-3-031-36938-4_4

1 Introduction

Machine learning communities are growing tremendously every day, as many real-world problems can be solved through machine learning. Several fields such as computer vision, natural language understanding, time series analysis, and healthcare analytics have adopted machine learning as the primary technology for data-driven solutions. The availability of extensive patient information in electronic formats makes it possible to utilize them and develop machine learning-based models for data analysis [1]. Still, most of the ongoing development of solutions is mostly in an academic context with less detail on how to automate the deployment, monitoring, or maintaining the performance of a model in production [2].

Once a trained model is deployed in the real world, several new challenges arise. A key challenge is the degradation of the model [3] and therefore its reliability. This must be recognized by monitoring. While traditional monitoring utilizes performance metrics such as request latency, memory, and serving frequency, machine learning applications require specialized metrics. These metrics include model performance, data-related metrics, drift metrics, and explainability metrics [4]. However, the assessment of a model's performance mainly depends on the ground truth of the incoming production data. Given a model deployed in the real world, it is therefore often not possible to evaluate the model's performance until the ground truth is collected. However, drift detection can help predicting poor model performance by detecting changes in the incoming data. This enables monitoring of the model independently of collecting the true labels [4].

In the following we present a case study of how drift detection can be applied in a text classification scenario.

Since the Kernel Maximum Mean Discrepancy (KMMD) [5] and Kolmogorov–Smirnov (KS) [6] have proven to be reliable distance measures for statistical hypothesis testing, we have chosen these as the basis for our drift detection. In detail, we will perform drift detection experiments based on the Drug Review Dataset [7] and evaluate the performance of the drift detectors. We then modify the drift monitoring process by introducing sub-sampling to improve detection performance. In addition, we will evaluate the performance of our drift detectors using different sample size configurations. Finally, we will compare our drift detectors in terms of time performance.

The structure of this paper is as follows: We will first discuss related work on drift detection in Sect. 1. After that, in Sect. 2, we present the methodology used, which focuses on how we perform statistical hypothesis testing. Then we describe our case study and in particular the experimental setup in Sect. 3. We then present the execution of our experiments as well as their results and evaluation in Sect. 4. Finally, we summarize the lessons learned and discuss future work in Sect. 5. It is worth mentioning that the work in this paper is based on applied research from [8].

Related Work

In this section we highlight previous research on drift detection. Before using the term drift in the context of machine learning-based pipelines, the problem of detecting drifts has been widely and explicitly discussed in the field of data mining. According to Barddal et al., the most common learning problem when it comes to streaming is classification. In a classification problem, a learner learns from features extracted from the data streams. Therefore, the drift detection is defined by the changes in the extracted features from the streaming data (feature drift) [9]. In addition, Jose G. Moreno-Torres et al. defined the term drift as the change in the joint statistical distribution between the training data and the test data [10]. They categorize drifts into covariate drifts for inputs' distribution, prior drifts for labels' distribution, and concept drifts for predictions' distributions.

According to Hu, Hanqing et al. in [11], the methods of detecting drifts can be broken down into performance-based and distribution-based methods. The performance-based detection methods monitors the drifts of the performance metrics of a machine learning model such as accuracy, precision, and F-score. Examples of these methods are Early Drift Detection Method (EDDM) and Adaptive Windowing (ADWIN) which are used to detect drifts in event streams [12, 13]. This category requires obtaining the ground truth of the prediction to fire a signal in case a drift occurs. Hence, it is defined as a supervised approach. In contrast, distribution-based detection methods are defined as unsupervised approaches in which the change in a data distribution is monitored to identify drifts. Examples of these methods are the Semi-supervised Adaptive Novel class detection (SAND), 1-norm Support Vector Machine (SVM), and Kolmogorov–Smirnov (KS) [14–16]. These categories are explained in depth in [11].

Furthermore, in [10] Jose G. Moreno-Torres et al. investigate neural network-based approaches for high-dimensional data using deep learning for feature extraction and training the model. Utilizing the extracted features for deep learning models has pushed recent advances in different fields such as computer vision, natural language processing and understanding, and reinforcement learning. Nevertheless, a proper drift detection technique can mitigate the drop in performance of neural network-based models. The underlying concept in drift detection for deep learning-based solutions is that the closed-world assumption does not apply in the real world [17]. An example of using neural networks for drift detection is the use of Auto-Encoders to investigate prediction quality by using the encoder part to regenerate features that are passed to a multilayer perceptron for prediction. The output is then passed to a decoder to forecast the quality of the prediction [18].

In addition to the mentioned solutions, uncertainty-based techniques have been investigated to detect concept drifts under the condition of the true label being unavailable or scarce. According to Baier et al., Monte-Carlo Dropout is used such that the model can provide predictions with high entropy in case of unknown inputs by randomly switching the neurons of the neural networks on and off [19]. Similarly,

Monte-Carlo DropConnect is used to identify unknown inputs, but by randomly switching the weights of the neural networks on and off [20].

Finally, to detect drift occurrences in a machine learning-based system before obtaining the ground truth, Rabanser et al. investigated using two-sample statistical hypothesis testing [21]. In their paper, they investigate several drift detection methods including the Auto-Encoder, the label classifier, the domain classifier, the Chi-Square test, Maximum Mean Discrepancy, and the Kolmogorov-Smirnov test. According to their results, Maximum Mean Discrepancy and the Kolmogorov–Smirnov test give comparable performances and can be used to detect drifts from images such as the MNIST dataset. Therefore, we use these techniques as they work with high-dimensional data. Besides, these statistical-based approaches can perform independently from any model we deploy and don't require training, but rather a proper data handling for drift monitoring. It is worth mentioning that in the context of this paper, we focus on covariate drifts as we desire independence from collecting labels by focusing on the incoming data. Moreover, we will be using the term drift to indicate covariate drift as defined in [10].

2 Two-Sample Statistical Hypothesis Testing

In the following, we describe the process of the two-sample test which we implement to investigate drifts. The two-sample test is a statistical hypothesis testing process that mainly focuses on assessing the similarity of statistical distributions from which the samples are observed [22]. Therefore, the two-sample test enables us to assess the similarity between the distribution of the data we monitor for drifts and a reference data distribution.

In a two-sample test two hypotheses are defined as follows: let P be the distribution from which the first sample is drawn, while Q is drawn from the second. Then the Null hypothesis states that $P = Q$. In contrast, the Alternative hypothesis states that the two samples observed are drawn from different distributions, $P \neq Q$. The decision on whether these two samples are drawn from the same distribution or not relies on rejecting the Null hypothesis.

Following the standard hypothesis testing method, we use the p-value to estimate the probability of the correctness of the test statistics. The p-value is expressed as the probability of multiple test statistics measures T taken at different situations being greater or equal to a calculated threshold measure $\widehat{T}$ as demonstrated in Eq. 1. To calculate the measures T and $\widehat{T}$, and estimate the p-value, the following steps are implemented.

- The measure T is estimated as the initial calculation taken by the test statistics.
- The set of measures T will then be taken after permuting the sample sets.
- After that, a probability of compliance with the null hypothesis is estimated, that is the p-value.

- This p-value is then compared to the significance level α. The significance level determines whether the Null hypothesis is rejected or not.

$$p\text{-}value = P(T \geq \widehat{T}) \geq \alpha \tag{1}$$

As discussed in Sect. 1, we select and compare Kernel-based Maximum Mean Discrepancy and Kolmogorov–Smirnov to calculate the distance T. Both algorithms are able to detect drifts for high-dimensional features which enables testing them on different drift scenarios [21].

2.1 Kernel Maximum Mean Discrepancy

The Kernel-based Maximum Mean Discrepancy (KMMD) is a distance measure used to differentiate between two statistical distributions. In this paper, we refer to it also as (MMD). For high-dimensional data, kernel functions such as the squared exponential kernel are used to calculate what is called the "mean embeddings" of the distributions in a reproducing kernel Hilbert space H [5]. For the two distributions of high-dimensional data p and q, and a function F, the MMD is defined in the Hilbert space by the following Equation.

$$\text{KMMD}[F, p, q] = ||\mu[p] - \mu[q]||_H \tag{2}$$

To calculate the MMD using a kernel function k as the function F, we use the following Eq. 3 which calculates a squared MMD with samples x from distribution p and samples y from the distribution q [5].

$$\text{MMD}[F, p, q] = \left[\frac{1}{n} \sum_{i,j=1}^{n} k(x_i, x_j) - \frac{2}{nm} \sum_{i,j=1}^{n,m} k(x_i, y_j) + \frac{1}{m^2} \sum_{i,j=1}^{m} k(y_i, y_j) \right]^{\frac{1}{2}} \tag{3}$$

To calculate the p-value using the MMD distance, we use the bootstrapping technique. In bootstrapping, we shuffle the values from both distributions and remeasure the MMD distance multiple times (e.g 1000 times). Then we calculate the p-value according to the steps explained in Sect. 2.

2.2 Kolmogorov–Smirnov Test

The Kolmogorov–Smirnov (KS) test is fundamentally a non-parametric test that computes the largest distance between the Cumulative Density Functions (CDFs) over all the values within the distributions [23]. For the distributions p and q which contain the values x and y respectively, the distance D can be calculated as follows.

$$D = \sup |F(X) - F(Y)| \tag{4}$$

where F is the CDF for samples x in X and the samples y in Y. For high-dimensional data such as high-dimensional features extracted from images, KS calculates the distances dimensional-wise. Namely, if we have features of K dimensions, then K distances will run. Then only the highest distance is considered. To calculate the p-value using KS, we use the inner (or inside) method which is a numerical technique that uses graphical topologies to estimate the p-value [6, 24]. This method is claimed to perform fast for high-dimensional data [6, 25].

3 The Clinical Case Study

In the following, we first present the monitoring schema that explains the components of the drift monitor as part of machine learning-based architecture. After that, we show how we explore the data and prepare it to test drift detectors by simulating different drift scenarios. Finally, we carry out the experiments through several experimental phases.

For the text-based case study we use the Drug Review Dataset [7]. We take the patient reviews from this dataset as input text and their corresponding conditions as labels. For the input, we calculated an average token length of 126. We extracted the 10 most frequent classes for text classification to demonstrate several drift scenarios. The labels are ADHD, Acne, Anxiety, Bipolar Disorder, Birth Control, Depression, Insomnia, Obesity, Pain, and Weight Loss. To classify these reviews to the given conditions we finetuned a pre-trained BERT Model [26]. We also use this pre-trained BERT Model as the feature extractor by omitting its last layer. This produces the document embeddings that will later be used as input to the drift detectors.

Our goal is for the Drift Detector to indicate whether the input data differs from the training data and whether these changes affect the model performance.

3.1 The Drift Monitoring Schema

For the given text classification use case, the experimental drift monitor is developed as shown in Fig. 1. The Training Data Batch is taken from the data our model used for training. The Input Batch represents the production data. Drift detection starts when both batches enter the monitoring system. The feature extractor takes the raw text as input and produces the corresponding embeddings. Then the hypothesis testing is conducted to compare the Training Data Batch and the Input Batch. Based on the result of the statistical hypothesis testing a drift alarm is triggered.

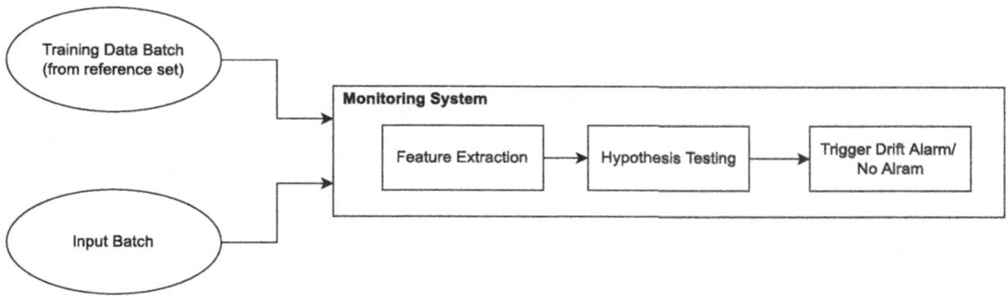

Fig. 1 A schema of the monitoring pipeline for drift detection

3.2 Drift Data Setup

By utilizing KMMD and KS as distance measures, we can implement statistical hypothesis testing to inspect drifts. Here, it is important to specify the reference set with which we compare the test data. Therefore we take samples as a batch from the training set with a specified size of 50. Here, we select the sample by random selection from the training set. Simulating drift data for testing is challenging because it is difficult to include all kinds of drifts that can appear in real-world scenarios. Therefore, we limit the drift detection experiments to three scenarios:

- **Scenario 1: No Drift**: In this scenario, the test text reviews are defined with similar labels as the reviews in the reference set. Namely, the incoming text reviews and the reference reviews both describe symptoms of ADHD, Acne, Anxiety, Bipolar Disorder, Birth Control, Depression, Insomnia, Obesity, Pain, and Weight Loss. Hence, the monitor should not define these reviews as drifts.
- **Scenario 2: Total Drift**: In this scenario, the reference text reviews describe symptoms of ADHD, Bipolar Disorder, Birth Control, Depression, and Obesity. On the other hand, the test reviews describe symptoms of Acne, Anxiety, Insomnia, Pain, and Weight Loss. Since the symptoms of these conditions are not present in the reference set, the monitor should define these reviews as drifts.
- **Scenario 3: Partial Drift**: In this scenario, the reference and test reviews share partially similar conditions. Namely, the reference reviews include symptoms of ADHD, Bipolar Disorder, Birth Control, Depression, Insomnia, Obesity, and Pain. The test reviews include also Insomnia, Obesity, and Pain, with additional symptoms about Weight Loss, Acne, and Anxiety which are not present in the reference reviews.

Finally, we also investigate the needed amount of drifting data within a batch to detect a drift with the given drift detectors. For this, we adjusted the Drift Data setup to different batch sizes of 10, 50, 100, 500, and 1000 texts.

3.3 *Sub-sampling*

To get around the problem of figuring out the sample sizes for statistical hypothesis testing, we are comparing the test batch to multiple reference batch samples rather than just one batch. By doing so, we try to cover as many relevant examples from the reference reviews as possible. We call this sub-sampling. The idea is based on the ensembling method. The resultant p-value will be the average p-values from all the conducted drift tests and the multiple batches are randomly selected from the reference set.

4 Analysis

In this section, we demonstrate the results of the run experiments and analyze the behavior of the drift detectors throughout the experiment's different phases. Furthermore, we show and compare the performance of the drift detectors using the sub-sampling technique. Additionally, we investigate the influence of the number of drift texts within the input batch. Namely, we investigate what is the minimum amount of drift data needed to fire a drift alarm by the drift detector. To examine the impact of drifts, we divide each experiment into 3 phases of 5 runs each. The experiment indices are categorized into the 3 phases, namely No Drift, Total Drift, and Partial Drift. A single run is called a drift test. With this setup, we measure the ability of the drift detector to detect drifts by observing if the p-value falls below the predefined threshold α (0.05). Additionally, we observe the model's accuracy, the monitor's p-values, and the measured distribution distance. In each of the 3 phases, we test the drift detector 5 times using different text reviews as batches from the production data. On the associated graphs, we plot 15 results per experiment in this way. It is worth mentioning that each batch included in these experiments has 50 texts, namely a sample size of 50. If the drift detector's p-value drops below the threshold, this indicates that the text reviews might include information that was not included in the reference text reviews, e.g. symptoms. For predictions linked to such a low p-value, it is possible that the classification model has predicted false conditions.

The results of the executed experiments are demonstrated in the Figs. 2 and 3. On the horizontal axis of each figure, we show the drift test index such that each index represents an execution of a drift test within the experiment. Moreover, every 5 indices cover one of the categories of input batches to be tested. The green background represents the No Drift phase, while the second and third backgrounds represent the Total Drift and Partial Drift phases respectively. On the vertical axis the drift detector's measured distances, the p-values as well as the p-value threshold, and the model's classification accuracy are displayed. For each experiment, we include the model's accuracy to visualize the model's behavior through the input batches during the experiments.

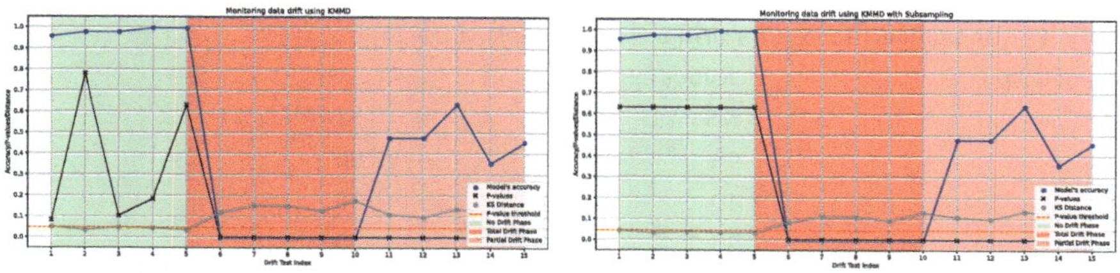

(a) KMMD before sub-sampling. (b) KMMD after sub-sampling.

Fig. 2 Performance of the KMMD drift detectors beside the model through the experiment's phases before and after sub-sampling

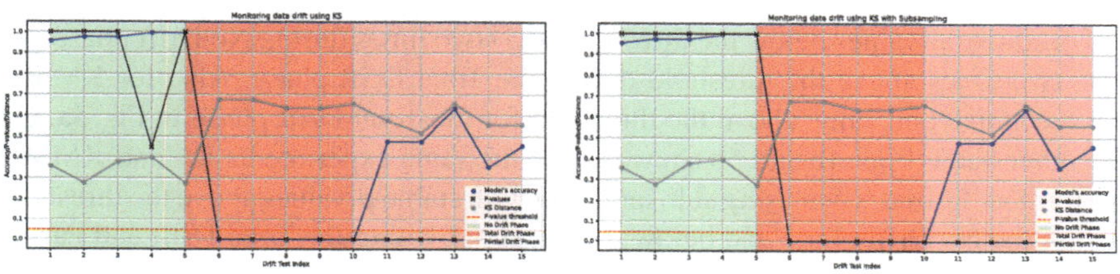

(a) KS before sub-sampling. (b) KS after sub-sampling.

Fig. 3 Performance of the KS drift detectors beside the model before and after sub-sampling

In Figs. 2a and 3a we display the results of using KMMD and KS respectively. We see that in the No Drift phase, the p-values for KMMD fluctuate highly within a range from around 0.1 and 0.8. For KS we have p-values that only fluctuate once at drift test index 4 to a value of around 0.45. However, for both detectors, the p-values stay above the set threshold α. For the Total Drift and the Partial Drift phases, we find that the p-values drop below the threshold to 0.0 for both detectors and stay there for all drift tests. Moreover, we see that the distribution distances for KMMD and KS rise higher for the Total Drift phase in comparison with the Partial Drift phase. Finally, we see that the model's classification accuracy stays high between 0.9 and 1.0 for the No Drift phase and drops to 0.0 for the Total Drift phase while fluctuating between 0.35 and 0.65 for the Partial Drift phase. The distance values for KS are higher compared to the KMMD distance values. For implementing the sub-sampling technique we observe in Figs. 2b and 3b that the p-values remain above the threshold without fluctuation for the No Drift phase. For the Total Drift phase and Partial Drift phase, the p-values still drop below the threshold and stay at 0.0 as before. In general, we notice for all experiments that the distribution distances rise higher for the Partial Drift phase and Total Drift phase in comparison to the No Drift phase. From these results, we observe that both KMMD and KS detectors perform as desired without falsely detecting drifts.

When applying the other batch sizes 10, 50, 100, 500, and 1000 to the experiment, we only experienced significant deviations with a batch size of 10. For the other batch sizes, the behavior was comparable with a batch size of 50. In Fig. 5a we notice that

Table 1 Minimum drift ratios for drift tests per sample sizes

Sample size	Ratios	
	KS (%)	KMMD (%)
10	90	90
50	40	30
100	30	20
500	10	10
1000	10	10

KS produces p-values below the threshold (0.5) two times during No Drift phase and also two times above the threshold during the Total Drift phase. Also for KMMD we recognize one value above the threshold for the Total Drift phase in Fig. 5a. On the other hand, Figs. 4b and 5b show that on the same experiments KS and KMMD perform as desired when using sub-sampling with stable p-values above the threshold within the No Drift phase and below the threshold within the Total Drift and Partial Drift phases.

4.1 Analysis with Different Drift Ratios

The results from investigating the drift ratio are summarized in Table 1. We observe that for using 10 samples from the Drug Review dataset for drift testing, the KMMD and KS detectors can detect drifts if there exist at least 90% drifting samples within the 10 test samples. Furthermore, if we increase the sample size up to 100, both the KMMD and KS detectors can detect drifts with less drifting samples down to 20% and 30% drifting samples respectively. Finally, if we increase the sample size to 1000, we observe that the KMMD can fire drift alarms with having 10% drifting samples just like the KS. By doubling the sample size from 500 to 1000 texts, no reduction of the required drift data can be achieved, so it remains at 10%.

4.2 Analysis of Time Performance

In Table 2 we show the average time taken to calculate the distances and the p-values by KMMD and KS using sample sizes of 10, 50, 100, 500, and 1000. From the table, we see that KMMD takes an average time between 0.135 for a batch of 10 samples and 2.715 seconds for a batch of 1000 samples while KS takes between 0.001 for a batch of 10 samples and 0.149 seconds for a batch of 1000 samples.

Table 2 Time Performance for drift tests per sample sizes

Sample size	Time (s)	
	KS	KMMD
10	0.001	0.135
50	0.008	0.191
100	0.013	0.202
500	0.061	0.845
1000	0.149	2.715

4.3 Discussion

By visualizing the results in Figs. 2a and 3a we notice that for the No Drift phase, both KMMD and KS detectors without sub-sampling indicate that the test batches are not drifting from the reference (training) batches. We derive this evidence by finding the p-values being above the predefined threshold and the model's accuracy remaining between 0.9 and 1.0. The opposite holds for the Total Drift and Partial Drift phases. In Figs. 2b and 3b we observe that using multiple sub-samples enables us to calculate p-values that are entirely below the threshold for the Total Drift and the Partial Drift phase and above it for the No Drift phase. In addition, we observe that both KMMD and KS's p-values do not drop below the threshold for the No Drift phase. As a result, we find that using more than one sub-sample from the reference set results in more consistent performance. This remark is more decisive for batch sizes of 10. Figure 5a shows unstable performance with false negatives at drift test indices 1 and 4 and false positives at drift test indices 6 and 8 for KS. Additionally, Fig. 4a shows a false positive at drift test index 6. The fluctuations until index 9 can be resolved by using sub-sampling that also leads to the elimination of the false negatives and false positives for both KS and KMMD and therefore to better overall performance. Furthermore, we can see from Table 1 that the performance of KMMD and KS depends on the sample sizes used for drift detection. Given small sample sizes, a high amount of the input batch has to be drifting to detect this drift. For big sample sizes it is sufficient if only a small amount of the input batch is drifting for detecting the drift. Finally it seems that even for big sample sizes at least 10% of the data has to be drifting to detect the drift. Although this seems as the detector can detect drifts perfectly, it also indicates the possibility of false drift alarms since KMMD and KS fire drift alarms even when no drifts occur for small sample sizes of 10. Finally, we can see from Table 2 that with all selected sample sizes, KS outperforms KMMD in terms of time performance which indicates that the inner method used to calculate the p-value in KS is faster than the bootstrapping method used in KMMD. From the table, we notice that KMMD responds to changes earlier than KS.

The discussion shows that drift detection is feasible for similar use cases in healthcare and highlights levers for adjusting to a specific use case, such as the required amount of drifting data in the production batch or the approach of subsampling for

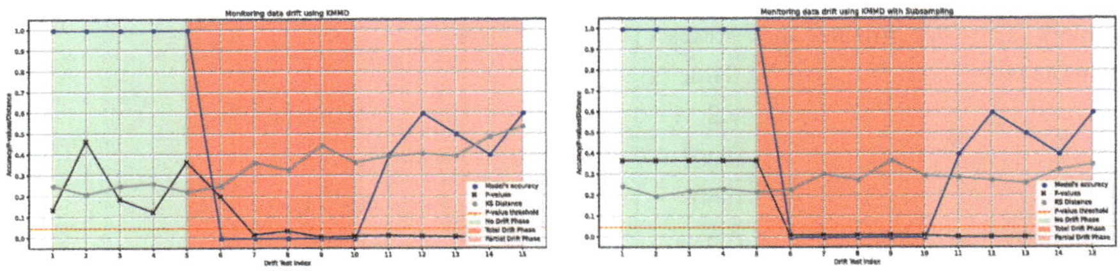

(a) KMMD before sub-sampling. (b) KMMD after sub-sampling.

Fig. 4 Performance of the KMMD drift detectors beside the model before and after sub-sampling (batch size = 10)

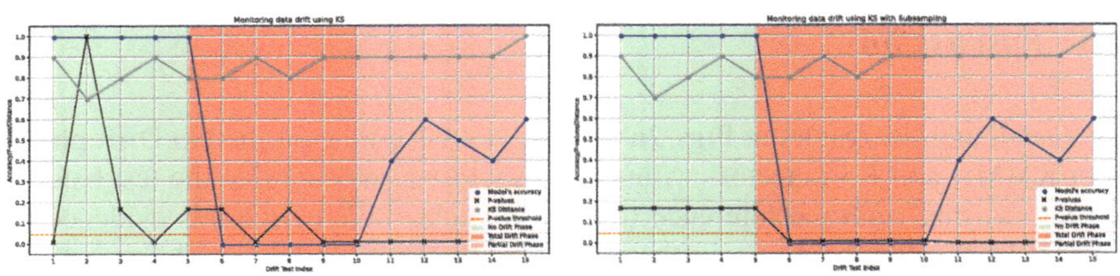

(a) KS before sub-sampling. (b) KS after sub-sampling.

Fig. 5 Performance of the KS drift detectors beside the model before and after sub-sampling (batch size = 10)

more stable performance. It is quite conceivable that this approach can be used to detect new symptoms, conditions, and also other features in text-based healthcare applications during operation.

5 Conclusion and Future Work

In this paper, we examined the ability to monitor drifts for clinical text classification models. In our case study, we used the Drug Review dataset to detect changes in the reviews. Based on the investigation of state-of-the-art techniques, we focused on the distribution-based approaches using statistical hypothesis testing on the features extracted from the input data. By following [21], we showed that we can use Kernel Maximum Mean Discrepancy (KMMD) and Kolmogorov–Smirnov (KS) in different drift experiments involving different types of drifts with different sample sizes of the test data. Finally, we showed that we can implement sub-sampling to cover more reference data from the reviews for better comparison with fewer false alarms.

For the given clinical case study, we learned that both KMMD and KS were able to detect drifts successfully when reviews describing new conditions were used. We have also found that the unstable performance of both detectors can be circumvented by using the sub-sampling method so that false positives and false negatives are mitigated. Furthermore, we found that it is challenging to generalize the minimum amount of drift data needed to guarantee firing a drift alarm. The reason lies within the dependency of the performance on the sample size used to implement drift detection. From the results, we have shown that for larger batch sizes, the required number of drifting text in a batch decreases to detect drifts. We also found that KS works faster than KMMD for the given implementation.

In future work, finding the right sample size to implement accurate drift detection needs to be investigated. Since our results showed that using multiple sub-samples yield better performance than using a single sub-sample, we can improve the use of sub-sampling by determining the optimal number of sub-sample groups to draw from the reference data. Furthermore, an appropriate feature extraction technique should be used—based on the nature of the input data. Therefore, the development of a feature extractor that optimally facilitates drift detection from text reviews is required. In addition, although we used the p-values as an indicator of drifts, we need to investigate how to choose a metric that better describes the severity of the drift, e.g., using the measured distance itself. In addition, we need to investigate approaches to select the reference data from the reference dataset instead of the random selection approach so that the extracted features can facilitate drift detection. For text data, in particular, one can select tokens from which the extracted features help detecting drifts in the incoming data easily. Also we can explore, if the token input length affects the detection results. Finally, we need to take a closer look at how to explain the results of a drift detector, e.g., why drifts occurred or in which parts of the incoming data the drift occurred.

The approach presented can be used to determine drifts regardless of the presence of labels for production data and can be mapped to a variety of similar use cases to enhance healthcare applications by adding a drift detection component for operation.

Acknowledgements This research has been funded by the Federal Ministry of Education and Research of Germany and the state of North-Rhine Westphalia as part of the Lamarr-Institute for Machine Learning and Artificial Intelligence, LAMARR22B. This work was partially done within the SmartHospital. NRW project, funded by the Ministry of Economic Affairs, Industry, Climate Action and Energy of the State of North Rhine-Westphalia, Germany.

References

1. Y. Wang, S. Sohn, S. Liu, F. Shen, L. Wang, E.J. Atkinson, S. Amin, H. Liu, BMC Med. Inf. Decis. Mak. **19**(1), 1 (2019)
2. T. Diethe, T. Borchert, E. Thereska, B. Balle, N. Lawrence, arXiv:1903.05202 (2019)
3. A. Shafaei, M. Schmidt, J.J. Little, arXiv:1809.04729 (2018)
4. J. Klaise, A. Van Looveren, C. Cox, G. Vacanti, A. Coca, arXiv:2007.06299 (2020)
5. A. Gretton, K.M. Borgwardt, M.J. Rasch, B. Schölkopf, A. Smola, J. Mach. Learn. Res. **13**(1), 723 (2012)
6. T. Viehmann, arXiv:2102.08037 (2021)
7. F. Gräßer, S. Kallumadi, H. Malberg, S. Zaunseder, in *Proceedings of the 2018 International Conference on Digital Health* (2018), pp. 121–125

8. H. Abdelwahab, Evaluation of drift detection techniques for automated machine learning pipelines. Master's thesis, Hochschule Bonn-Rhein-Sieg, Grantham-Allee 20, 53757 St. Augustin, Germany (2022). Joined project with Fraunhofer. Alexander Asteroth, Nico Hochgeschwender, and Claudio Martens supervising
9. J.P. Barddal, H.M. Gomes, F. Enembreck, in *2015 IEEE 27th International Conference on Tools with Artificial Intelligence (ICTAI)* (IEEE, 2015), pp. 1053–1060
10. J.G. Moreno-Torres, T. Raeder, R. Alaiz-Rodríguez, N.V. Chawla, F. Herrera, Pattern Recognit. **45**(1), 521 (2012). https://doi.org/10.1016/j.patcog.2011.06.019
11. H. Hu, M. Kantardzic, T.S. Sethi, Wiley interdisciplinary reviews. Data Min. Knowl. Discovery **10**(2), e1327 (2020)
12. M. Baena-Garcıa, J. del Campo-Ávila, R. Fidalgo, A. Bifet, R. Gavalda, R. Morales-Bueno, in *Fourth International Workshop on Knowledge Discovery from Data Streams*, vol. 6 (2006), pp. 77–86
13. M. Hassani, in *ECMS* (2019), pp. 230–239
14. A. Haque, L. Khan, M. Baron, in *Proceedings of the AAAI Conference on Artificial Intelligence*, vol. 30 (2016)
15. J. Zhu, S. Rosset, R. Tibshirani, T.J. Hastie, in *Advances in Neural Information Processing Systems* (Citeseer, 2003), p. None
16. N.M. Razali, Y.B. Wah et al., J. Stat. Model. Anal. **2**(1), 21 (2011)
17. M. Mundt, Y.W. Hong, I. Pliushch, V. Ramesh, arXiv:2009.01797 (2020)
18. G. Wang, A. Ledwoch, R.M. Hasani, R. Grosu, A. Brintrup, Appl. Soft Comput. **85**, 105683 (2019)
19. L. Baier, T. Schlör, J. Schöffer, N. Kühl, arXiv:2107.01873 (2021)
20. L. Wan, M. Zeiler, S. Zhang, Y. Le Cun, R. Fergus, in *International Conference on Machine Learning* (PMLR, 2013), pp. 1058–1066
21. S. Rabanser, S. Günnemann, Z.C. Lipton, arXiv:1810.11953 (2018)
22. E.L. Lehmann, J.P. Romano, G. Casella, *Testing Statistical Hypotheses*, vol. 3 (Springer, 2005)
23. W.J. Faithfull, J.J. Rodríguez, L.I. Kuncheva, Inf. Fusion **45**, 202 (2019)
24. J.L. Hodges, Arkiv för Matematik **3**(5), 469 (1958)
25. T. Viehmann, L. Antiga, D. Cortinovis, L. Lozza, (2020). https://github.com/TorchDrift/TorchDrift
26. S. Targ, D. Almeida, K. Lyman, arXiv:1603.08029 (2016)

Neural Bandits for Data Mining: Searching for Dangerous Polypharmacy

Alexandre Larouche, Audrey Durand, Richard Khoury, and Caroline Sirois

Abstract Polypharmacy, most often defined as the simultaneous consumption of five or more drugs at once, is a prevalent phenomenon in the older population. Some of these polypharmacies, deemed inappropriate, may be associated with adverse health outcomes such as death or hospitalization. Considering the combinatorial nature of the problem as well as the size of claims database and the cost to compute an exact association measure for a given drug combination, it is impossible to investigate every possible combination of drugs. Therefore, we propose to optimize the search for potentially inappropriate polypharmacies (PIPs). To this end, we propose the OptimNeuralTS strategy, based on Neural Thompson Sampling and differential evolution, to efficiently mine claims datasets and build a predictive model of the association between drug combinations and health outcomes. We benchmark our method using two datasets generated by an internally developed simulator of polypharmacy data containing 500 drugs and 100 000 distinct combinations. Empirically, our method can detect up to 72% of PIPs while maintaining an average precision score of 99% using 30 000 time steps.

Keywords Bandit · Neural network · Data mining · Polypharmacy

---

A. Larouche (✉)
Université Laval, 2325 Rue de l'Université, Quebec, QC G1V 0A6, Canada
e-mail: alexandre.larouche.7@ulaval.ca

A. Durand
Université Laval, 2325 Rue de l'Université, Quebec, QC G1V 0A6, Canada
e-mail: audrey.durand@ift.ulaval.ca

R. Khoury
Université Laval, 2325 Rue de l'Université, Quebec, QC G1V 0A6, Canada
e-mail: richard.khoury@ift.ulaval.ca

C. Sirois
Centre d'excellence sur le vieillissement de Québec 1050 Chemin Ste-Foy, Quebec, QC G1S 4L8, Canada
e-mail: caroline.sirois@pha.ulaval.ca

A. Shaban-Nejad et al. (eds.), *Artificial Intelligence for Personalized Medicine*, Studies in Computational Intelligence 1106,
https://doi.org/10.1007/978-3-031-36938-4_5

1 Introduction

Polypharmacy is most often defined as the simultaneous consumption of five or more drugs at once by a patient [18] and is a prevalent phenomenon in the older population. In the USA, 65.1% of older adults experience polypharmacy, with most of them using more than 5 medications at once [26]. In Canada, older adults in long term care facilities use on average 9.9 drug classes, while older adults living outside of these facilities use on average 6.7 drug classes [2]. Some polypharmacies can be dangerous, in the sense that they can lead to to negative health outcomes, like death or hospitalization. Fortunately, screening tools exists to avoid prescription of potentially inappropriate drugs (e.g., opioids and benzodiazepines [21]). These tools essentially consist of finite lists of individual drugs and drug pairs that have been identified as dangerous by experts during pharmaco-epidemiological studies as well as from experience. By definition, this means that potentially dangerous combination resulting from more than two drugs interacting together cannot be identified using the screening tools, in addition to other currently unknown dangerous drug combinations. In order to prevent the prescription of potentially harmful polypharmarcies, it would be important to expand the screening tools until they ideally contain all potentially dangerous combinations. Unfortunately, given all the possible drug combinations as well as their varying effects depending on different patient characteristics, pharmaco-epidemiologists cannot investigate all of them.

The goal of this work is therefore to build a predictive model able to identify drug combinations at risk of being harmful, so that they can be investigated further. We propose to achieve this by leveraging neural networks to predict an association measure to a health outcome given any input describing an arbitrary number of drugs. In practice, such model would be trained using historical data on drugs prescribed to patients, their clinical and sociodemographic characteristics, and their health outcomes. These datasets are typically very large, which makes the association measure expensive to compute, and highly unbalanced.

We therefore tackle the general problem of efficiently mining historical data to train a generalizable and useful model. To achieve this, we formulate the problem under the neural bandit setting so that it can be addressed with the Neural Thompson Sampling (NeuralTS) [27] strategy. However, using this strategy on a very large action space (such as the one considered here) also raises challenges, which we address by combining NeuralTS with differential evolution (DE) [24]. The proposed OptimNeuralTS approach finally results in an ensemble predictor made of the evolving sequence of models trained on all the intermediate data subsets. We evaluate the potential of OptimNeuralTS in simulated experiments. Our results show that our approach can be used to iteratively build an information-rich dataset that can in turn be used for training a predictive model, resulting in an ensemble model capable of extracting new potentially inappropriate polypharmacies (PIPs). We finally provide an overview of related work in machine learning (in general) applied to polypharmacy discovery and bandit strategies applied for data mining. We highlight two contributions:

1. Tackling the problem of efficient creation of information-rich datasets under the contextual bandit setting.
2. Introducing the OptimNeuralTS approach to learn predictive models by mining relevant data from very large unbalanced datasets.

2 Problem Formulation

Let $\mathcal{D}$ denote a historical dataset containing information about drug combinations and health outcomes. Table 1 shows a simple example of such a historical dataset, where each line corresponds to a drug combination identified in a binary format, (i.e. 0/1 indicate whether a drug was taken or not by an individual) along with a binary variable indicating whether the individual developed (or not) a given health outcome while consuming the drug combination. In the targeted application, the measure of association is called the relative risk (RR) and is described as the risks for a given health outcome in the exposed population over the risks in the unexposed population [23]. An exposed patient is a patient which takes a given drug combination. The exposed population is simply the portion of the population which consumes the combination of drugs. Mathematically, given Table 2, the RR is defined as:

$$\mathrm{RR} = \frac{a(c+d)}{c(a+b)}.$$

RR has the advantage of having an implicit threshold: if RR > 1 then a drug combination is associated with a given health outcome, therefore it is potentially

Table 1 Simple example of historical dataset

ID	Drug 1	Drug 2	...	Drug N	Outcome
1	1	0	...	1	0
1	1	0	...	1	1
2	0	1	...	1	1
⋮	⋮	⋮	⋱	⋮	⋮

Table 2 Example of contingency table. An exposed patient is a patient taking a given combination of drugs. A health outcome is computed as a yes if the health outcome occurred during the duration of the prescription of the drug combination (1 in the corresponding line in Table 1)

	Health outcome	
	Yes	No
Exposed	a	b
Not exposed	c	d

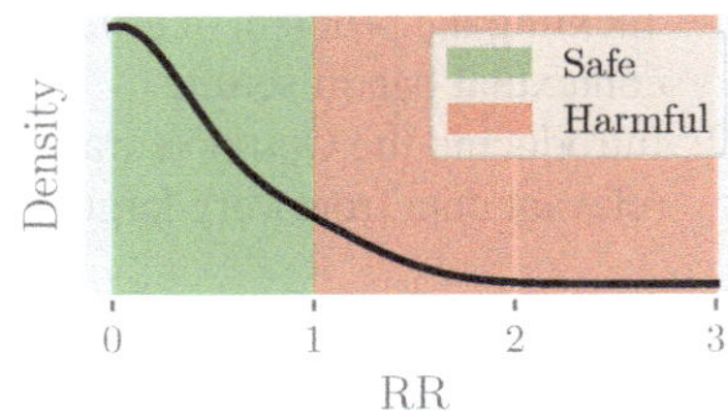

Fig. 1 Example of a typical distribution the estimated RR for historical data. A drug combination with a RR > 1 is considered harmful while if RR < 1, the combination is considered safe

harmful, while RR < 1 implies that a drug combination protects against a given health outcome, therefore it is safe. RR = 1 means that a drug combination is neither protective against nor associated with a given health outcome.

In practice, historical datasets are extracted from medico-administrative databases. Therefore, medication claims data typically contain hundreds of columns related to drug usage and millions of rows enumerating all possible simultaneous drug combinations taken by all patients in a large population, but no precomputed association measure. Unfortunately, computing the association measure for a given drug combination is computationally expensive since it requires enumerating every row in the dataset. Therefore, as the size of the data grows, it becomes harder to compute. This raises a challenge when aiming to train a model capable of predicting an association measure to the considered health outcome for any drug combination provided as input because it is not computationally realistic to compute the RR for the whole dataset $\mathcal{D}$. The training of a predictive model must therefore be performed on a subset of at most T samples from $\mathcal{D}$, where $T \ll |\mathcal{D}|$. This opens the question of how to sample such subsets from $\mathcal{D}$.

In addition, one must note that the historical dataset $\mathcal{D}$ is highly unbalanced. Indeed, prescribers usually prescribe drugs which do not interact together in a harmful way, thus most drug combinations have a low measure of association to health outcomes such as hospitalization and death. Figure 1 displays a simulated but typical distribution of the estimated association measures observed in the real data. Therefore, randomly sampling from $\mathcal{D}$ would yield a training dataset containing mostly combinations of drugs with low measures of association as PIPs associated with adverse health outcomes are rare. However, it is well known that the performance of a predictive model highly depends on the quality of the underlying data [10]. In other words, if the training dataset contains few to no PIPs associated with adverse health outcomes, the predictive model is very unlikely to learn to identify such PIPs. Therefore we need a strategy capable of sampling a training dataset with higher odds of containing PIPs associated with adverse health outcomes.

3 Proposed Approach

We tackle this challenge by formulating the training dataset creation problem as a contextual bandit problem [14], where we leverage the NeuralTS [27] action selection strategy combined with DE [24] to select drug combinations for which to compute the RR.

3.1 Neural Contextual Bandits

A contextual bandit environment is described by a collection of actions $\mathcal{A}$, a feature space $\mathcal{X}$, and an unknown reward function $h : \mathcal{X} \mapsto \mathbb{R}$, such that each action $a \in \mathcal{A}$ is associated with a feature vector, or context, $x_a \in \mathcal{X}$. At each time step $t = 1, 2, \ldots, T$ of the contextual bandit game, the player (agent interacting with the environment) is presented with a subset of actions $\mathcal{A}_t \subset \mathcal{A}$. It then selects an action $a_t \in \mathcal{A}_t$ to play and observes a noisy reward $r_t = h(x_{a_t}) + \xi_t$, where ξ_t is a σ-sub-Gaussian noise (e.g., $\xi_t \sim \mathcal{N}(0, \sigma)$). The goal of an agent playing this game is to maximize the cumulative reward, defined as:

$$\sum_{t=1}^{T} r_t \tag{1}$$

The obvious solution is to simply play the optimal action a_t^* at every round, which has the highest expected reward for this round. That is $a_t^* = \arg\max_{a \in \mathcal{A}_t} h(x_a)$. However, function h is not known a priori, therefore, the agent must learn by trial-and-error in order to improve its behavior over time.

In the training dataset construction problem from historical data, $\mathcal{A}$ corresponds to the set of all possible drug combinations, $\mathcal{A}_t$ corresponds to the set of drug combinations that are available to explore at time t, the features x_a correspond to a multi-hot representation of drug combination a, and the reward function h corresponds to the measure of association between a drug combination and a given health outcome, i.e., the RR. At each time step t, the agent selects a drug combination a_t for which to compute the RR. Since computing the RR on the historical dataset is computationally expensive, it is instead computed on a subset of the data, hence the noisy reward r_t.

Neural bandit strategies, such as NeuralUCB [28] and NeuralTS [27], rely on a neural network $f(\cdot; \theta) : \mathcal{X} \mapsto \mathbb{R}$ to model the reward function h in order to predict the expected reward given any feature $x \in \mathcal{X}$. More importantly, these approaches can estimate the confidence interval around the prediction of the neural network to guide the exploration. They achieve this by using the gradient on the activation, $g(\cdot; \theta) : \mathcal{X} \mapsto \mathbb{R}^{|\theta|}$.

3.1.1 NeuralTS

NeuralTS uses the gradient to estimate the distribution of reward for an action. Indeed, at step t, the parameters of a normal prior $\mathcal{N}(f_t, s_t)$ are estimated as follows:

$$f_t(\cdot) = f(\cdot; \theta_t) \quad \text{and} \tag{2}$$

$$s_t(\cdot) = \sqrt{\lambda g(\cdot; \theta_t)^\top U_t^{-1} g(\cdot; \theta_t)/m}, \tag{3}$$

where $U_t = \lambda \mathbf{I}_m + \sum_{i=1}^{t} g(x_{a_i}, \theta_{i-1}) g(x_{a_i}, \theta_{i-1})^\top$ is a design matrix containing the gradient computed for the inputs (selected actions) up to time t, λ is a regularization parameter and m is the number of parameters in the neural network. NeuralTS selects the action a_t to play by sampling a value from $\mathcal{N}(f_t(x_a), s_t(x_a)), \forall a \in \mathcal{A}_t$ and picking the action with the highest value. After an action a_t is played, the associated context x_{a_t} as well as the observed reward r_t are added to the training dataset $\mathcal{D}_t = \mathcal{D}_{t-1} \cup \{(x_{a_t}, r_t)\}$ used for computing the new network parameters θ_t, before moving on to the next time step.

Since the bandit strategy will seek to play actions which yield high rewards, this should lead to a dataset $\mathcal{D}_T$ containing a reasonable amount of drug combinations with high measures of association to a given health outcome. We therefore hypothesize that such a dataset will make it possible to train a predictive model capable of identifying PIPs with high precision. However, for the bandit strategy to be able to recommend actions with high RR, such actions need to be contained in the set of available actions $\mathcal{A}_t$.

3.2 *Generating Relevant Available Action Sets*

From the neural contextual bandit problem formulation, action a_t is selected from the subset $\mathcal{A}_t \subset \mathcal{A}$ containing all available actions at time t. This is due to the fact that the bandit strategy must consider each action $a \in \mathcal{A}_t$ in order to recommend a_t, and that the complete action set $\mathcal{A}$ is typically too large to be entirely considered at every time step. This is also the case in the considered application due to the combinatorial nature of polypharmacy. For the same reason that the predictive model training dataset cannot be sampled at random from $\mathcal{D}$, we cannot generate $\mathcal{A}_t$ by randomly sampling from $\mathcal{A}$ due to the highly skewed distribution of RRs. We must therefore generate subsets $\mathcal{A}_t$ such that the presence of potentially harmful drug combinations is favored.

To achieve this, we propose to generate subsets of available actions $\mathcal{A}_t$ using differential evolution (DE) [24], an evolutionary optimization algorithm which does not rely on a gradient signal to converge to a solution. The general principle behind DE is to maintain a population and mutate its members, which are feature vectors, according to a strategy. Here, we consider the best/1/bin strategy [24] described in

Algorithm 1 DE best/1/bin

Input: Population size N, crossover rate C, differential weight F, number of optimization steps S, objective function $q(\cdot)$ to maximize
Output: Best member $b_\star$ in the final step

Initialize population $\mathcal{W}$ with N feature vectors sampled from the domain $\mathcal{X}$.
for $s \leftarrow 1 \ldots S$ **do**
  Let $b \leftarrow \arg\max_{w_i \in \mathcal{W}} q(w_i)$
  for $w_i \in \mathcal{W}$ **do**
    Sample $v \sim \mathcal{U}(0, 1)$
    Randomly select indices l, r_1 and r_2 in $[N] - [i]$
    Generate a new feature vector:

$$m_i \leftarrow b + F(w_{r_1} - w_{r_2})$$

    Generate a mutated feature vector where components j are computed as follows:

$$u_{i,j} \leftarrow \begin{cases} m_{i,j} & \textbf{if } j = l \textbf{ or } v \leq C \\ w_{i,j} & \textbf{otherwise} \end{cases}$$

$$w_i \leftarrow \begin{cases} u_i & \textbf{if } q(u_i) \leq q(w_i) \\ w_i & \textbf{otherwise} \end{cases}$$

  end for
end for
$b_\star \leftarrow \arg\max_{w_i \in \mathcal{W}} q(w_i)$
return $b_\star$

Algorithm 1, where the objective function $q : \mathcal{X} \mapsto \mathbb{R}$ corresponds to an action value function sampled from a neural network. The DE optimization process is therefore conducted *on* function q.

DE with best/1/bin therefore corresponds to considering $|\mathcal{A}_t| = N \times S$ available actions at each time step t. Parameters N and S are typically chosen such that $N \times S \ll |\mathcal{A}|$. The best member returned after the S steps of DE corresponds to the action features maximizing q, which is a value function given by the neural network model. Therefore the best member would correspond to a_t. For example, with NeuralTS, $q(\cdot)$ corresponds to a sample from the distribution $\mathcal{N}(f_{t-1}(\cdot), s_{t-1}(\cdot))$ at time step t (see Eqs. 2 and 3). Now, computing r_t on-the-fly on $\mathcal{D}$ requires the selected action to be contained in $\mathcal{D}$. However, DE (best/1/bin) is not constrained to $\mathcal{D}$, so this condition may not be fulfilled. In order to account for this situation, we propose to select the action a_t as being the 1-nearest-neighbor in $\mathcal{D}$ to the action returned by DE (best/1/bin) with ties broken arbitrarily. This ensures that the association measure for the drug combination can be computed.

Algorithm 2 OptimNeuralTS

Input: Dimension of feature vectors d, number of time step T, number of warm-up steps τ, regularization term λ, exploration factor ν, number of training epochs J, learning rate η, DE population size N, DE crossover rate C, DE differential weight F, number of DE steps S, historical dataset $\mathcal{D}$

Output: Generated dataset $\mathcal{D}_T = \{(x_{a_i}, r_i)\}_{i=1}^{T}$ and ensemble neural network models $\{\theta_i\}_{i=1}^{T}$

1: Initialize neural network parameters θ_0
2: $U_0 \leftarrow \lambda I_{|\theta_0|}$ and $\mathcal{D}_0 \leftarrow \{\}$
3: **for** $t \leftarrow 1 \ldots \tau$ **do**
4: Randomly play action $a_t \in \mathcal{D}$, observe r_t
5: $U_t \leftarrow U_{t-1} + g(x_{a_t}; \theta_0) g(x_{a_t}; \theta_0)^\top$
6: $\mathcal{D}_t \leftarrow \mathcal{D}_{t-1} \cup \{(x_{a_t}, r_t)\}$
7: **end for**
8: $\theta_\tau \leftarrow$ Train network with parameters η, J, $\mathcal{D}_\tau$, θ_0
9: **for** $t \leftarrow \tau + 1 \ldots T$ **do**
10: $\hat{a}_t \leftarrow \mathrm{DE}(N, C, F, S, \mathcal{N}(f_{t-1}, \nu s_{t-1}))$
11: $U_t \leftarrow U_{t-1} + g(x_{\hat{a}_t}, \theta_{t-1}) g(x_{\hat{a}_t}, \theta_{t-1})^\top$
12: Set a_t as the 1-nearest-neighbor of $\hat{a}_t$ in $\mathcal{D}$
13: Play a_t, observe r_t
14: $\mathcal{D}_t \leftarrow \mathcal{D}_{t-1} \cup \{(x_{a_t}, r_t)\}$
15: $\theta_t \leftarrow$ Train network with parameters η, J, $\mathcal{D}_t$, θ_{t-1}
16: **end for**
17: **return** $\mathcal{D}_T, \{\theta_i\}_{i=\tau}^{T}$

3.3 OptimNeuralTS

Algorithm 2 describes the resulting OptimNeuralTS. The agent warms up by randomly sampling actions during the first τ steps, observing their rewards and updating the internal parameters U (lines 3–7). The neural network is then trained for the first time on the random data using a standard gradient descent with the L2 regularization scheme of NeuralTS (line 8), before the agent starts playing actions according to the NeuralTS and DE strategy. We slightly abuse the notation when calling DE (line 10) to indicate that the objective function q evaluated at features x consists in a normal distribution centered at $f_{t-1}(x)$ with standard deviation $s_{t-1}(x)$ (see Eqs. 2 and 3). The design matrix U of the agent is then updated (line 11), the agent plays the transformed action a_t, observes the reward r_t, and updates the dataset $\mathcal{D}_t$ before updating the neural network parameters with the same procedure as previously stated (lines 13–15). OptimNeuralTS finally returns the dataset generated by the algorithm as well as the ensemble model corresponding to all the intermediate models $(\theta_\tau \ldots \theta_T)$ encountered along the search (line 17). Indeed, as experiments will show, the subsets of data encountered along the neural contextual bandit game will result in neural network models that are specialized in different relevant regions of PIPs. Combining these models in an ensemble therefore results in a strong predictive model with a good coverage of the input space, which can then be used to predict a RR for any given drug combination. Detecting new PIPs is then only a matter of applying a threshold over the predicted RR's lower confidence bound, which can be computed from $f_t(\cdot)$ and $s_t(\cdot)$.

3.3.1 Warming-Up

Previous work showed that non-informative priors can impact performance [17]. To mitigate this, we allow the agent to warm-up by selecting τ random actions $a \in \mathcal{D}$ and observe their rewards as a way to initialize its belief about the data. This effectively creates a small randomly sampled dataset composed of the seen contexts and the observed rewards. This is helpful, as the data gathered after this point is dependent on the previously gathered data, therefore breaking the i.i.d. assumption of data in supervised learning. This in turn can lead to a failure in learning as an agent without any representation of the relationship in the data may sample it poorly when playing, leading to a poor representation and so on.

3.3.2 Transforming Recommended Actions into Playable Actions

As previously mentioned, our problem requires that we transform $\hat{a}_t$ into a_t. However, U_t is updated with $g(x_{\hat{a}_t}; \theta_{t-1})$ instead of $g(x_{a_t}; \theta_{t-1})$ (line 12). Two facts motivate this choice of update. The first is that if the entirety of $\mathcal{A}$ was present in $\mathcal{D}$, then no transformation would be needed. Indeed, the transformation is only implemented in order to train the neural network on a relationship existing in the historical data and so $\hat{a}_t$ is the recommended action. Secondly, due to the distribution of RRs on our data, the gradient of a_t often contains very little information. This is due to the fact that the RR is concentrated around the mean, which leads to the last bias term of the neural network being almost the only participating term in the prediction. The gradient vector $g(x_{a_t}; \theta_t)$ is then almost barren, which in turn leads to a design matrix containing little information. In practice, updating U_t with $g(x_{a_t}; \theta_t)$ works, but we have found it leads to much less new PIPs detected much later during the bandit algorithm's training.

4 Experiments

Evaluating the proposed approach requires a dataset with a ground-truth and a structure similar to the real world data. As no such dataset is readily available, we first develop a simulator to generate synthetic data.

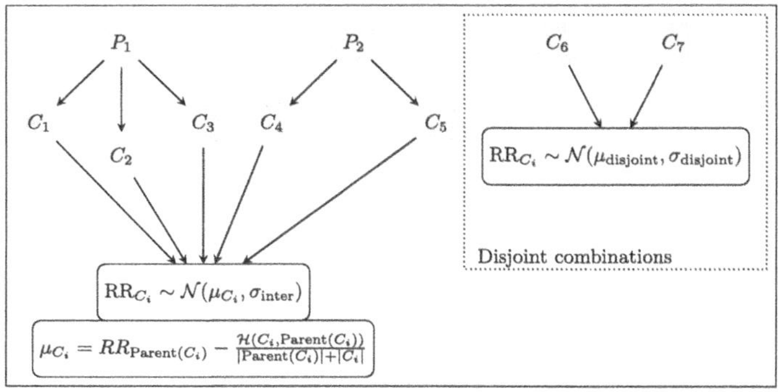

Fig. 2 Overview of the simulated assignment of RR to drug combinations. The RR attributed to a combination C is proportional to its similarity to its nearest dangerous pattern P, shown as the parent in the resulting tree. σ_{inter}, $\sigma_{\mathrm{disjoint}}$ and μ_{disjoint} are user-defined parameters

4.1 Synthetic Data Generation

Our main hypothesis guiding data generation is that a drug combination similar to a drug combination with a high RR should have a similarly high RR. Consequently, we generate by sampling from binomial distributions the set $\mathcal{P}$ of what we call "dangerous patterns", which are characterized by a high RR. Likewise, we randomly generate the set of distinct drug combinations $\mathcal{C}$ without attributing them a RR. The similarity between each drug combination $C \in \mathcal{C}$ and each dangerous pattern $P \in \mathcal{P}$ are then computed using the Hamming distance. Two cases can arise from this: (1) either drug combination C has some drug(s) in common with the nearest pattern or (2) it does not. If it does, then the drug combination is said to *intersect with the pattern* and is attributed a RR proportional to its similarity to its nearest pattern P. Alternatively, if a combination is disjoint of the nearest pattern, then its RR is sampled from a normal distribution $\mathcal{N}(\mu_{\mathrm{disjoint}}, \sigma_{\mathrm{disjoint}})$. This procedure results in a dataset with only distinct combinations with a precomputed RR. The Hamming distance, by definition, favors dangerous patterns containing smaller subsets of drugs during the nearest pattern search. As a result, a combination's nearest dangerous pattern is not always the one with the biggest overlap in terms of drugs. This results in a dataset with a RR not necessarily increasing proportionally to the size of the intersection between combinations and patterns, which adds difficulty to the problem. Figure 2 gives the overview of the RR generation process.

4.2 Experimental Setup

We devise one experiment on two datasets generated by the simulator. The goal of the experiment is to detect drug combinations with an RR above a certain threshold.

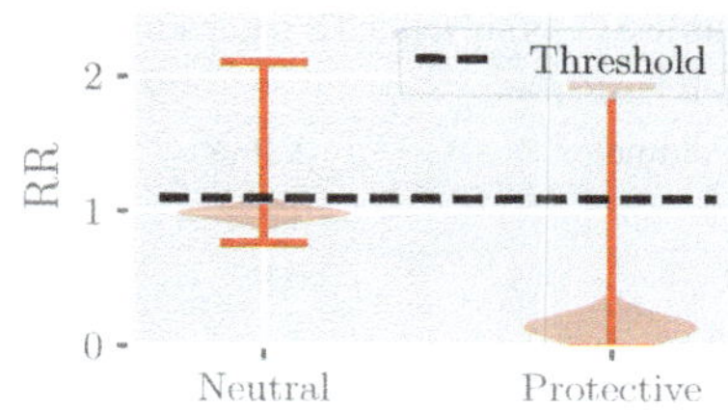

Fig. 3 Distribution of RRs for the neutral and protective datasets

In our work, we consider a threshold of 1.1 for the RR since we consider a RR between 1.0 and 1.1 to be too low to be significant. **The most important aspect here is not to find every PIP but to detect them with as few false positives as possible.** This is crucial, as in practice these findings need to be further studied by healthcare experts and it is laborious to do so. Furthermore, since we plan on using samples of the real dataset in our practical application, we also add a noise term $\xi_t \sim \mathcal{N}(0, 0.1)$ to the observed RR for a drug combination to simulate sample noise. Therefore, to ensure a low false positive rate, a drug combination a is only classified as potentially harmful if $f_t(x_a) - 3s_t(x_a) > 1.1$, to emulate a pessimistic 99% lower confidence bound. The choice of $\sigma = 0.1$ for the normal distribution is so the noise does not dominate the reward signal while still resulting in a challenging instance in the datasets described below.

We generate two datasets each representing a different hypothesis on the effect of the consumption of drugs: a neutral effect instance,where most RRs are near 1, and a protective effect instance (where most RRs are concentrated near 0). The average RR in the latter is well below the dangerous RR threshold, as would be expected in a real dataset. However, to study the robustness of the proposed approach, we also consider a neutral effect dataset where the distribution of RRs is concentrated around the threshold. This leads to a more challenging instance as the noise ξ_t is more likely to make a safe combination appear potentially harmful. Figure 3 shows the distribution of RRs in both datasets. Both datasets contain 100 k distinct combinations of 500 possible drugs, with RRs computed from 10 random dangerous patterns that do not appear in the distinct combinations. The two datasets are highly unbalanced, with the neutral and protective datasets respectively containing 2082 and 7805 distinct dangerous drug combinations (RR > 1.1).

Table 3 shows the OptimNeuralTS parameters used in this experiment. The number of time steps T is chosen such that only a small fraction of the entire drug combination space can be investigated during the bandit game. Indeed, in real life applications, millions of drug combinations are typically available. Setting T to a small number compared to the number of possible combinations thus requires OptimNeuralTS to be efficient in its choice of drugs to investigate at every round in order to succeed. The results shown here are for a warm-up duration of $\tau = 10$ k samples and an exploration factor of $\nu = 1$ taken as the best configuration from a grid search of the space $\tau \in \{1\,\text{k}, 10\,\text{k}, 20\,\text{k}, 30\,\text{k}\}$ and $\nu \in \{1, 10\}$.[1] All the configurations in

[1] The total number of configurations tried is thus 7, as $\tau = 30$ k is the same for any ν.

Table 3 OptimNeuralTS parameter values; left side are specific to the DE component

DE best/1/bin		OptimNeuralTS	
Parameter	Value	Parameter	Value
N	32	d	500
C	0.9	T	30 k
F	1	τ	10 k
S	16	λ	1
		ν	1
		J	100
		η	0.01*

*η is reduced when the loss reaches a plateau during training

the grid search succeed in finding variable amounts of PIPs with high precision, except when the warm-up phase is too long (e.g. $\tau = 30$ k), highlighting the need for a bandit strategy. Furthermore, the considered space for the grid search of ν is never below 1 to discourage greedy action (i.e. exploiting the already known PIPs). Furthermore, we schedule the learning rate $\eta = 0.01$ to decrease as the loss reaches a plateau during training. This parameter was found by a hyperparameter search guided by OpTuna [1] and Tune [15]. As for DE, the parameters were selected manually to maximize the mutation of the population at each optimization step while still maintaining a very small population and a quick runtime.

As previously mentioned, by applying a threshold over the lower bounds on neural network predictions, the regression problem of learning a mapping from drug combinations to a RR can be turned into a binary classification problem where a prediction lower bound over a threshold (e.g. 1.1) corresponds to a PIP, else to a safe drug combination. Therefore, classical classification performance metrics such as precision and recall are used here to evaluate the model trained by OptimNeuralTS. In addition to these two metrics, we also report the ratio of dangerous patterns used to generate the data that were found (Ratio $\mathcal{P}$), as well as the ratio of PIPs detected in $\mathcal{D}$ that are not in $\mathcal{D}_T$ (Ratio $\notin \mathcal{D}_T$). These last two metrics aim to evaluate the generalization to PIPs unseen during training as the dangerous patterns $\mathcal{P}$ are not in $\mathcal{D}$ but are still known to have a high RR. All the results are reported for 25 repetitions of the experiments and every evaluation metric is computed at every 200 time steps of training. Furthermore, in order to quantify the benefits of using an ensemble model we compare the metrics of the ensemble approach to that of the latest trained model (i.e. the single model trained by OptimNeuralTS before the evaluation).

4.2.1 Implementation Details

Since feature vectors are multi-hot vectors, the recommended drug combination ($\hat{a}_t$) is transformed into the most similar one in $\mathcal{D}$ (a_t) using the Hamming distance. Furthermore, unlike the original NeuralTS training routine that trains for a set amount of gradient steps and then returns the last parameters θ computed with the gradient step, our implementation keeps the parameters θ associated with the lowest loss on the training dataset as this maximizes likelihood [9]. As training is time consuming, the neural network is retrained every 10 steps and uses the Adam [12] optimizer due to its faster convergence than regular SGD. Finally, in order to simplify and accelerate computation, we rely on the same tricks as the original NeuralTS implementation [27]: we approximate the matrix U by taking only its diagonal, remove the division by m (see Eq. 3) and only compute the L2 penalty on the current weights θ_t.

5 Results

Figures 4 and 5 respectively display the precision and recall of the ensemble predictive model produced using the OptimNeuralTS training procedure over the growing time horizon. We observe a monotonic improvement of the recall over time in both settings, while managing to keep the precision excellent. The lower recall and slightly lower precision in the neutral instance is expected and is due to the possibility of the observed RR crossing the threshold because of the noise term ξ_t. Indeed, as time progresses, the agent must focus on drug combinations near the RR threshold, thus leading to slightly more false positives.

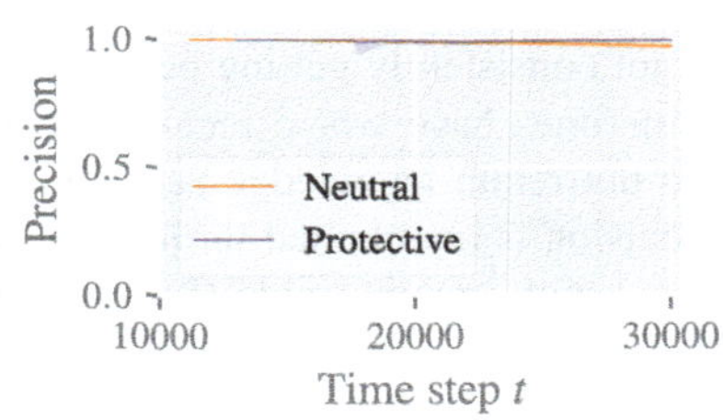

Fig. 4 Precision on the neutral and protective instances using the ensemble predictive model

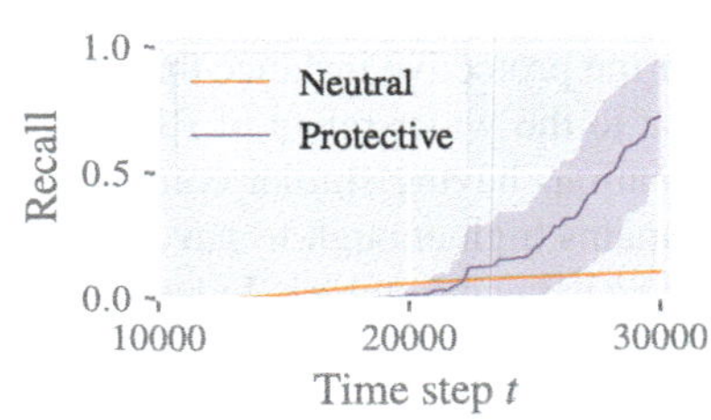

Fig. 5 Recall on the neutral and protective instances using the ensemble predictive model

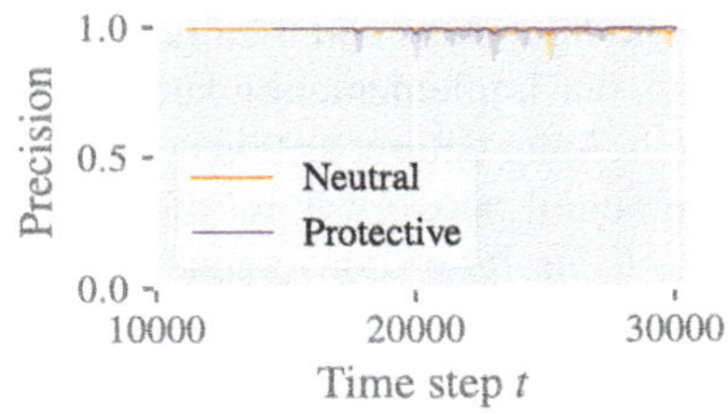

Fig. 6 Precision on the neutral and protective instances using the single latest model

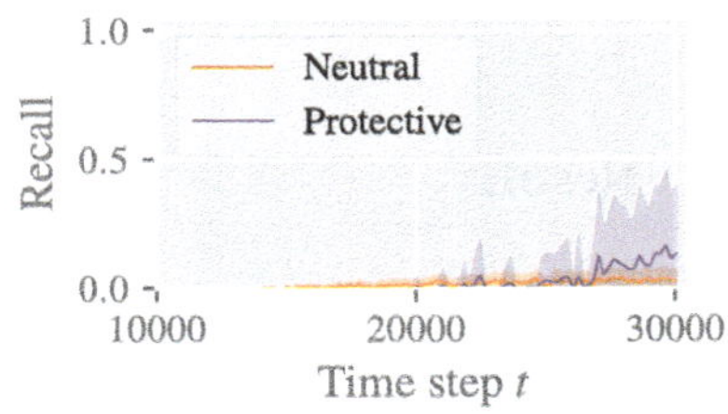

Fig. 7 Recall on the neutral and protective instances using the single latest model

5.1 Impact of Ensemble

To highlight the benefits of using the history of generated models as an ensemble rather than simply relying on the most recent model, Figs. 6 and 7 respectively show the precision and recall obtained at each time step by using only the most recent model instead of the ensemble. We first observe (Fig. 7) that the most recent model is not consistently getting better at detecting PIPs on its own. Indeed, later models sometimes have worse recall than those of earlier iterations, suggesting that they did not retain knowledge acquired earlier. However, we observe (Fig. 6) that high precision is maintained throughout the time steps, although with more noise compared with the ensemble (Fig. 4). That is, every individual neural network trained by OptimNeuralTS has a low false positive rate. The ensemble leverages this fact efficiently by using a single vote to classify a combination as potentially harmful. Moreover, we observe in general that the precision for single models fluctuates more for the protective instance than for the neutral instance. This behavior is most likely due to the wider range of RR of the protective dataset which results in drug combinations having similar components but very different RR. Even so, the precision remains high enough to have very little false positives in practice for both datasets when using ensembles. Indeed, the ensemble approach results in 819 ± 47 correctly detected PIPs for 20 ± 9 false positives on the neutral instance while it yields on the 1504 ± 481 correctly detected PIPs for 1 ± 0 false positives on the protective instance. The bigger fluctuation in the number of true positives for the protective instance can also be attributed to the bigger range of RR to cover.

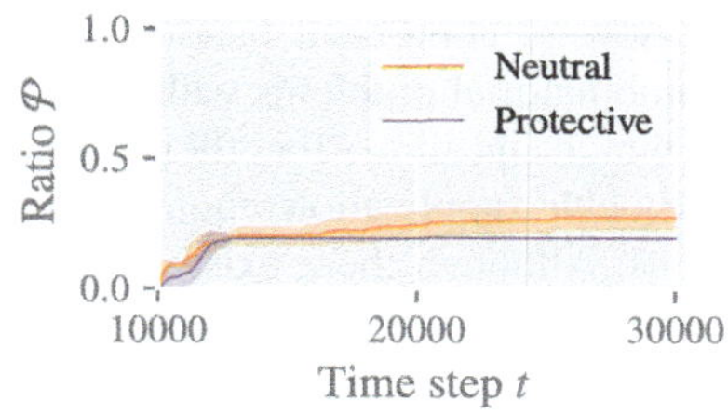

Fig. 8 Ratio $\mathcal{P}$ on the neutral and protective instances using the ensemble predictive model

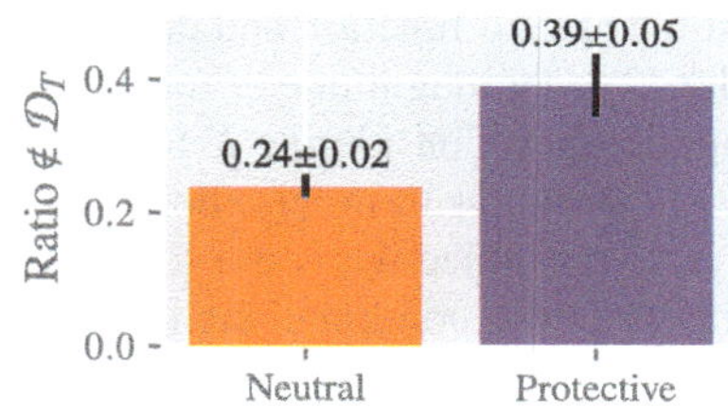

Fig. 9 Ratio $\notin \mathcal{D}_T$ on the neutral and protective instances using the ensemble predictive model

5.2 *Generalization*

Figure 8 shows the ratio of dangerous patterns $\mathcal{P}$ identified as PIPs by the resulting model. As previously mentioned, the dangerous patterns used to generate data are not in $\mathcal{D}$. However, although they can never be observed directly, the ensemble is capable of detecting dangerous patterns. Furthermore, we observe that a good percentage of the detections made by the ensemble are not contained in the training data $\mathcal{D}_T$.

We observe (Fig. 9) that on average up to 39% (on the protective instance) of detected PIPs were not even in $\mathcal{D}_T$ upon the completion of training. This is promising, as it indicates that the resulting ensemble can pick up on observed trends to predict unseen patterns. In practice, this would represent drug combinations that have never been prescribed together before, but whose combination could be dangerous, and so detecting them will prevent health risks for patients.

6 Related Work

There exists prior work on data mining in the polypharmacy context. Methods have been proposed to detect new potentially inappropriate medications or model the association of polypharmacy to side effects using small datasets [11, 25]. General machine learning techniques have also been used to model polypharmacy and its side effects [13, 19, 30] by using complex drug data that are usually not contained in claims database. Therefore, efficiently learning from large datasets such as those considered in the current work require new approaches, hence motivating the proposed bandit angle.

Several contextual bandit strategies have been proposed previously to handle combinatorial problems with linear rewards [3, 7]. The linear reward assumption is common and allows for efficient computations of action recommendations. However, the tackled application requires the estimation of a (possibly) non-linear reward functions. Although there exists non-linear reward combinatorial bandit strategies [6], they rely on oracles to recommend an action to play. Such oracles are typically designed for specific problems and unfortunately, our problem is not one of them. Alternatively, strategies to extract the top-K best actions [22] do not assume linearity of the reward function and do not rely on an oracle. However, they require to preset the K-order magnitude of relevant actions, which is a priori unknown in the current application. Considering K too low would result in missing PIPs, while setting it too high would result in false positives.

As our objective is not to maximize the cumulative rewards (Eq. 1), but rather explore drug combinations in order to detect potentially dangerous ones, the pure exploration setting would also be a natural formulation for the current application. Several combinatorial pure exploration bandit strategies have been proposed [4, 5, 8] previously. However, due to their combinatorial nature, they all exhibit a dependency on an oracle, which makes them then unusable in the tackled problem. Pure exploration neural bandits have also been studied previously [29]. Although theoretically relevant, these methods bear important implementation challenges that prevent them from being used efficiently. The proposed OptimNeuralTS strategy is simpler to implement while still maintaining high precision and a capacity to generalize to unseen data.

Finally, thresholding bandits [16] is another relevant setting where the objective is to extract actions with a mean value estimate over a certain threshold. However, proposed approaches for this setting [16, 20] only maintain mean estimates of every actions encountered during the game and could therefore not lead to a model that can predict the association measure for any new drug combination. Without such a model it becomes impossible to generalize to unseen actions as required by the tackled application.

7 Conclusion

This paper introduces the OptimNeuralTS approach combining NeuralTS [27] and differential evolution [24] to data mine relevant data from very large unbalanced datasets. This method leverages the neural contextual bandit formulation to create an information-rich dataset on which to learn an ensemble predictive model. OptimNeuralTS is a general method for data mining that can be applied to any unlabelled dataset with a combinatorial structure. We conduct experiments using simulated datasets representing both protective and neutral settings. Results show that the predictive model learned with OptimNeuralTS is empirically capable of detecting PIPs with high precision. More importantly, the model is able to identify underlying dangerous patterns that are not observed directly in the data. These encouraging results suggest that OptimNeuralTS is a promising approach for guiding pharmaceutical research by recommending potentially dangerous drug combinations to investigate further and therefore contribute to safer prescriptions.

In future work, one could attempt to improve the sample efficiency using pure exploration neural bandits methods. Furthermore, while we do not take them into account, other important factors other than the presence of drugs (e.g. sex, age, medical conditions) contribute to whether a combination should be considered a PIP. Therefore, our simulation data can still be improved to portray a more complete setting.

Acknowledgements This project is supported by the Canadian Institute of Health Research and the Natural Sciences and Engineering Research Council of Canada, grant number CPG-170621. Caroline Sirois receives a Junior 2 salary award from the Fonds de recherche du Québec-Santé. We would also like to thank CIFAR for the CCAI Chair funding.

References

1. T. Akiba, S. Sano, T. Yanase, T. Ohta, M. Koyama, Optuna: a next-generation hyperparameter optimization framework, in *Proceedings of the 25rd ACM SIGKDD International Conference on Knowledge Discovery and Data Mining* (2019)
2. Canadian Institute for Health Information. Drug use among seniors in Canada, 2016 (2018)
3. N. Cesa-Bianchi, G. Lugosi, Combinatorial bandits. J. Comput. Syst. Sci. **78**(5), 1404–1422 (2012)
4. L. Chen, A. Gupta, J. Li, M. Qiao, R. Wang, Nearly optimal sampling algorithms for combinatorial pure exploration, in *Proceedings of the 2017 Conference on Learning Theory*, ed. by S. Kale, O. Shamir. *Proceedings of Machine Learning Research*, vol. 65, PMLR, pp. 482–534. Accessed 07–10 July 2017
5. S. Chen, T. Lin, I. King, M.R. Lyu, W. Chen, Combinatorial pure exploration of multi-armed bandits. Adv. Neural. Inf. Process. Syst. **27**, 379–387 (2014)
6. W. Chen, Y. Wang, Y. Yuan, Combinatorial multi-armed bandit: general framework and applications, in *International Conference on Machine Learning* (PMLR, 2013), pp. 151–159
7. R. Combes, M.S. Talebi Mazraeh Shahi, A. Proutiere, et al., Combinatorial bandits revisited, in *Advances in Neural Information Processing Systems*, vol. 28 (2015)
8. Y. Du, Y. Kuroki, W. Chen, Combinatorial pure exploration with full-bandit or partial linear feedback, in *Proceedings of the AAAI Conference on Artificial Intelligence*, vol. 35 (2021), pp. 7262–7270
9. I. Goodfellow, Y. Bengio, A. Courville, *Deep Learning* (MIT Press, 2016)
10. V. Gudivada, A. Apon, J. Ding, Data quality considerations for big data and machine learning: going beyond data cleaning and transformations. Int. J. Adv. Softw. **10**(1), 1–20 (2017)
11. F. Held, D.G. Le Couteur, F.M. Blyth, V. Hirani, V. Naganathan, L.M. Waite, M.J. Seibel, D.J. Handelsman, R.G. Cumming, H.G. Allore et al., Polypharmacy in older adults: association rule and frequent-set analysis to evaluate concomitant medication use. Pharmacol. Res. **116**, 39–44 (2017)
12. D.P. Kingma, J. Ba, Adam: a method for stochastic optimization (2014). arXiv:1412.6980
13. A.Lakizadeh, M. Babaei, Detection of polypharmacy side effects by integrating multiple data sources and convolutional neural networks. Mol. Divers. 1–11 (2022)
14. J. Langford, T. Zhang, The epoch-greedy algorithm for multi-armed bandits with side information, in *Advances in Neural Information Processing Systems*, vol. 20 (2007)

15. R. Liaw, E. Liang, R. Nishihara, P. Moritz, J.E. Gonzalez, I. Stoica, Tune: a research platform for distributed model selection and training (2018). arXiv:1807.05118
16. A. Locatelli, M. Gutzeit, A. Carpentier, An optimal algorithm for the thresholding bandit problem, in *Proceedings of The 33rd International Conference on Machine Learning*, ed. by M.F. Balcan, K.Q. Weinberger. *Proceedings of Machine Learning Research*, vol. 48 (PMLR, New York, USA), pp. 1690–1698. Accessed 20–22 June 2016
17. J. Mary, R. Gaudel, P. Philippe, Bandits warm-up cold recommender systems (2014). arXiv:1407.2806
18. N. Masnoon, S. Shakib, L. Kalisch-Ellett, G.E. Caughey, What is polypharmacy? A systematic review of definitions. BMC Geriatr. **17**(1), 1–10 (2017)
19. R. Masumshah, R. Aghdam, C. Eslahchi, A neural network-based method for polypharmacy side effects prediction. BMC Bioinf. **22**(1), 1–17 (2021)
20. S. Mukherjee, K.P. Naveen, N. Sudarsanam, B. Ravindran, Thresholding bandits with augmented ucb (2017). arXiv:1704.02281
21. A.G.S.B.C.U.E. Panel, D.M. Fick, T.P. Semla, M. Steinman, J. Beizer, N. Brandt, R. Dombrowski, C.E. DuBeau, L. Pezzullo, J.J. Epplin, et al., American geriatrics society 2019 updated ags beers criteria® for potentially inappropriate medication use in older adults. J. Amer. Geriatr. Soc. **67**(4), 674–694 (2019)
22. I. Rejwan, Y. Mansour, Top-k combinatorial bandits with full-bandit feedback, in *Algorithmic Learning Theory* (PMLR, 2020), pp. 752–776
23. C.L. Sistrom, C.W. Garvan, Proportions, odds, and risk. Radiology **230**(1), 12–19 (2004)
24. R. Storn, K. Price, Differential evolution-a simple and efficient heuristic for global optimization over continuous spaces. J. Global Optim. **11**(4), 341–359 (1997)
25. R.E. Thomas, L.T. Nguyen, D. Jackson, C. Naugler, Potentially inappropriate prescribing and potential prescribing omissions in 82,935 older hospitalised adults: association with hospital readmission and mortality within six months. Geriatrics **5**(2), 37 (2020)
26. E.H. Young, S. Pan, A.G. Yap, K.R. Reveles, K. Bhakta, Polypharmacy prevalence in older adults seen in united states physician offices from 2009 to 2016. PLoS ONE **16**(8), e0255642 (2021)
27. W. Zhang, D. Zhou, L. Li, Q. Gu, Neural Thompson sampling, in *International Conference on Learning Representation (ICLR)* (2021)
28. D. Zhou, L. Li, Q. Gu, Neural contextual bandits with ucb-based exploration, in *International Conference on Machine Learning* (PMLR, 2020), pp. 11492–11502
29. Y. Zhu, D. Zhou, R. Jiang, Q. Gu, R. Willett, R. Nowak, Pure exploration in kernel and neural bandits. Adv. Neural. Inf. Process. Syst. **34**, 11618–11630 (2021)
30. M. Zitnik, M. Agrawal, J. Leskovec, Modeling polypharmacy side effects with graph convolutional networks. Bioinformatics **34**(13), i457–i466 (2018)

Dynamic Outcomes-Based Clustering of Disease Trajectory in Mechanically Ventilated Patients

Emma Rocheteau, Ioana Bica, Pietro Liò, and Ari Ercole

Abstract The advancement of Electronic Health Records (EHRs) and machine learning have enabled a data-driven and personalised approach to healthcare. One step in this direction is to uncover patient sub-types with similar disease trajectories in a heterogeneous population. This is especially important in the context of mechanical ventilation in intensive care, where mortality is high and there is no consensus on treatment. In this work, we present a new approach to clustering mechanical ventilation episodes, using a multi-task combination of supervised, self-supervised and unsupervised learning techniques. Our dynamic clustering assignment is explicitly guided to reflect the *phenotype*, *trajectory* and *outcomes* of the patient. Experimentation on a real-world dataset is encouraging, and we hope that we could someday translate this into actionable insights in guiding future clinical research.

Keywords Electronic health records · Temporal clustering

1 Introduction and Related Work

Patients on mechanical ventilation are a highly heterogeneous group, with widely differing outcomes. Some have relatively healthy lungs e.g. if they are recovering from surgery on another organ; whereas others have varying degrees of pulmonary failure. Pulmonary failure can be acute e.g. Acute Respiratory Distress Syndrome (ARDS)

E. Rocheteau (✉) · P. Liò · A. Ercole
University of Cambridge, Cambridge, UK
e-mail: ecr38@cam.ac.uk

P. Liò
e-mail: pl219@cam.ac.uk

A. Ercole
e-mail: ae105@cam.ac.uk

I. Bica
University of Oxford, Oxford, UK
e-mail: ioana.bica@eng.ox.ac.uk

A. Shaban-Nejad et al. (eds.), *Artificial Intelligence for Personalized Medicine*,
Studies in Computational Intelligence 1106,
https://doi.org/10.1007/978-3-031-36938-4_6

and deteriorate rapidly, or chronic, typically evolving slowly. Unfortunately, patients on ventilators have high mortality [14, 17] and there is no established consensus on optimal treatment strategies from randomised controlled trials [2]. Therefore, there is *great potential benefit to be gained from phenotype discovery* in order to guide future clinical studies.

To this end, we have developed a dynamic clustering approach for mechanically ventilated patients. Previous work using simple clustering techniques has revealed *actionable* sub-phenotypes by secondary analysis of RCT data. For example, latent trajectory modelling of inflammatory biomarkers has revealed sub-types of ARDS [8]. Clustering of transcriptomic data has revealed patient populations in which steroid therapy may be beneficial in sepsis [1]. Routinely collected data has also been used to find trajectory clusters in sepsis based on physiological parameters [3].

We know that temporal neural network architectures can handle the heterogeneous population in the Intensive Care Unit (ICU), both using supervised [10, 19] and unsupervised [15, 27] approaches. Temporal clustering approaches have been applied successfully to other domains e.g. in Parkinson's [29], diabetes [20] and cystic fibrosis [12] and increasingly in intensive care as discussed above.

We have designed our clusters to share similarities in *phenotype*, *trajectory* and *outcomes*. We generate a cluster for each hour of a patient's stay, meaning that if an event happens which alters the predicted trajectory and outcomes, there will be a shift in the cluster assignment. This is interesting, not only because it can reveal which events are associated with these shifts, but also what might have happened if the ventilation strategy had been different. We hope that our work could someday translate into actionable insights in guiding future clinical research.

2 Methods

Broadly, our strategy was to train a temporal encoder to embed the patient data at every timestep (this is analogous to returning all the hidden states for an LSTM model). We used a mixture of supervised, unsupervised and self-supervised learning to do this (see Sect. 4 below). Once the encoder training was complete, we used an unsupervised method to cluster the embeddings, so that we get a cluster for every timestep in the patient's ventilation episode. The code can be found at: https://github.com/EmmaRocheteau/Mechanical-Ventilation-Clustering.

The data consisted of both timeseries and static features. The supervised tasks included two binary tasks: predicting hospital mortality and the risk of receiving a tracheostomy,[1] and two duration tasks: the remaining length of stay (LoS) from timestep t, and their remaining ventilation duration (VD). This ensured that the patient *outcomes* are stored within the embedding. In addition, we trained a decoder to reconstruct timestep t and the static data. This unsupervised approach encourages

[1] A tracheostomy is a procedure designed for long term mechanical ventilation of a patient.

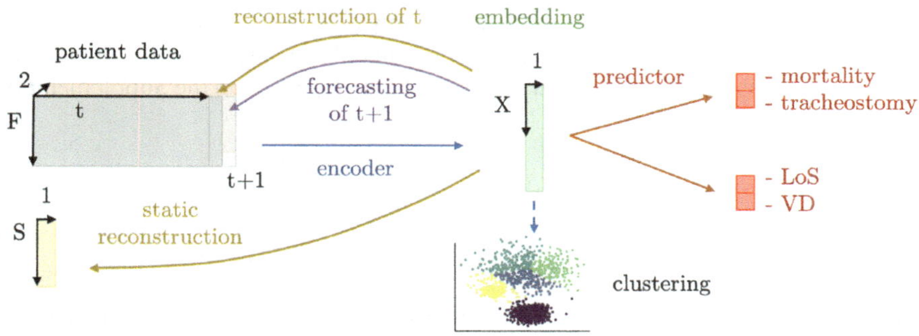

Fig. 1 Overview of our model. Only one timestep, t, is shown for simplicity. F and S are the number of time series and static variables respectively. At timestep t, the static variables (yellow) and preceding time series variables (grey) and their corresponding decay indicator variables (orange, explained under 'Time Series' in the Supplementary Material) are given to the encoder, which produces an embedding (green) for timestep t. This is then given to the decoder networks (yellow), forecasting network (purple) and the predictor network to obtain the four patient outcomes (red). After training is complete, the test embeddings are used for clustering

the embedding to retain the patient *phenotype*. Finally, we predicted timestep $t + 1$, a self-supervised approach designed to embed the patient *trajectory* (Fig. 1).

Encoder In recent years, LSTMs have been by far the most popular model for predicting clinical outcomes and have achieved state-of-the-art results [10, 18, 21, 25]. They have also been applied to other patient prediction tasks e.g. forecasting diagnoses and medications [6, 13], and mortality prediction [5, 10, 22]. More recently, the Transformer model [26] has marginally outperformed the LSTM when predicting LoS [23]. Rocheteau et al. [19] showed that Temporal Pointwise Convolution (TPC) outperformed both the LSTM and Transformer models on mortality and LoS. Therefore, we chose to investigate these three encoders. Details of their implementation are given under 'Additional Implementation Details' in the Supplementary Material.

K-Medoids Clustering We used k-medoids clustering to cluster the learned embeddings. K-medoids is similar to k-means, except that it operates with medoids rather than centroids. This means that the medoids will always be a true observation in the data, while that is not usually the case for centroids. The main advantage is that k-medoids are less sensitive to outliers than k-means, which is more suitable in this context where the data is noisy and heavily skewed.[2]

Both k-means and k-medoids operate on pairwise similarities. We decided to use Euclidean distance rather than cosine similarity. This is because intuitively, it is not only the direction that the patient is moving in that matters, but also the distance along that axis. For example, if a particular 'direction' represents acute decompensated heart failure, we also care how severe the decompensation is.

[2] Preliminary experiments revealed that k-means were more likely to produce small clusters which lay far away from the rest of the data, because it is more affected by outliers. This made the clustering process less reliable and reproducible.

We applied batch normalisation [11] to the embeddings, to ensure that the embedding distribution remained within a reasonable range. The value of k (5 for all models) was chosen using the elbow method (see 'Number of Clusters' in the Supplementary Material).

3 Data

We used the Amsterdam UMC database version 1.0.2 [24], which contains 23,106 ICU admissions from 20,109 patients admitted between 2003 and 2016. We selected all of the mechanical ventilation episodes with a minimum duration of 4 h, capping the maximum duration after 21 days to reduce computational costs. This corresponded to 14,836 episodes which occurred during 13,502 ICU admissions from a cohort of 12,597 unique patients. We selected 31 time series features and 14 static features. The data were split such that 70%, 15% and 15% were used for training, validation and testing respectively. These were split by *patient*, not ventilation episode, to avoid data leakage from the train set. Further details of the data are provided under 'Data Preprocessing' in the Supplementary Material.

4 Prediction Tasks

Remaining Length of Stay and Ventilation Duration We assigned a remaining length of stay (LoS) and remaining ventilation duration (VD) target to each hour of the ventilation episode, ending when the patient dies or is extubated. We only trained on data from the first 21 days of the ventilation episode to protect against batches becoming overly long and slowing down training.

The remaining LoS and VD each have a significant positive skew which makes the duration tasks more challenging. We partly circumvent this by replacing the commonly used mean squared error (MSE) loss with mean squared *log* error (MSLE), as in Rocheteau et al. [19]. We reported on 2 LoS and VD metrics: mean absolute deviation (MAD) and mean squared log error (MSLE). The MAD was used as the primary metric in Harutyunyan et al. [10] but MSLE is arguably the more holistic metric [19].

Mortality and Tracheostomy Unlike the duration tasks, these tasks are static, i.e. the labels do not change during the ventilation episode. Both tasks have significant class imbalance (only 14.6% and 7.4% of patients died or received a tracheostomy respectively). In order to encourage the model to prioritise learning these important outcomes, we applied class weighting to the task. We used binary crossentropy as the loss function. We report the area under the receiver operating characteristic curve (AUROC) and the area under the precision recall curve (AUPRC) as metrics.

Reconstruction and Forecasting As shown in Fig. 1, we use the embedding to reconstruct the timestep t, and forecast one timestep $(t + 1)$ ahead. For the reconstruction of t and forecast of $t + 1$, we apply the mean squared error. We also reconstruct the following static features: sex, urgency of admission, agegroup, weightgroup, and heightgroup. The first two are binary, and so we apply the binary crossentropy loss function. The other three are ordered categorical (as explained under 'Static Features' in the Supplementary Material), therefore we use the mean squared error loss function. Since these tasks are *auxiliary* (we are not interested in the performance as an outcome of the model), we reported their loss function values as 'metrics' since they do not need to be interpretable.

The relative weightings of all of these tasks are given under 'Hyperparameter Search Methodology' in the Supplementary Material.

5 Results

In this section, we highlight important performance differences between the three encoders, analyse an ablation study on the tasks, and provide a detailed analysis of the clusters produced by the TPC model. A deeper evaluation of the results can be found in the discussion.

(a)—Full Task Setting The TPC model performs significantly better than the LSTM and Transformer on the outcome tasks (Table 1a), which is in line with previous findings in MIMIC-IV and eICU [19]. The superiority of the TPC model is also evident in the variational and ablation experiments. Interestingly, the Transformer performs poorly on the binary tasks but better on the duration tasks with respect to the LSTM. Additionally, the LSTM performs the best on the reconstruction and forecasting tasks (Table 2a). Possible reasons for these findings are explored in the discussion.

(b)—Variational Embedding Spaces We experimented with making the embeddings 'variational', by representing the embedding as a set of means and standard deviations to allow sampling of embedding coordinates. The rationale was that by forcing the embedding space to be smoother, we might improve the quality of the clustering as the distances between patients in the embedding space become more reliable. However, this was found to universally hurt performance (Tables 1b and 2b) and it produced clusters which were more homogeneous in terms of outcomes and features, which was counter to the aim of producing clinically distinct clusters.

Ablation Study We performed an ablation study on the tasks used to train the representation space. The results are shown in Table 3. Firstly, we see that the best results for all tasks (except for the duration tasks) are achieved in the full multi-task setting. Not a single metric improves in the other ablation settings, and yet at least one metric showed a deterioration in performance (the exception in task setting (g) is discussed below). Overall this indicates that having multiple competing learning objectives has a stabilising effect on learning the representation.

Table 1 Encoder performance on the prediction tasks averaged over 5 independent training runs. The error margins are 95% confidence intervals. For mortality and tracheostomy, higher AUROC and AUPRC is better; for LoS and VD, lower MAD and MSLE is better. (a) shows the full multi-task setting as shown in Fig. 1, (b) is a variational alternative to the full task setting. Statistically significant differences are indicated by daggers († = $p < 0.05$, ‡ = $p < 0.001$). If the result is significantly better than the comparison models*, it is highlighted in blue, if it is significantly worse it is highlighted in pink. *In (a) the statistical testing compares the three model types, in (b) each model type is compared to its corresponding 'non-variational' model in table (a)

	Model	In-Hospital mortality		Tracheostomy		Length of stay		Vent. duration	
		AUROC	AUPRC	AUROC	AUPRC	MAD	MSLE	MAD	MSLE
	TPC	$0.833\pm0.010^{\dagger}$	$0.644\pm0.013^{\ddagger}$	$0.804\pm0.007^{\ddagger}$	$0.507\pm0.020^{\dagger}$	$7.20\pm0.13^{\ddagger}$	$0.359\pm0.010^{\ddagger}$	$3.24\pm0.07^{\ddagger}$	$0.210\pm0.008^{\ddagger}$
(a)	Transformer	0.697 ± 0.012	0.434 ± 0.019	0.760 ± 0.012	0.419 ± 0.033	8.46 ± 0.07	0.495 ± 0.007	3.95 ± 0.20	0.256 ± 0.016
	LSTM	0.823 ± 0.002	0.608 ± 0.008	0.774 ± 0.002	0.473 ± 0.015	9.16 ± 0.06	0.663 ± 0.008	5.57 ± 0.04	0.681 ± 0.011
	TPC	$0.807\pm0.006^{\ddagger}$	$0.584\pm0.014^{\ddagger}$	$0.775\pm0.008^{\ddagger}$	$0.437\pm0.012^{\ddagger}$	$9.06\pm0.10^{\ddagger}$	$0.555\pm0.018^{\ddagger}$	$4.42\pm0.03^{\ddagger}$	$0.347\pm0.006^{\ddagger}$
(b)	Transformer	$0.660\pm0.023^{\dagger}$	$0.373\pm0.039^{\dagger}$	$0.714\pm0.020^{\ddagger}$	$0.353\pm0.018^{\dagger}$	$9.42\pm0.27^{\ddagger}$	$0.623\pm0.020^{\ddagger}$	$4.63\pm0.27^{\ddagger}$	$0.359\pm0.030^{\ddagger}$
	LSTM	$0.803\pm0.004^{\ddagger}$	$0.555\pm0.006^{\ddagger}$	$0.748\pm0.005^{\ddagger}$	$0.411\pm0.010^{\ddagger}$	$10.2\pm0.1^{\ddagger}$	$0.813\pm0.016^{\ddagger}$	$5.95\pm0.04^{\ddagger}$	$0.775\pm0.007^{\ddagger}$

Table 2 Losses for the reconstruction tasks and forecasting task averaged over 5 independent training runs. The error margins are 95% confidence intervals. See Sect. 4 for explanations of the losses shown. The meaning of (a), (b), the colour scheme and statistical tests are defined in the legend to Table 1

	Model	Reconstruction tasks			Forecasting
		Last timestep	Static (Binary)	Static (Other)	
(a)	TPC	0.334±0.004	0.013±0.000	0.210±0.038	0.334±0.005
	Transformer	0.351±0.005	0.013±0.000	0.354±0.005	0.347±0.001
	LSTM	**0.297±0.006‡**	**0.012±0.001†**	**0.078±0.010‡**	**0.299±0.004‡**
(b)	TPC	0.345±0.002‡	0.013±0.000	0.332±0.006‡	0.345±0.003‡
	Transformer	0.355±0.006	0.013±0.000†	0.356±0.001	0.353±0.005†
	LSTM	0.322±0.003‡	0.012±0.000	0.266±0.004‡	0.323±0.003‡

(c)—No Forecasting Experiment (c) included all the tasks except forecasting one timestep ahead. When we compare experiment (c) to (a), we see that the results are mostly similar, but there is a consistent decrease in performance, which is statistically significant at the $p<0.05$ level on the tracheostomy task (AUPRC in the TPC model and AUROC in the Transformer model). On the reconstruction task, again the performance is similar but statistically worse in the last timestep reconstruction in the LSTM model. This means that the forecasting task is contributing slightly to the performance in (a), but the benefit is small.

(d)—No Reconstruction Experiment (d) removes both the timestep t reconstruction and the static data reconstruction tasks, but keeps the forecasting task. The effect size is larger than in (c), but again is only statistically significant on the tracheostomy task. The forecasting task performs significantly worse in the Transformer and LSTM models without the reconstruction.

(e)—Prediction Tasks Only Experiment (e) includes the binary and duration prediction tasks, but no reconstruction or forecasting. The performance again deteriorates, particularly on the tracheostomy task, we also start to see a more noticeable deterioration in the duration tasks, although this is not yet statistically significant.

(f)—Binary Tasks Only Experiment (f) also shows worsening performance as tasks are removed. This means that the mortality and tracheostomy tasks consistently benefit from supplementary tasks which help to distinguish signal from noise.

(g)—Duration Tasks Only Experiment (g) shows unexpected results; all of the models return better results when only predicting LoS and VD. This is not what has been observed previously in multitask settings ([10, 19]). This is explored further in the discussion.

However, overall the trend is such that the more tasks that are included, the better the average results across tasks.

Cluster Analysis As the best performing encoder, we have focused on analysing the clusters produced by the TPC model. In order to analyse the average differences between the patients in each cluster, it was necessary to flatten the clustering into one 'primary' cluster per patient. This was to prevent confusion, since patients can enter

Table 3 Prediction task results for the task ablation study. The full task setting from Table 1a has been repeated for ease of comparison. Various task ablations are compared to (a): (c) includes all tasks except for the forecasting task, (d) includes all tasks except for the reconstruction tasks, (e) includes only the prediction tasks, (f) is only the binary tasks, and (g) is only the duration tasks. The colour scheme, metrics and statistical test comparisons are explained in the legend to Table 1

	Model	In-Hospital mortality		Tracheostomy		Length of stay		Vent. duration	
		AUROC	AUPRC	AUROC	AUPRC	MAD	MSLE	MAD	MSLE
	TPC	0.833±0.010	0.644±0.013	0.804±0.007	0.507±0.020	7.20±0.13	0.359±0.010	3.24±0.07	0.210±0.008
(a)	Transformer	0.697±0.012	0.434±0.019	0.760±0.012	0.419±0.033	8.46±0.07	0.495±0.007	3.95±0.20	0.256±0.016
	LSTM	0.823±0.002	0.608±0.008	0.774±0.002	0.473±0.015	9.16±0.06	0.663±0.008	5.57±0.04	0.681±0.011
	TPC	0.831±0.006	0.645±0.009	0.796±0.006	**0.499±0.016**†	7.24±0.12	0.360±0.005	3.26±0.07	0.210±0.004
(c)	Transformer	0.675±0.052	0.399±0.079	**0.743±0.011**†	0.406±0.022	8.44±0.29	0.492±0.024	3.95±0.25	0.251±0.026
	LSTM	0.820±0.003	0.608±0.003	0.773±0.005	0.473±0.014	9.16±0.04	0.663±0.005	5.60±0.05	0.685±0.010
	TPC	0.832±0.005	0.645±0.016	0.796±0.007	**0.483±0.020**†	7.28±0.09	0.362±0.007	3.29±0.06	0.213±0.002
(d)	Transformer	0.698±0.017	0.431±0.041	**0.743±0.008**†	0.391±0.008	8.44±0.23	0.492±0.019	3.91±0.39	0.253±0.033
	LSTM	0.820±0.003	0.608±0.007	0.773±0.002	0.464±0.011	9.19±0.04	0.669±0.006	5.59±0.03	0.688±0.010
	TPC	0.828±0.004	0.643±0.010	0.798±0.005	**0.480±0.020**†	7.38±0.20	0.367±0.020	3.24±0.07	0.212±0.012
(e)	Transformer	**0.676±0.019**†	0.410±0.034	**0.736±0.021**†	0.383±0.026	8.67±0.27	0.509±0.024	4.12±0.22	0.268±0.017
	LSTM	0.819±0.005	0.604±0.013	0.773±0.002	0.475±0.008	9.20±0.04	0.669±0.008	5.61±0.04	0.691±0.012
	TPC	**0.823±0.006**†	**0.626±0.014**†	**0.793±0.002**†	**0.477±0.017**†	–	–	–	–
(f)	Transformer	0.669±0.036	**0.373±0.048**†	**0.737±0.021**†	0.400±0.038	–	–	–	–
	LSTM	**0.817±0.003**†	**0.597±0.007**†	**0.767±0.003**‡	0.458±0.016	–	–	–	–
	TPC	–	–	–	–	**6.99±0.10**†	**0.341±0.007**†	**3.08±0.09**†	**0.180±0.004**‡
(g)	Transformer	–	–	–	–	**8.18±0.12**‡	**0.472±0.012**†	**3.68±0.18**†	**0.224±0.009**†
	LSTM	–	–	–	–	**9.05±0.05**†	**0.644±0.006**‡	5.55±0.01	**0.668±0.003**†

Table 4 Average outcomes by cluster ± 95% confidence intervals for the TPC model. Each patient has been classified into a primary cluster, which is the cluster that they spent the majority of their time in. LoS and VD are shown in days

Cluster	Patients	Mortality (%)	Tracheostomy (%)	Length of stay	Vent. duration
1	232	72.0±5.8	1.3±1.5	3.8±0.8	2.4±0.3
2	133	34.6±8.2	38.3±8.4	30.0±3.6	21.4±2.2
3	1,292	1.9±0.7	1.5±0.7	2.8±0.3	0.7±0.0
4	347	4.0±2.1	31.1±4.9	22.0±1.8	7.4±0.9
5	227	26.0±5.7	8.4±3.6	13.0±1.6	7.2±0.9

Table 5 Key features averaged by cluster ± 95% confidence intervals. 'Urgency' is a flag given to the patient at admission. Mandatory Ventilation (MV) settings are provided in the Supplementary Material. The peak inspiratory pressure, P/F Ratio and PEEP are expressed in mmHg. A normal P/F ratio at sea level is ≈400–500 mmHg; whereas 200–300 mmHg is consistent with mild ARDS [9]. Lung compliance is expressed in ml/cmH_2O (normal for a mechanically ventilated patient is 50–100 ml/cmH_2O)

Cluster	Age 70+ (%)	Sex (% male)	Urgency (%)	MV (%)	Peak Insp. pressure	Lung Comp.	P/F Ratio	PEEP
1	52.2±6.5	59.7±6.3	63.4±6.3	68.3±0.8	25.3±0.2	32.7±0.5	217±2	10.09±0.07
2	54.1±8.6	65.8±8.1	39.1±8.4	43.2±0.4	23.2±0.1	36.8±0.3	220±1	9.97±0.03
3	39.8±2.7	69.7±2.5	14.9±1.9	38.6±0.6	16.1±0.1	58.8±0.7	260±1	6.78±0.03
4	25.9±4.6	68.4±4.9	41.5±5.3	22.1±0.4	17.8±0.1	57.5±0.4	237±1	8.19±0.28
5	40.1±6.4	69.6±5.9	43.2±6.5	41.8±0.5	20.3±0.1	47.1±0.4	243±1	8.83±0.38

multiple clusters during their ICU stay (sometimes only for one or two timepoints), and this is disproportionately true of the long stay patients. The cluster in which each patient spent the majority of their time in was assigned its primary cluster. If there were multiple modes, then the mode experienced later in the sequence was chosen. The next two sections characterise the behaviour of the primary clusters. Subsequently, we analyse the dynamic aspects of the clustering from multiple different perspectives.

Differences in Phenotype and Outcomes Table 4 shows the mean outcomes for each cluster. We also analysed some key features in the original data, to visualise differences in patient *phenotype* that the model identified. The average values of key features in patients divided by primary cluster are shown in Table 5. Broadly we can say that:

- Cluster 1 contains the sickest patients, with an average mortality of 72.0%. They are short stay patients with low rates of tracheostomy as most do not survive or stay long enough to require complex respiratory weaning. Table 5 shows they are primarily ventilated with 'mandatory' ventilation settings, meaning the machine is breathing for the patient. Furthermore, they have evidence of mechanical and functional damage to the lung parenchyma. This is in keeping with severe respiratory distress. We could describe this phenotype as a '*early, life-threatening pulmonary injury*' patient group.

- Cluster 2 display substantial mortality and severe pulmonary dysfunction like cluster 1. However this phenotype is characterised by very long LoS and VD, with consequent high rates of tracheostomy: this represents patients who are difficult to wean from mechanical ventilation. This might be described as a '*pulmonary critical illness*' phenotype.
- Cluster 3 have the best outcomes, with short LoS and low mortality. They are extubated without tracheostomy. This appears to be a '*short stay*' phenotype who require a brief period of organ support, perhaps after significant surgery.
- Cluster 4 have relatively low mortality but high rates of tracheostomy. Table 5 shows modest levels of respiratory failure and good lung compliance. Thus, whilst these patients are difficult to wean from mechanical ventilation (like cluster 2), this is due to factors that are not primarily related to pulmonary pathology. We could therefore describe them as a '*general critical illness*' phenotype.
- Cluster 5 shows a moderate to severe group, who are not as acutely unwell as cluster 1, but are still high-risk. From Table 5 we see that pulmonary injury is not a prominent feature so we could characterise these patients as '*early, life-threatening non-pulmonary injury*' patients.

Overall, the findings from Tables 4 and 5 show that there are clinically meaningful differences between the clusters. These can be visualised in Fig. 2.

Medoid Analysis The medoids produced by the clustering algorithm are shown in Fig. 3 and give a description of a representative patient in each cluster. Note that each medoid corresponds to a specific time-point in their ventilation episode.

- The medoid patient for cluster 1 (female, age 60–69) died 4 h after the episode shown without a tracheostomy. Infection (high WBC) and pulmonary dysfunction are particularly noteworthy.
- The typical medoid patient representing cluster 2 (male, 80+ years old) received a tracheostomy 19 days after the episode shown, and was discharged at 23 days. This patient required late as well as early mandatory ventilation suggesting possible infectious complications (his CRP is also high).
- The medoid patient in cluster 3 (female, age 60–69) was discharged from hospital the day after her brief window of ventilation. She does not display substantial physiological derangement.
- The patient in cluster 4 (female, age 60–69) received a tracheostomy 3 days after the sequence shown. Her lung compliance and P/F ratio are both high, indicating good lung function. Therefore, we can conclude that she needed a tracheostomy for reasons other than lung injury.
- Lastly, the patient in cluster 5 (female, 80+ years old) stayed for 9 further days in hospital before being discharged. The short duration of ventilation and relatively normal pulmonary physiology is consistent with a non-pulmonary phenotype.

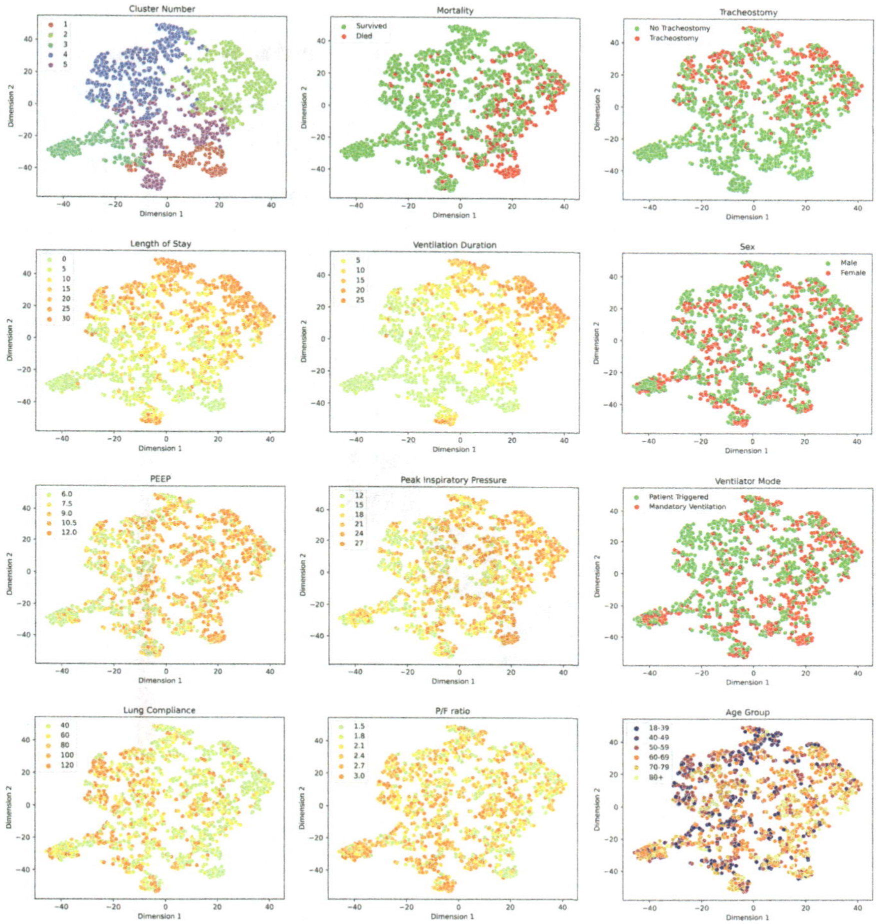

Fig. 2 t-SNE plots for the embeddings produced by the TPC model. For these figures, 1500 random samples were selected from the test set and projected. In each plot, a different attribute is highlighted

Temporal Analysis Broadly, there are two perspectives when evaluating the dynamic aspects of this clustering.

One is the 'Markovian' perspective, where we can examine the transition function between clusters. This is shown in Fig. 4. Unsurprisingly, this reveals that the patient is always most likely to remain in the same cluster. However the most common inter-cluster transitions are from cluster 5 to cluster 4, and cluster 1 to cluster 5. Note that these clusters are next to one another and share lengthy borders in Fig. 2. Most of the patients who transition to 'Died' come from cluster 1, and most of the 'Discharged' patients come from cluster 3.

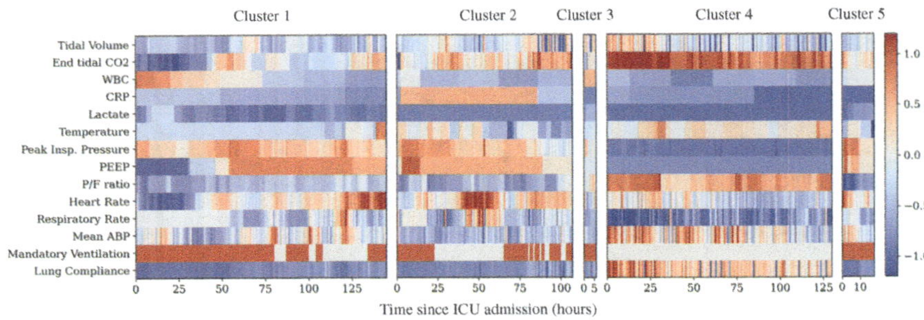

Fig. 3 Raw data from each of the medoids. The data have been standardised around the mean value for each feature. Red means the value is high and blue means low. We can see that each medoid largely follows the average pattern for the cluster shown in Table 5. WBC is white blood count, CRP is C Reactive Protein, ABP is arterial blood pressure

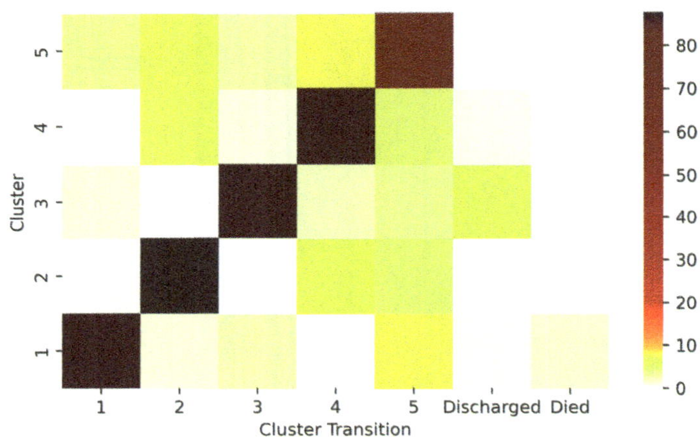

Fig. 4 A transition matrix for the TPC model, showing the probability of entering each cluster at time $t + 1$, plus the categories 'discharged' or 'died', given their cluster at time t

The other perspective is to look at the number of patients in each cluster at different time points after admission, and observe the transitions between them (Fig. 5). Transitions from cluster 3 to 'extubated' are very common within the first day, but then they almost disappear by 3 days. This cannot be seen with the Markovian perspective in Fig. 4. Cluster 2 contains patients with the longest ventilation episodes, which can be seen by its low rate of attrition over time.

Number of Clusters per Patient Figure 6 shows that most patients remain in only one cluster during their ventilation episode. However, when the distribution is broken down by primary cluster, we can see that this is heavily driven by the behaviour of cluster 3 patients, which tend to remain in cluster 3 for their entire ventilation duration (note that they tend to have short VDs so this is not so surprising). In contrast, clusters 2 and 5 most commonly appear alongside other clusters during a single ventilation

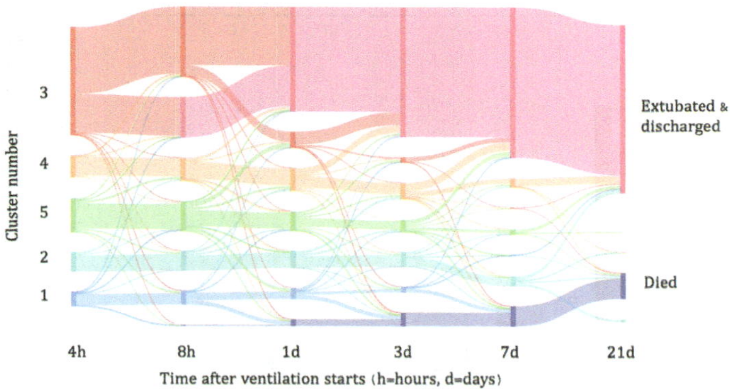

Fig. 5 A sankey plot showing the evolution of the clustering across time. We begin at 4 h to allow the clustering to stabilise at the start of the time series. At 21 days there are still some patients without a final outcome (mostly from cluster 2) but this is because they are ventilated for longer than 21 days and have been right censored

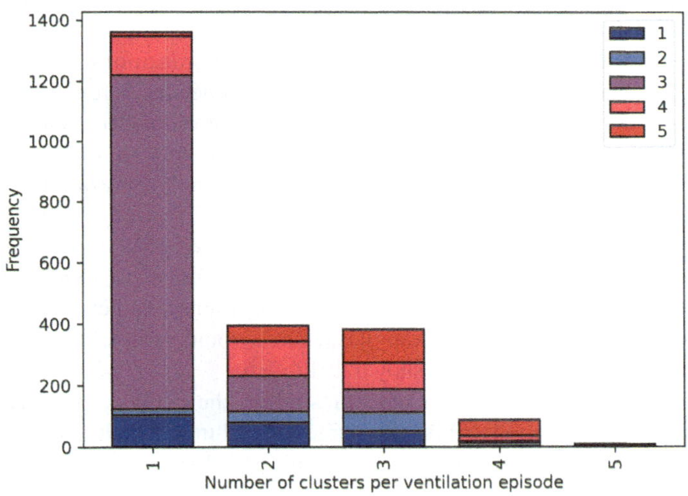

Fig. 6 Distribution of the number of clusters that the patient enters during their ventilation episode, separated by *primary* cluster (shown by the colour key). For example, cluster 3 (purple) mainly appears on its own i.e. the patient starts the episode in cluster 3 and remains in cluster 3 for the whole duration, whereas cluster 5 (red) rarely appears on its own

episode. This means that for most episodes attributed to cluster 2 or 5, there are transitions either into or out of these clusters. These are explored next.

Cluster Transitions The clusters produced by the TPC model are remarkably stable over time, given that there is no explicit loss incentive to constrain the representation to behave in this way. Figure 7 shows the distribution of timepoints that the patients first enter their primary cluster. Clusters 2 and 3 are particularly likely to accurately

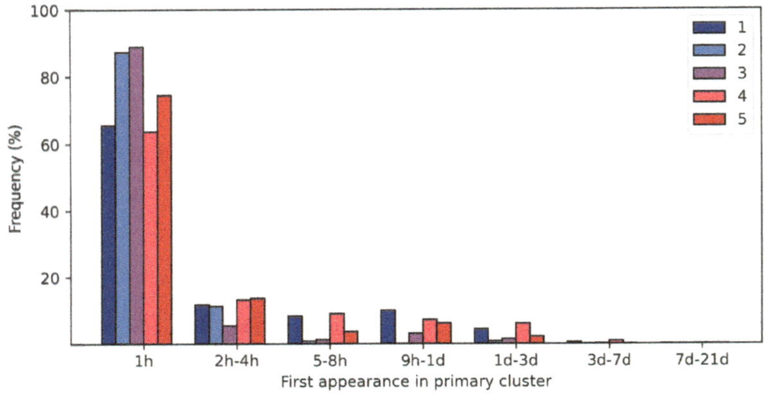

Fig. 7 Percentage of patients who have entered their primary cluster against time

assigned during the first hour of ventilation (87 and 89% respectively), while cluster 4 is the least likely to be identified early (64%).

Next, we investigated what we will refer to as 'stable' transitions between clusters. In order to be characterised as stable, the origin cluster needed to remain stable in the 5 h preceding the transition, and the patient was not permitted to re-enter the origin cluster for 5 h following the transition. This was primarily to screen out patients who were at the boundary between two clusters, continually crossing back and forth but not representing a true transition from one to the other. Before screening, there were 22,036 cluster transitions, corresponding to 870 separate ventilation episodes (39% of the total in the test set). Of these transitions, only 291 represented stable movement between clusters. We further removed any transitions between two clusters that had fewer than 15 transition examples, as this would be insufficient to analyse. The remaining 230 transitions are shown in Table 6.

Firstly, it is noteworthy that the outcomes reflect the *destination* cluster, not the origin cluster. The exception to this is the 'urgency' column, which is not an outcome, but a label assigned at *admission* and hence is more likely to reflect the *origin* cluster.

Cluster 5 stands out as being disproportionately involved in inter-cluster transitions. Of these, the most common is 5→3, which occurs when the model overestimates the risk to the patient early on in the ventilation episode. Not shown in Table 6, is that the average predicted risk of death drops from 56.4% 5 h prior to the transition, to 41.7% at the point of transition. There is also a corresponding reduction in tracheostomy risk (–13%), LoS (–17.1% after adjustment[3]) and VD (–26.4% after adjustment) as predicted by the model, and dramatic improvements in physiological parameters such as lung compliance (+35%) and P/F ratio (+15%).

Another interesting transition is 3→1, which happens when the model initially believes the patient to be relatively healthy, but then quickly re-adjusts to predict

[3] There is a 5 h gap between these predictions, therefore this time difference needs to be removed from the first prediction.

Table 6 Stable cluster transitions (origin cluster → destination cluster) with a count of ≥15, sorted by destination cluster. The median rather than the mean time is displayed to show a more representative time of transition (as there is positive skew)

Transition	Count	Median time	Mortality (%)	Tracheostomy (%)	Urgency (%)	VD	LoS
3→1	17	3	76.5	0.0	47.1	0.5	0.7
5→1	29	16	51.7	10.3	55.2	4.3	5.3
1→3	28	11	10.7	0.0	67.9	1.0	2.6
5→3	46	9	15.2	4.3	41.3	1.2	6.5
2→4	28	17	10.7	21.4	42.9	6.2	12.8
5→4	27	10	11.1	7.4	48.1	3.4	9.1
1→5	25	3	44.0	4.0	68.0	3.9	6.5
3→5	15	4	13.3	13.3	53.3	1.9	4.6
4→5	15	56	26.7	26.7	46.7	6.6	11.5

poor outcomes. Looking in more detail at the raw data, we discovered that these patients are younger (only 23.5% are 70+), which could explain why the model was initially optimistic and why the deterioration is so rapid.[4] We also observed a deterioration in the lung compliance (–26.3%) and P/F ratios (–12.8%), and a change in the ventilator settings—namely higher PEEP and peak inspiratory pressure and lower tidal volumes—reflecting a drop in lung compliance of the patients. Most of these patients died within 12 h of the transition to cluster 1.

Reliability We investigated the reproducibility of these phenotypes. We chose to analyse the clusters in the following settings: (i) alternative encoder models, (ii) retraining the TPC model with different random seeds and (iii) varying the value of k. The clusters were found to be surprisingly stable, with key features of the extracted phenotypes remaining similar between models. With increasing value of k, we noticed that rather than completely rearranging the position of the clusters, increasing k progressively subdivides existing clusters, hinting that the clusters are hierarchically organised. The full analysis is included in the Supplementary Material.

6 Discussion

We evaluated the use of TPC model, trained using supervised, unsupervised and self-supervised learning techniques, for the purposes of *phenotype* discovery in mechanically ventilated patients. We discuss the most important findings in turn.

Firstly, we reaffirmed that the TPC model performs better than alternative encoders on EHR data for patient outcome prediction.

---

[4] This is because younger patients can mask a problem by compensating deceptively well, until they reach a point where the homeostatic mechanisms can no longer cope.

Secondly, we found that the Transformer outperformed LSTM on LoS and VD, but performed much worse on the mortality task, and slightly worse on the tracheostomy task. This may be because the task weighting was more favourouble to the LSTM and TPC models, whereas the Transformer would have benefited from greater weighting towards the binary tasks. Another possibility is that the binary tasks benefit from biases in the LSTM and TPC encoders, because these models naturally emphasise recent timepoints (and these are especially important for solving the mortality task). As for the reason that the Transformer performs better on tracheostomy than mortality, it could be because there is positive correlation between the LoS, VD and tracheostomy tasks. Solving the duration tasks makes the tracheostomy task easier, whereas the relation to mortality is more complex (Fig. 2).

To briefly comment on the reconstruction results in Table 2; it may seem surprising that the LSTM model performs best on the reconstruction and forecasting tasks. However, this could be explained if the LSTM is creating 'lower level' representations that are easier to translate back to the original data with the decoder networks.

Thirdly, Table 3 reveals a general trend that the more tasks that are added, the better results across all the tasks, with particular benefits to the tracheostomy task. The exception to this was the duration only setting. There are two possible explanations:

1. The weighting of the duration task was not sufficient.
2. The tracheostomy task reduces the performance on the duration tasks.

The former does not seem likely, because the Transformer is probably over-weighting the duration tasks, and yet, it follows the same trend as the LSTM and TPC. The latter may appear to be counter-intuitive, because the duration tasks are correlated with tracheostomy. Usually this is an advantage in multitask learning, because it can enhance the signal:noise ratio. However, looking closely at Fig. 2, we can see that there is an area of patients in cluster 5 who have long VD and LoS but have been separated from the other long stay patients in clusters 2 and 4. The separation can be attributed to these patients never receiving tracheostomies, therefore the tracheostomy task forces the representation space to separate these groups when they would be otherwise be aligned. Given the simple nature of the predictor networks, this may harm the performance on the duration tasks because the predictor cannot effectively map these patients to appropriately long LoS. This theory could be formally tested by accompanying the duration tasks with the mortality task only.

Finally, regarding the repeatability of the clustering with different encoders and TPC instances, we demonstrated that key aspects of the representations are consistently recognised. Whereas the separation between the sickest patients and moderately ill was more malleable. This suggests that there may not be a clear distinction but rather a scale of deterioration through which an arbitrary line can be drawn.

7 Limitations and Future Work

Hierarchical Clustering It is evident that certain clusters are more related than others. A tree based hierarchy of clusters seems more natural than a flat structure. We are particularly interested in modifying an approach for genetics data [4, 7, 16].

Contrastive Learning Currently, there is no explicit loss to enforce relative positioning of the embeddings. Despite this, we have empirically found the clusters to be very stable, both temporally and to encoder type. This is likely because our predictor and decoder networks are very simple, which already constrains the network to place similar patient trajectories in similar parts of the representation space. Nevertheless, contrastive learning (e.g. Yèche et al. [28]) may provide further regularisation.

8 Summary

While we acknowledge important limitations in our work, we have shown that:

1. The TPC model outperforms alternative encoders on patient outcome prediction.
2. We can generate clinically meaningful and interpretable clusters.
3. The phenotypes are similar across choices of encoder and number of clusters.
4. The cluster assignment is remarkably stable over time, and membership is determined early on. This is particularly encouraging as a substrate for future intervention studies, because they rely on phenotyping before any intervention.
5. Stable cluster transitions do occur but they are infrequent. Studying these transitions is an important avenue for future work.

Supplementary Material

These can be accessed at: https://emmarocheteau.com/publication/cluster_supplementary.pdf.

Acknowledgements The authors would like to thank Petar Veličković, Sophie Xhonneux, Stephanie Hyland and Mihaela van der Schaar for helpful discussions and advice. We also thank the Armstrong Fund, the Frank Edward Elmore Fund, and the School of Clinical Medicine at the University of Cambridge for their generous funding.

References

1. D.B. Antcliffe, K.L. Burnham, F. Al-Beidh, S. Santhakumaran, S.J. Brett, C.J. Hinds, D. Ashby, J.C. Knight, A.C. Gordon, Transcriptomic signatures in sepsis and a differential response to steroids. From the VANISH randomized trial. Am. J. Respir. Crit. Care. Med. **199**(8), 980–986 (2019)
2. T. Bein, S. Grasso, O. Moerer, M. Quintel, C. Guerin, M. Deja, A. Brondani, S. Mehta, The standard of care of patients with ARDS: ventilatory settings and rescue therapies for refractory hypoxemia. Intensiv. Care Med. **42**(5), 699–711 (2016)
3. S.V. Bhavani, M. Semler, E.T. Qian, P.A. Verhoef, C. Robichaux, M.M. Churpek, C.M. Coopersmith, Development and validation of novel sepsis subphenotypes using trajectories of vital signs. Intensiv. Care Med. **48**(11), 1582–1592 (2022)
4. I. Chami, A. Gu, V. Chatziafratis, C. Ré, From trees to continuous embeddings and back: hyperbolic hierarchical clustering (2020). CoRR arXiv.org/abs/2010.00402
5. Z. Che, S. Purushotham, K. Cho, D. Sontag, Y. Liu, Recurrent neural networks for multivariate time series with missing values. Sci. Rep. **8**(1), 6085 (2018)
6. E. Choi, M.T. Bahadori, A. Schuetz, W.F. Stewart, J. Sun, Doctor AI: predicting clinical events via recurrent neural networks, in *JMLR Workshop and Conference*, vol. 56 (2015), pp. 301–318
7. G. Corso, R. Ying, M. Pándy, P. Veličković, J. Leskovec, P. Liò, Neural distance embeddings for biological sequences (2021). https://doi.org/10.48550/ARXIV.2109.09740, https://arxiv.org/abs/2109.09740
8. K.R. Famous, K. Delucchi, L.B. Ware, K.N. Kangelaris, K.D. Liu, B.T. Thompson, C.S. Calfee, A. Network, Acute respiratory distress syndrome subphenotypes respond differently to randomized fluid management strategy. Am. J. Respir. Crit. Care Med. **195**(3), 331–338 (2017)
9. A.D.T. Force, V.M. Ranieri, G.D. Rubenfeld, B.T. Thompson, N.D. Ferguson, E. Caldwell, E. Fan, L. Camporota, A.S. Slutsky, Acute respiratory distress syndrome: the Berlin definition. JAMA **307**(23), 2526–2533 (2012). https://jamanetwork.com/journals/jama/articlepdf/1160659/jsc120003_2526_2533.pdf
10. H. Harutyunyan, H. Khachatrian, D.C. Kale, G. Ver Steeg, A. Galstyan, Multitask learning and benchmarking with clinical time series data. Sci. Data **6**(96) (2019)
11. S. Ioffe, C. Szegedy, Batch normalization: accelerating deep network training by reducing internal covariate shift, in *Proceedings of the 32nd International Conference on International Conference on Machine Learning* vol. 37, JMLR, ICML'15 (2015), pp. 448–456
12. C. Lee, M. van der Schaar, Temporal phenotyping using deep predictive clustering of disease progression (2020). arXiv.org/abs/2006.08600
13. Z.C. Lipton, D.C. Kale, C. Elkan, R.C. Wetzel, Learning to diagnose with LSTM recurrent neural networks (2015). CoRR. arXiv:1511.03677
14. J. Máca, O. Jor, M. Holub, P. Sklienka, F. Burša, M. Burda, V. Janout, P. Ševčík, Past and present ARDS mortality rates: a systematic review. Respir. Care **62**(1), 113–122 (2017). http://rc.rcjournal.com/content/62/1/113.full.pdf
15. R. Miotto, L. Li, B.A. Kidd, J.T. Dudley, Deep patient: an unsupervised representation to predict the future of patients from the electronic health records. Sci. Rep. **6**(1), 26094 (2016)
16. A. Patel, D.M. Montserrat, C. Bustamante, A. Ioannidis, Hyperbolic geometry-based deep learning methods to produce population trees from genotype data (2022). bioRxiv
17. J. Poole, C. McDowell, R. Lall, G. Perkins, D.F. McAuley, F. Gao, D. Young, Individual patient data analysis of tidal volumes used in three large randomized control trials involving patients with acute respiratory distress syndrome. BJA: Br. J. Anaesth. **118**(4), 570–575 (2017). http://www.oup/backfile/content_public/journal/bja/118/4/10.1093_bja_aew465/1/aew465.pdf
18. A. Rajkomar, E. Oren, K. Chen et al., Scalable and accurate deep learning with electronic health records. Nature **1**(1), 18 (2018)
19. E Rocheteau, P Liò, S Hyland, Temporal pointwise convolutional networks for length of stay prediction in the intensive care unit, in *Proceedings of the Conference on Health, Inference, and Learning, Association for Computing Machinery*, New York, NY, USA, CHIL'21, (2021), pp. 58–68

20. A. Rusanov, P.V. Prado, C. Weng, Unsupervised time-series clustering over lab data for automatic identification of uncontrolled diabetes, in *2016 IEEE International Conference on Healthcare Informatics (ICHI)* (2016), pp. 72–80
21. S. Sheikhalishahi, V. Balaraman, V. Osmani, Benchmarking machine learning models on eICU critical care dataset (2019). 1910.00964
22. B. Shickel, T.J. Loftus, L. Adhikari, T. Ozrazgat-Baslanti, A. Bihorac, P. Rashidi, DeepSOFA: a continuous acuity score for critically Ill patients using clinically interpretable deep learning. Sci. Rep. (2019)
23. H. Song, D. Rajan, J. Thiagarajan, A. Spanias, Attend and diagnose: clinical time series analysis using attention models, in *32nd AAAI Conference on Artificial Intelligence, AAAI 2018* (2018), pp. 4091–4098
24. P.J. Thoral, J.M. Peppink, R.H. Driessen, E.J.G. Sijbrands, E.J.O. Kompanje, L. Kaplan, H. Bailey, J. Kesecioglu, M. Cecconi, M. Churpek, G. Clermont, M. van der Schaar, A. Ercole, A.R.J. Girbes, P.W.G. Elbers, Sharing ICU patient data responsibly under the society of critical care medicine/European Society of Intensive Care Medicine Joint Data Science Collaboration: The Amsterdam University Medical Centers Database (AmsterdamUMCdb) Example. Crit. Care Med. **49**(6) (2021)
25. C. Tong, E. Rocheteau, P. Veličković, N. Lane, P. Liò, *Predicting patient outcomes with graph representation learning* (Springer International Publishing, Cham, 2022), pp.281–293
26. A. Vaswani, N. Shazeer, N. Parmar, J. Uszkoreit, L. Jones, A.N. Gomez, U. Kaiser, I. Polosukhin, Attention is all you need, in *Proceedings of the 31st International Conference on Neural Information Processing Systems*, Curran Associates Inc., NIPS'17 (2017), pp. 6000–6010
27. Y. Wang, Y. Zhao, T.M. Therneau, E.J. Atkinson, A.P. Tafti, N. Zhang, S. Amin, A.H. Limper, S. Khosla, H. Liu, Unsupervised machine learning for the discovery of latent disease clusters and patient subgroups using electronic health records. J. Biomed. Inf. **102**, 103364 (2020)
28. H. Yèche, G. Dresdner, F. Locatello, M. Hüser, G. Rätsch, Neighborhood contrastive learning applied to online patient monitoring, in *Proceedings of the 38th International Conference on Machine Learning, PMLR, Proceedings of Machine Learning Research*, vol. 139, ed. by M. Meila, T. Zhang (2021), pp. 11964–11974
29. X. Zhang, J. Chou, J. Liang, C. Xiao, Y. Zhao, H. Sarva, C. Henchcliffe, F. Wang, Data-driven subtyping of Parkinson's disease using longitudinal clinical records: a cohort study. Sci. Rep. **9**(1), 797 (2019)

Bayesian-Based Parameter Estimation to Quantify Trust in Medical Devices

Mini Thomas, Omar Boursalie, Reza Samavi, and Thomas E. Doyle

Abstract In this paper, we propose a data-driven approach to estimate Bayesian parameters when trust needs to be quantified in the domain of wearable medical devices (WMD). Our approach extracts the probability of a trust determinant (e.g., reliability or robustness) being in a specific state from the data. Then, we use the Bayesian approach to estimate the parameters for the intermediate nodes in the network and ultimately compute the trust score. The trust score we compute is used as a *relative measure* of trustworthiness between different WMDs evaluated in the same test conditions and with the same Bayesian network (BN). To evaluate our approach, we develop a BN for the trust quantification of similar wearable medical devices from two manufacturers under identical test conditions. The results demonstrate the learnability and generalizability of our data-driven parameter estimation approach.

Keywords Bayesian parameter estimation · Trust quantification · Trustworthy AI · Wearable devices

---

M. Thomas (✉)
Department of Computing and Software, McMaster University, Hamilton, ON, Canada
e-mail: thomam21@mcmaster.ca

O. Boursalie · R. Samavi
Department of Electrical, Computer and Biomedical Engineering, Toronto Metropolitan University, Toronto, ON, Canada
e-mail: boursalie@torontomu.ca

R. Samavi
e-mail: samavi@torontomu.ca

T. E. Doyle
Department of Electrical and Computer Engineering, McMaster University, Hamilton, ON, Canada
e-mail: doylet@mcmaster.ca

School of Biomedical Engineering, McMaster University, Hamilton, ON, Canada

R. Samavi · T. E. Doyle
Vector Institute, Toronto, ON, Canada

A. Shaban-Nejad et al. (eds.), *Artificial Intelligence for Personalized Medicine*, Studies in Computational Intelligence 1106,
https://doi.org/10.1007/978-3-031-36938-4_8

1 Introduction

Advances in sensor technology have revolutionized wearable medical devices (WMD) capability to monitor patients' health remotely. In addition to technical advances, trust in WMD is essential for their continued acceptance and adoption by patients, health professionals, and institutions. The trust depends on multiple factors, including the secured and reliable collection of data by the network of sensors and the flow of data through several layers and units of a WMD. If the data generated by WMD cannot be trusted, it could jeopardize the trustworthiness of the entire system. However, trust is stochastic and subjective, as demonstrated in the following motivating scenario. Assume Alex, an athlete, and his family, trainer and physician (primary stakeholders) are considering different WMDs to monitor Alex's health during indoor and outdoor activities. Alex and his family's trust in the WMD may depend on the device's safety and privacy. At the same time, Alex's trainer and physician may select a WMD depending on the device's accuracy and quality of collected data. The important question in this scenario is how these stakeholders, each with possibly different and subjective views of trustworthiness, select the suitable device.

There is a growing interest in using BN [29] to quantify trust in WMD. In a recent study, [31] proposed a BN to quantify trust where each node in the network represents a trust determinant, as shown in Fig. 1a. The BN is structured so that the nodes in the first layer ($X_1 - X_4$) capture trust factors that are directly measurable (e.g., quality of the heart rate signals as a trust determinant) while the remainder of the network represents the probability of subjective belief on the state of each refined attributes such as reliability (X_5) and robustness (X_6). Then, the trustworthiness of a WMD from a stakeholder's perspective is represented in terms of the probability of a subjective belief in trust to be in a specific state (e.g., low or high). Computing this probability requires access to all intermediary prior and posterior probabilities of the BN parameters, which appear to be a major challenge [11]. As a result, usually arbitrary values from experts [11] or random values assuming some Gaussian distribution have been used [31]. The latter can lead to over-fitting and biased objective quantification. The former requires experts to provide an exponential number of prior and posterior probabilities, which is impractical and may lead to expensive, biased, or contradictory estimates.

This paper addresses this gap by proposing a data-driven parameter estimation approach for BN to quantify trust. As shown in Fig. 1a, our goal is to compute the probability of X_7, which represents the trustworthiness of a WMD, from a stakeholder's perspective, to be in a specific state (e.g., low or high). We extract prior probabilities of the measurable attributes ($X_1 - X_4$) directly from the data collected from the WMD sensors. For example, in Fig. 1b, the probability that the quality of the heart rate sensor is *high* is calculated by comparing our data to the manufacturer's specifications. We capture the subjectivity of stakeholders' perspectives by asking the experts to provide the impact of each measurable attribute on its immediate intermediary node. For example, how X_1 and X_2 impacts reliability (X_5). This is a known method in software engineering to investigate the impact of a specific

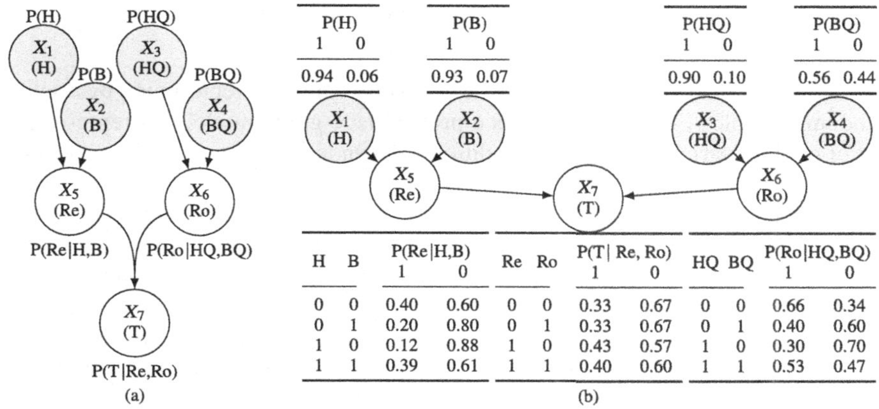

P(H) 1	P(H) 0
0.94	0.06

P(B) 1	P(B) 0
0.93	0.07

P(HQ) 1	P(HQ) 0
0.90	0.10

P(BQ) 1	P(BQ) 0
0.56	0.44

H	B	P(Re|H,B) 1	0
0	0	0.40	0.60
0	1	0.20	0.80
1	0	0.12	0.88
1	1	0.39	0.61

Re	Ro	P(T| Re, Ro) 1	0
0	0	0.33	0.67
0	1	0.33	0.67
1	0	0.43	0.57
1	1	0.40	0.60

HQ	BQ	P(Ro|HQ,BQ) 1	0
0	0	0.66	0.34
0	1	0.40	0.60
1	0	0.30	0.70
1	1	0.53	0.47

Fig. 1 BN to quantify trust (T) in WMD by analyzing heart (H) and breathing (B) rate sensor validation, heart (HQ) and breathing (BQ) rate quality, reliability (Re), and robustness (Ro)

software architecture on quality attributes such as privacy, security and trust [24] or in understanding the alignment of different business models with a corporate strategy [25]. In this way, instead of expecting experts to estimate numerous probabilities, their subjective views are captured by a minimal number of qualitative labels. Then, we use the Bayesian approach [14] to estimate the parameters by calculating the joint probability distributions. Since trust is subjective, the computed probability of trustworthiness for one device may not be interpretable. Instead, the trust probability provides a relative measure of trustworthiness when probabilities are computed for different WMDs with the same test conditions, i.e. same BN and subjective measures.

To evaluate our approach, we developed a hierarchical four-layered BN. We identified trust factors from the literature and mapped them to the regulatory standards' requirements for trustworthy medical devices [4, 6–8, 10]. We learned the BN parameters, including the probability of trust on each device, using our data-driven approach with the data collected from two WMDs. We evaluated the learnability and generalizability behaviour of our BN. For learnability, we measured the similarity of the quantified trust probability for the two WMDs under the same test conditions, such as sitting, standing, and walking. For generalizability, we assessed how decreasing the WMDs' sampling rates impacts our quantified trust probability. We successfully trained a BN using our Bayesian parameter estimation approach. The quantified trust score for the two WMDs was similar for predefined activities, demonstrating high learnability. The results also showed that the trust scores did not depend on the sampling rate, demonstrating high generalizability.

We make the following contributions. First, we present a data-driven approach for estimating the parameters of the BN with data collected from the WMDs. Second, we present a compact BN structure for computing the probability of trust using factors from the literature mapped to regulatory guidelines. Finally, we present a proof-of-concept BN to compare the trust score between two WMDs.

2 Parameter Estimation of Bayesian Network

The stochastic nature of trust can be viewed as probabilities represented by a set of random variables [31]. A BN [29] is a probabilistic graph model that represents a set of random variables and their conditional dependencies in a compact way using *nodes* and *edges*. Formally, a BN is represented as $G = (X, E)$, where X is a finite set of n discrete random variables as nodes and E is a finite set of directed edges. For each random variable i, the prior probability is defined as $P(X_i)$, capturing the node's aleatory uncertainty and edges representing a causal relationship between the nodes. Following terminologies in [16], we have two types of nodes: non-descendant (have no parents) and descendant (have parents), with the assumption of independence between the non-descendant nodes. Let Pa_{X_i} denote the set of parents of a node X_i. Figure 1b shows an example of a BN with seven nodes (X_1 to X_7) and six edges. Nodes X_1 to X_4 are non-descendant and X_5 to X_7 descendant nodes, $Pa_{X_5} = \{X_1, X_2\}$, $Pa_{X_6} = \{X_3, X_4\}$, and $Pa_{X_7} = \{X_5, X_6\}$. We can see a BN as a structure (G) with a set of prior probabilities ($P(X_i)$) for non-descendant and a set of parameters ($\boldsymbol{\theta}$) for descendant nodes. The structure encodes the probability density of the random variables with their conditional dependencies in the form of a directed acyclic graph (DAG) [16]. The prior probabilities are the given evidence, and the BN parameters are the intermediate (descendant) nodes' conditional probabilities computed as:

$$P(X_i|Pa_{X_i}) = \frac{P(Pa_{X_i}|X_i)P(X_i)}{P(Pa_{X_i})}, \quad \forall i = descendant, \tag{1}$$

where $P(Pa_{X_i}|X_i)$ is the likelihood probability of the evidence for a particular data based on $\boldsymbol{\theta}$, $P(X_i)$ is the prior probability before the evidence is considered, and $P(Pa_{X_i})$ is the marginal probability of the evidence under any circumstance [16]. Then the joint probability distribution for the BN can then be expressed as:

$$P(X_1, ..., X_n) = \prod_{i=1}^{n} P(X_i|Pa_{X_i}). \tag{2}$$

2.1 Data-Oriented Bayesian Parameter Estimation

With its simple structure, a BN can capture many random phenomena in the presence of multiple interrelated aspects that relate to a specific reasoning task. For example, we might be interested to know the probability of a patient having flu given several interrelated pieces of evidence, including the season and symptoms where the same symptoms might also indicate another diagnosis, such as hay fever ([16]-Chap. 1).

In our approach, we define the trustworthiness of a WMD in a similar way and in terms of the probability of the *trust* node in the BN (X_7) to be in a specific state (e.g., low, medium, or high). The first step is to define the BN structure to capture how trust is interrelated to other phenomena, whether they are directly measurable or not. For example, in Fig. 1b, trust (X_7) cannot be measured directly and also it is not directly related to measurable observations such as the heart rate sensor validation (X_1) or the quality of heart rate signal (X_3). However, we know the relationship between the relative reliability of two devices (X_5), the measured observations in X_1 and X_2, and the relationship between reliability (X_5) and trust (X_7). Undoubtedly, this structure is subjective both in terms of structure and also the qualitative levels that we define to indicate the relative trustworthiness of a WMD. However, we will show that if these subjective aspects are given by the domain experts, then the parameters of the entire structure can be estimated, i.e. the conditional probability of all descendant nodes, including the trustworthiness captured in the ultimate node of the BN. In this way, instead of domain experts subjectively guessing the trustworthiness of a device, they use observed data from the device's behaviour, which are informative but not fully indicative of the device's trustworthiness, to quantify and measure trust.

Before formalizing our notion of parameter estimation for the trust network, we should point to a specific constraint. The parameter estimation approaches such as maximum likelihood estimator (MLE) [14, 16] rely on the availability of observations for all nodes in the network. However, in our trust network, the data is available only for the non-descendant nodes. Therefore, our formalism should include methods for generating data for all intermediary nodes, e.g., generating indirect observations for reliability (X_5) using direct observations of X_1 and X_2.

Let y be a constant integer that represents the number of qualitative levels of an observed or inferred evidence in the BN, i.e. values of every node in the BN are always mapped to a fixed set of discrete values, $k = 1, ..., y$. Therefore, after the BN structure is developed, our first step is to discretize the data collected by the WMD to y mutually exclusive levels to perform parameter estimation. For every i representing a non-descendant node, let $S_i = \{S_i^1, ..., S_i^m\}$ represent the raw samples of m number of observations. Then the equivalent discretized values of S will be computed as:

$$D^q_{i(nondescendant)} = \begin{cases} 1, & l_{i,1} \leq S_i^q < u_{i,1} \\ 2, & l_{i,2} \leq S_i^q < u_{i,2} \\ \dots & \\ y, & l_{i,y} \leq S_i^q < u_{i,y} \\ 0, & \text{otherwise} \end{cases}, \tag{3}$$

where D_i^q is the discrete data with operational range values for $q = 1, ..., m$, which is equal to the number of non-discretized observed values of S_i^q, and $l_{i,k}$ and $u_{i,k}$, are the lower and upper thresholds for each discrete level of S_i^q, for $k = 1, ..., y$ discrete levels, respectively. These two values are defined as:

$$l_{i,k} = min(O_i) + \Delta H * k, \quad u_{i,k} = l_{i,k} + \Delta H, \tag{4}$$

where $O_i = \{min, max\}$ is the set of the minimum and maximum operating values with user-defined thresholds based on the manufacturer's datasheet or a validation device, $min(O_i)$ is the minimum threshold value, $max(O_i)$ is the maximum threshold value, and $\Delta H = (max(O_i) - min(O_i))/ y$ is the step size between the levels. For example, in Fig. 1b, if the raw heart rate sensor validation data collected for X_1 is $S_1 = \{40, 78, 100, 115, 150\}$ the discrete sample observation for y=3 is D_1={0, 1, 1, 2, 0} with $l_{1,1} = 76.66$ and $u_{1,1} = 113.32$.

After the values of non-descendant nodes are discretized, we generate sample data for each immediate descendent node. As part of the BN structure, we assume the domain experts provide information on the impact of each parent of an intermediate node. The impact is defined as positive (+) or inverse positive (–). For example, in Fig. 1b, an expert may (subjectively) interpret that measurement of X_1 and X_2 has synergy with the measurement of reliability (X_5). Thus the discrete values of X_1 and X_2 for all observed instances in the dataset will be added to generate the instances of a random variable (X_5). On the other hand, if the two parents were inversely impacting the descendant node, then the values of instances will be subtracted. This approach can be extended to more than two parents with different impacts as follows:

$$S^q_{i(descendant)} = \sum_{j=1}^{t_+}(D^q_{Pa(i),j}) - \sum_{j=1}^{t_-}(D^q_{Pa(i),j}), \quad \forall q = 1, ..., m, \tag{5}$$

where t_+ and t_- are the number of parents of a descendant node i which positively or inverse positively impact the node i and $D_{Pa(i)}$ is the discrete value of the instance q of the parent of node i. The result of Eq. 5 will be a generated sample data for each intermediate node. To get $D^q_{i(descendant)}$, the generated data needs to be discretized to y level following Eq. 3. Following our example, the generated data for X_5 can be computed based on the impact of the immediate parent nodes X_1 and X_2, which in this case, both impacts are considered positive. Assume D_2={1, 0, 1, 1, 2}, then following Eq. 5 $S_5 = \{1, 1, 2, 3, 2\}$ and if we follow Eq. 3 to discretize these values to $y = 3$ levels, $D_5 = \{0, 0, 1, 2, 1\}$.

Now that we have data (observed and generated) for all nodes of our BN, our next step is to compute prior and posterior probabilities for all nodes. The prior probability for node X_i in our BN with multinomial data set D_i that takes y discrete levels is computed as:

$$P(X^k_i) = \sum_{q=0}^{m}(D^q_i = k)/m, \quad \forall k = 1, ..., y, \tag{6}$$

where $P(X^k_i)$ is the short form for $P(X_i = k)$. For example, the prior probabilities of nodes X_1, X_2, and X_5 for three levels of y are $P(D^0_1) = P(D^1_1) = P(D^0_5) = P(D^1_5) = 0.4$, $P(D^2_1) = P(D^0_2) = P(D^2_2) = P(D^2_5) = 0.2$ and $P(D^1_2) = 0.6$.

Next, we estimate the parameter θ_i for each descendant node. Note that we currently have discrete data set for all nodes, i.e. $\{D_i[1], ..., D_i[m]\}$ for $i = 1, ..., n$. The structure of BN allows us to reduce the parameter estimation to a set of unrelated (disjoint) problems. Let $P(D_i^k[1], ..., D_i^k[m])$ represent the joint probability of instances of node i to have value k. The joint distribution for the data set of D_i^k and θ_i is then given as:

$$\begin{aligned} P(D_i^k[1], ..., D_i^k[m], \theta_i) &= P(D_i^k[1], ..., D_i^k[m] | \theta_i) P(\theta_i), \\ &= P(\theta_i) \prod_{q=1}^{m} P(D_i^k[m] | \theta_i), \quad \forall i = descendant, \end{aligned} \tag{7}$$

where $\prod_{q=1}^{m} P(D_i^k[m]|\theta_i)$ is the likelihood function $L(\theta_i : D_i^k)$ of the parameters, and $P(\theta_i)$ is the prior probability of the descendant node in concern [16]. Then the posterior probability of θ_i given the instances of D_i^k is computed as:

$$P(\theta_i | D_i^k[1], ..., D_i^k[m]) = \frac{P(D_i^k[1], ..., D_i^k[m] | \theta_i) P(\theta_i)}{P(D_i^k[1], ..., D_i^k[m])}, \tag{8}$$

where $P(D_i^k[1], ..., D_i^k[m])$ is the marginal probability [16]. If we follow these equations, in our example, we can compute θ_5 to be in state 0 for the descendant node X_5 as: $P(\theta_5^0 | D_1^0[1], ..., D_1^0[5]) = 0.04$.

2.2 Constructing the Trust Network

In this section, we describe a proof-of-concept BN by extracting trust factors from the American (USA) [6, 7], Canadian [8–10], and European Union (EU) [4] regulatory requirements for trustworthy medical devices. To quantify the regulatory requirements in our BN, we mapped the requirements to more granular trust factors from the literature until we reached factors that can be directly measured on a WMD, as shown in Fig. 2. The trust-determining factors in this study are not exhaustive due to domain dependency. Nevertheless, our proposed mapping enables stakeholders to define their own trust network for WMDs.

We identified four main trust-determining factors: reliability, operations, security and privacy. Reliability is the measurement of the capacity of the WMD network [1, 21]. We refined reliability in terms of network quality and loss, to evaluate the efficiency [15, 18] and effectiveness [3, 26] of the network, respectively. For network quality, we measure network latency [18], power consumption [19]. For network loss, we measure received signal strength [31], memory consumption [12] and the signal loss [1]. Operations is the measurement of the sensor performance [26, 27]. We refined operations in terms of sensor accuracy to assess if the sensor behaves according to the manufacturer's specifications [19, 22], and sensor quality to evaluate the recorded signal data quality compared to a baseline [19]. We further refined sensor accuracy to measurable factors such as bio-signal data from multiple sensors [18], time since the last calibration [27], age of the sensor [27] and sensor profile [27]. For

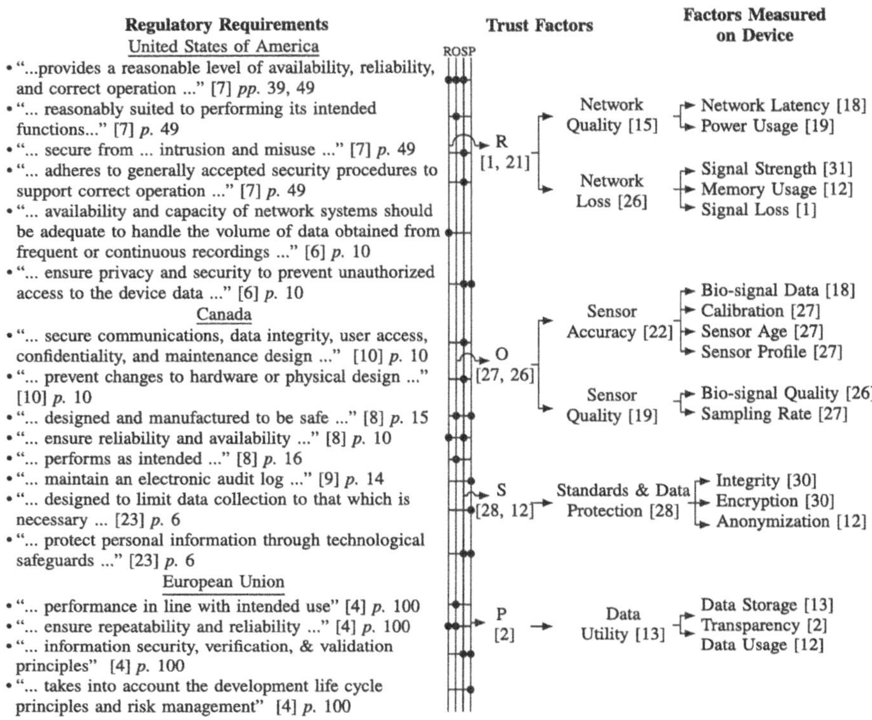

Fig. 2 Mapping USA, Canada, and European Union regulatory requirements to trust factors from the literature that can be directly measured on the devices. Dots denote mapping between regulatory requirements and reliability (R), operations (O), security (S), and privacy (P) trust factors

sensor quality, we measure the bio-signal quality data [26] and the sampling rate of the sensors [27]. Security is the measurement of how safe and protected communication is between different parts of the system [12, 28]. We refined security in terms of standards and data protection to evaluate compliance with security protocols and confidentiality [28]. Standards and data protection were further refined to evaluate system encryption [30], integrity [30] (unauthorized parties do not change sensor data), and anonymization [12]. Privacy assesses that personal data in the system is treated safely and securely [2, 31]. We refined privacy in terms of data utility to evaluate if the data is used only by authorized users and for its approved purpose [13]. Data utility was further refined to safe data storage [13], transparency [2] (data flow is visible to stakeholders) and authorized data usage [12].

Figure 3 presents an instantiation of a subset of the trust factors mapped in Fig. 2. For our proof-of-concept BN, we considered two main trust-determining factors: reliability and operations. The refined measurable factors forming the non-descendant nodes are the heart (X_1) and breathing rate sensor validation (X_2) heart rate (X_3) and breathing rate quality (X_4); memory consumption and signal loss (X_5, X_6), and

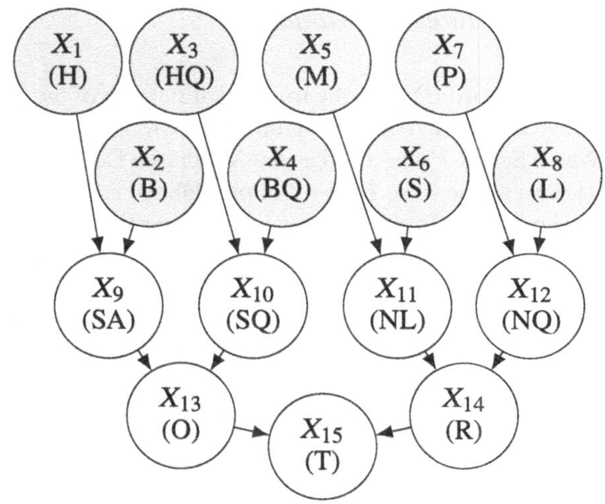

Fig. 3 Bayesian network to quantify trust (T) in WMD analyzing heart (H) and breathing (B) rate sensor validation, heart (HQ) and breathing (BQ) rate quality, signal loss (S), memory (M) and power (P) usage, latency (L), sensor accuracy (SA) and quality (SQ), network loss (NL) and quality (NQ), operations (O), and reliability (R)

power consumption and latency (X_7, X_8). The instantiated trust factors forming the intermediary nodes are sensor accuracy, sensor quality, network loss, and network quality ($X_9 - X_{12}$). The intermediary nodes contribute to the trust-determining factors reliability and operations ($X_{13} - X_{14}$), which determine the overall trust (X_{15}) of a WMD. Although some trust factors may be dependent (e.g., network loss and quality), we assume the dependency may distribute evenly among the trust determinants.

3 Experimental Evaluation

In this section, we evaluate our data-driven approach to estimate Bayesian parameters in terms of learnability and generalizability. For learnability, we determine how similar the trust probability is for the two WMDs under the same test conditions. We hypothesize that the probability of trust to be in a specific state will be similar across devices in the same usage conditions (e.g., walking in the evening). We also compare the BN quantified scores for individual nodes (e.g., reliability) between the wearable devices. We expect the inferences made on the same node to be similar for different devices under identical use cases. For generalizability, we assess how decreasing the WMD's sampling rates in orders of ten impacts the trust scores. We expect the trust score will be similar with reduced sample sizes.

3.1 Parameter Estimation

We developed BNs for the trust quantification of the Zephyr BioHarness 3.0 device[1] (Device 1) and Astroskin Vital Signs Monitoring System[2] (Device 2). The Apple Watch Series 7[3] was used as our Validation Device. We evaluated the devices in two use cases. Use Case 1 is a set of pre-defined activities (sitting, standing, and walking at a speed of 4 MPH) performed by the participants for 30 minutes each indoors and outdoors. An *indoor condition* is when the activities were performed when two or fewer people were within a room of 50 m by 50 m to investigate the impact on sensor accuracy. An outdoor condition is when the activities are performed with at least ten people and wireless devices (e.g., phones, smartwatches, and laptops) in a park to investigate the effect on signal quality, loss, and latency of the devices. The outdoor activity was performed in the morning (6–8 AM) and evening (6–8 PM) when the park had 10–12 and 40–50 people with wireless devices, respectively. In Use Case 2, activities (e.g., sleeping, sitting, and walking) were performed by the participants wearing the devices for up to 24 h in hybrid (indoors and outdoors) conditions.

The study involved a cohort of three subjects, one female (45-55 years) and two males (20–30 years). The recruited participants wore the Astroskin (right side of the lower abdomen), Zephyr (left side of the upper abdomen) and Apple Watch (left wrist) at the same time. We collected data $\{S_1^1, \ldots, S_8^m\}$ for the non-descendant nodes (X_1 to X_8) of our BN from each wearable device under different conditions. The raw samples of the data collected are converted to discrete-valued data set using Eqs. 3–4. Data set D_1^q and D_2^q for nodes X_1, X_2 are obtained by comparing the heart and breathing sensor validation data with user defined thresholds $l_{i,k}$ and $u_{i,k}$[1,2,3] [5]. For D_3^q and D_4^q (X_3 and X_4), we generate the data set by comparing the heart and breathing rate quality level with a pre-determined noise threshold[4,5]. For D_5^q (X_5), the memory consumed was compared with user-defined thresholds. For D_6^q (X_6), we consider the null or missing data [17, 32]. For D_7^q (X_7), we compare the battery level with a pre-determined threshold.[6] For D_8^q (X_8), we compare the latency with the acceptable threshold for the application. During data collection, both devices' orientations were kept constant and the devices were synchronized at the beginning of each recording session. Sixteen sets of data (S_i^q) for the use cases and conditions were collected. Priors and parameters were then estimated using our methodology described in Sect. 2. The experimental analysis was conducted on a 64-bit Windows 10 laptop with a 2.5 GHz Intel Core i9 CPU. The study was approved by the local research ethics committee, and all subjects gave consent.

---

[1] Zephyr Bioharness 3 User Manual.

[2] Hexoskin 3.0 User Manual.

[3] Apple Watch 7 User Manual.

[4] Hexoskin Datatype.

[5] Zephyrn Bioharness 3 Log Data Description.

[6] https://docs.rs-online.com/585c/0900766b81249809.pdf.

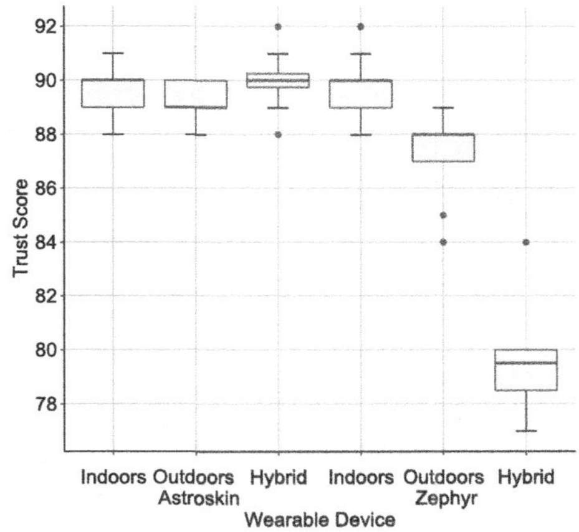

Fig. 4 Trust scores for WMDs in indoor, outdoor, and hybrid use cases. The lines within the boxplot boxes denote medians; bottom and top border denote the 25 and 75th percentiles, respectively. Dots represent outliers

Table 1 Query-based inference results on trust

Inference	Astroskin	Zephyr	Use Case
$P(X_{15}^1 \vert X_{13}^0, X_{14}^0)$	0.88	0.86	1 (Indoors)
$P(X_9^1 \vert X_1^1, X_2^0)$	0.90	0.90	1 (Indoors)
$P(X_{15}^1 \vert X_{13}^1, X_{14}^0)$	0.88	0.82	1 (Outdoors)
$P(X_{11}^1 \vert X_5^0, X_6^1)$	0.88	0.86	1 (Outdoors)
$P(X_{14}^1 \vert X_{11}^0, X_{12}^0)$	0.87	0.68	2 (Hybrid)
$P(X_{13}^1 \vert X_9^0, X_{10}^1)$	0.90	0.79	2 (Hybrid)

3.2 Results and Discussion

Figure 4 and Table 1 present our learnability results. Figure 4 shows the trust probability for Use Cases 1 and 2. The learned parameters for our BN resulted in similar trust probabilities on both devices under the same use cases. Our results indicate that though the data collected from the two WMDs differed, the parameter estimation approaches were generalized to produce similar results under the same conditions. Table 1 shows the individual scores in the BN by querying the nodes in high and low states. The inferences of individual nodes for both devices were also similar under the same condition. However, the results for Use Case 2 (Fig. 4) did not align with our expectations. The average trust value for Astroskin is higher than Zephyr. This was surprising and may be due to the difference in parameter estimation with the data collected under hybrid conditions for normal daily life activities for a long duration. Our BN may also be capturing actual behaviour differences between the devices. More data collection with refined use cases is required to investigate the results for Use Case 2 thoroughly. Our results to assess generalizability are shown in Table 2. Our results demonstrate the suitability of the Bayesian-based parameter estimation with fewer training samples as compared with discriminative models [20]. Reducing

Table 2 Trust scores for different sample sizes for Use Case 1 (indoors) with Astroskin and Zephyr

Samples	Astroskin	Zephyr
10,000,000	0.88	0.87
1,000,000	0.88	0.87
10,000	0.86	0.87
1,000	0.85	0.84
100	0.59	0.53
10	0.52	0.50

the sampling had little impact on Bayesian parameter estimation because it is based on distributions, not individual samples, for learning the parameters.

Our proposed BN provides a mechanism for the stakeholders in our motivating scenario to select a trustworthy WMD. The relative trust helps Alex choose the WMD based on trustworthiness factors such as operation. Alex's doctor and the trainer also decide which WMD can be used for monitoring Alex's health during indoor and outdoor activities based on the WMD's sensor accuracy and reliability.

Our trust probabilities provide a relative measure of trustworthiness between WMDs. We presented a BN structure for trust quantification identified from our mapping of regulatory requirements, trust factors from the literature and measurements from WMD. Using a BN structure enables stakeholders to explicitly define the trustworthiness factors used to assess the WMD. Our data-driven approach can then estimate the parameters for any BN structure using data collected with the WMD instead of defining the parameters based on Gaussian distributions or expert knowledge.

Given these results, we are currently assessing our proposed data-driven approach by comparing the trust probabilities of the WMD with data collected under normal working conditions with the scores obtained by artificially injected noisy data. The noise represents missing data due to signal loss by connectivity problems, data overwritten due to memory shortage, or battery failure due to excessive power consumption. We expect that the trust probabilities will decrease with increased noise.

Our study exhibits some limitations: First, our BN is based on a fixed structure. Second, our evaluation is based on limited data sets. Third, we evaluated the comparative behaviour of our network on two WMDs.

4 Conclusion and Future Work

In this paper, we present a data-driven approach to estimate Bayesian parameters when subjective and stochastic concepts such as trust need to be quantified. To assess our data-driven parameter estimation approach, we developed a proof-of-concept BN structure based on trust factors in the literature we mapped to regulatory standards. We then learned the BN parameters using our data-driven approach for two WMDs under identical use cases. Our results demonstrate the learnability and generalizability of our data-driven approach. Future work will investigate our data-driven approach to estimate Bayesian parameters for more WMDs, use cases, and participants. We will also extend our data-driven parameter estimation approach to concepts such as ethical machine learning, security, and privacy. Finally, we will investigate learning the BN structure from data directly instead of using a fixed, user-defined network.

Acknowledgements This work is supported by the Natural Sciences and Engineering Research Council of Canada Discovery and Innovation for Defence Excellence and Security grants.

References

1. S. Al-Sarawi, et al., IoT communication protocols, in *ICIT* (IEEE, 2017)
2. M. Chen, et al., Privacy protection and intrusion avoidance for cloudlet-based medical data sharing. IEEE Trans. Cloud Comput **8**(4), 1274–1283 (2016)
3. J.K. Devine, L.P. Schwartz, S.R. Hursh, Technical, regulatory, economic, and trust issues preventing successful integration of sensors into the mainstream consumer wearables market. Sensors **22**(7), 2731–2739 (2022)
4. EU: Medical device regulation (2017). https://eur-lex.europa.eu/legal-content/EN/TXT/?uri=celex:32020R0561
5. M. Falter, et al., Accuracy of Apple Watch measurements for heart rate and energy expenditure in patients with cardiovascular disease: cross-sectional study. JMIR mHealth uHealth **7**(3), 1–9 (2019)
6. FDA: Guidance for industry, digital health technologies for remote data acquisition in clinical investigations (2021). https://www.fda.gov/media/155022/download
7. FDA: Health FDA draft for cybersecurity in medical devices (2022). https://www.fda.gov/media/119933/download
8. Government of Canada: Medical devices regulations (2022). https://laws-lois.justice.gc.ca/eng/regulations/sor-98-282/
9. Government of Ontario: Personal health information protection act (2022). https://www.ontario.ca/laws/statute/04p03
10. Health Canada: Pre-market requirements for medical device cybersecurity (2019). https://www.canada.ca/content/dam/hc-sc/documents/services/drugs-health-products/medical-devices/application-information/guidance-documents/cybersecurity-guidance.pdf
11. M. Holden, M. Pereyra, K.C. Zygalakis, Bayesian imaging with data-driven priors encoded by neural networks: theory, methods, and algorithms (2021). arXiv
12. F.T. Jaigirdar, C. Rudolph, C. Bain, Can I Trust the Data I See? A physician's concern on medical data in IoT health architectures, in *ACSW* (ACM, 2019)
13. U. Jayasinghe, et al., Data centric trust evaluation and prediction framework for IoT, in *ITU* (IEEE, 2017)
14. Z. Ji, Q. Xia, G. Meng, A review of parameter learning methods in BN, in *ICIC* (Springer, 2015)
15. J. Kim, Energy-efficient dynamic packet downloading for medical IoT platforms. IEEE Trans. Ind. Inf. **11**(6), 1653–1661 (2015)
16. D. Koller, N. Friedman, Probabilistic Graphical Models: Principles and Techniques (MIT Press, 2009)
17. L. Kong, et al., Data loss and reconstruction in sensor networks, in *INFOCOM* (IEEE, 2013)

18. S. Maitra, K. Yelamarthi, Rapidly deployable IoT architecture with data security: implementation and experimental evaluation. Sensors **19**(11), 2484–2492 (2019)
19. D.K. Ming, et al., Continuous physiological monitoring using wearable technology to inform individual management of infectious diseases, public health and outbreak responses. Int. J. Inf. Dis. **96**, 648–656 (2020)
20. A. Ng, M. Jordan, On discriminative vs. generative classifiers: a comparison of logistic regression and Naive Bayes, in *NeurIPS* (MIT Press, 2001)
21. D. Pal, S. Funilkul, B. Papasratorn, Antecedents of trust and the continuance intention in IoT-based smart products: the case of consumer wearables. IEEE Access **7**, 184160–184171 (2019)
22. M. Pobiruchin, et al., Accuracy and adoption of wearable technology used by active citizens: a marathon event field study. JMIR mHealth uHealth **5**(2), 1–9 (2017)
23. Privacy Commissioner of Canada: Guidance for manufacturers of IoT devices (2020). https://priv.gc.ca/en/privacy-topics/technology/gd_iot_man
24. R. Samavi, T. Topaloglou, Designing privacy-aware personal health record systems, in *ER* (Springer, 2008)
25. R. Samavi, E. Yu, T. Topaloglou, Strategic reasoning about business models: a conceptual modeling approach. Inf. Syst. E-Bus. Manag. **7**, 171–198 (2009)
26. A. Sawand, S. Djahel, Z. Zhang et al., Multidisciplinary approaches to achieving efficient and trustworthy ehealth monitoring systems, in *ICCC* (IEEE, 2014)
27. M. Shin, Secure monitoring with unreliable mobile devices. J. Biomed. Biotech. 1–5 (2012)
28. S. Sicari et al., Security, privacy and trust in IoT: the road ahead. Comput. Netw. **76**, 146–154 (2015)
29. L. Sucar, Probabilistic graphical models: principles and application (Springer, 2015)
30. W. Sun, et al., Security and privacy in the medical IoT: a review. Secur. Commun. Netw. 1–9 (2018)
31. M. Thomas, R. Samavi, T.E. Doyle, Trust quantification for autonomous medical advisory systems, in *PST* (IEEE, 2021)
32. M. Zhang, A. Raghunathan, N. Jha, Trustworthiness of medical devices and BAN. Proc. IEEE **102**(8), 1–9 (2014)

EEG Analysis of Neurodevelopmental Disorders by Integrating Wavelet Transform and Visual Analysis

Soo-Yeon Ji, Sampath Jayarathna, Anne M. Perrotti, Katrina Kardiasmenos, and Dong H. Jeong

Abstract Identifying neurodevelopmental disorders, ADHD, autism spectrum disorder, and other disorders (e.g., depression and mental health diseases), are important for planning appropriate treatments and early intervention. EEG is a commonly used method that measures the electrical activity of a brain to examine such disorders. This study introduced an approach to understanding the disorders by integrating wavelet transform and visual analysis on EEG signals. Wavelet-based features are extracted to find informative information associated with any changes in the EEG signals to differentiate them from healthy subjects. The effectiveness of the features is determineed by proposing two different feature selection methods (DWT-PCA and DWT-ANOVA) and evaluated by applying ML classification algorithms, such as KNN and Naive Bayes. Also, visual analysis is conducted to assess the features and to enhance the understanding of the features. We found that the classification with DWT-PCA features provided better performances. Although there was no clear distinction between normal (i.e., healthy) and abnormal (i.e., disorders), similarities and differences between them were identified through visualization. Overall, the integration of using both wavelet-based feature extraction and visual analysis was effective in identifying diagnostic neurodevelopmental disorders.

S.-Y. Ji (✉) · K. Kardiasmenos
Bowie State University, Bowie, MD, USA
e-mail: sji@bowiestate.edu

K. Kardiasmenos
e-mail: kkardiasmenos@bowiestate.edu

S. Jayarathna · A. M. Perrotti
Old Dominion University, Norfolk, VA, USA
e-mail: sampath@cs.odu.edu

A. M. Perrotti
e-mail: aperrott@odu.edu

D. H. Jeong
University of the District of Columbia, Washington, DC, USA
e-mail: djeong@udc.edu

A. Shaban-Nejad et al. (eds.), *Artificial Intelligence for Personalized Medicine*,
Studies in Computational Intelligence 1106,
https://doi.org/10.1007/978-3-031-36938-4_9

Keywords Neurodevelopmental disorders · Wavelet transform · Visual analysis · Feature selection · EEG

1 Introduction

Working memory (WM) stores and manipulates data needed to perform complex cognitive tasks in our daily lives [2]. As a functional brain system, the deficiency of the WM can affect our lives due to limited cognitive abilities and, eventually, reduced life satisfaction. In addition, it could cause neurodevelopmental disorders such as autism and attention deficit hyperactivity disorder (ADHD) or mental illnesses such as depression [6], major depressive disorder [19], and obsessive-compulsive disorder [1]. Thus, early detection of such disorders has gained increasing attention from researchers across different disciplines in recent years. Working memory (WM) stores and manipulates data needed to perform complex cognitive tasks in our daily lives [2]. As a functional brain system, the deficiency of the WM can affect our lives due to limited cognitive abilities and, eventually, reduced life satisfaction. In addition, it could cause neurodevelopmental disorders such as autism and attention deficit hyperactivity disorder (ADHD) or mental illnesses such as depression [6], major depressive disorder [19], and obsessive-compulsive disorder [1]. Thus, early detection of such disorders has gained increasing attention from researchers across different disciplines in recent years.

ADHD affects approximately 10.0% of children and 6.5% of adolescents in the United States (U.S.) [21]. It may impact educational performance because it makes people have disruptive behaviors, such as difficulty remaining seated and organizing tasks, playing noisily, and sustaining attention during schoolwork or play activities. Guevara et al. [5] addressed the cost-effectiveness of ADHD in the U.S. by comparing children with and without ADHD [5]. They found that additional cost for health care is needed for children with ADHD due to the chance of the coexisting mental health disorders. Early diagnosis of neurodevelopmental disorders is crucial for planning appropriate early intervention and providing caregivers with accurate information. It would reduce the high financial costs associated with patients themselves or their caregivers [4]. Electroencephalography (EEG) is a widely used technique for capturing brain activity and identifying brain disorders because it provides low-cost, non-invasive, and portable capability [7]. Therefore, EEG analysis is used in predicting neurodevelopmental disorders [20]. There are well-known brain waves classified into five frequency band components associated with each type of brain wave, called delta (δ : 0.5$-$4Hz), theta (θ : 4$-$8Hz), alpha (α : 8$-$13Hz), beta (β : 13$-$30Hz), and gamma (γ : 30$-$45Hz) [18]. Although neurological or psychiatric dysfunction is one of the public health concerns, limited EEG analyses associated with brain disorders have been performed. Loh et al. [9] conducted an in-depth literature review on ADHD detection using either machine learning (ML) or deep learning (DL) approaches utilizing Magnetic Resonance Imaging (MRI) and physiological signals. They identified 69 studies using ML and 23 studies utilizing DL techniques. Among various ML studies, support vector machine (SVM) was determined as the most commonly used ML technique, and the convolutional neural network (CNN) was the most broadly used model in DL for ADHD research. MRI and EEG signals are the most widely used data in ML- or DL-based ADHD research. They also identified that EEG is the most prevalent data used in ADHD research. Although the average accuracy reported by numerous ML and DL studies was high ($80\% \sim 90\%$ since 2013), they emphasized that much improvement should be performed prior to being used in clinical use for ADHD diagnosis.

In this study, we aim to explore the feasibility of wavelet feature-based machine learning to predict neurodevelopmental conditions. We propose an integration of wavelet transform and visual analysis to identify the similarities and differences associated with neurodevelopmental and neurological disorders. In detail, extracting essential features representing the characteristics of the disorders is performed. A performance evaluation with ML classification algorithms is conducted to validate the effectiveness of the features. Since data often consists of imbalanced cases, classification performance may not be a perfect solution for finding their distinctive patterns. Thus, we utilized visualization techniques to represent all data instances and support a visual analysis to identify trends and patterns of the diseases compared to healthy subjects.

2 Previous Work

Researchers have performed various studies to understand neurodevelopmental disorders, specifically ADHD. Musser and Nigg [13] performed a study to identify the coherence of facial and autonomic behaviors of emotion reactivity and regulation in children with ADHD. They performed an experiment with 100 children aged 7–11 years old, where 50 of them had ADHD (62% male, 78% white) to understand emotion induction and suppression for negative and positive emotion-provoking task conditions by collecting electrocardiogram and impedance cardiography data. They found similar behavior for facial affect between children with and without ADHD. The children with ADHD exhibited reduced coherence between facial affect behavior and an index of parasympathetic functioning. They also found that children with ADHD may receive conflicting emotional signals compared to children without ADHD. Mohamed et al. [11] proposed an approach to detect two different cognitive skills, focused attention and working memory, to understand the effect of the outcomes of subjects' learning processes using EEG signals. They performed an assessment test to measure the complete cognitive profile of the subjects. By analyzing the EEG signals, a total of 280 domain features associated with time and frequency were extracted and used to predict three levels (i.e., low, average, and high) of the cognitive skills of learners. With the approach, they identified classification accuracies for the skills as 84% and 81%, respectively. Singh et al. [17] conducted a review on the diagnosis of ADHD in both children and adults. Specifically, they focused on understanding the process of how it develops, what associated problems might be, and how many other children and adults might be affected. They defined that ADHD diagnosis might rely on a combination of neuropsychological tests, teacher rating scales, direct observations of behavior, examinations of genetic factors, and evaluations of the impact of treatment trials and more. Although researchers have agreed upon a strong contribution to the occurrence of ADHD by genetic factors, they emphasized that numerous studies should be performed to understand symptoms and proper medical treatments clearly.

Koh et al. [8] developed a software tool to categorize ADHD or conduct disorder (CD) automatically by analyzing ECG signals. After transforming ECG signal data with empirical wavelet transform (EWT), entropy features were used. Then, analysis of variance (ANOVA) was applied to determine the variability and highly significant features. With three classification ML algorithms: SVM, decision tree (DT), and k-nearest neighbor (KNN), they showed the capability of diagnosing behavioral disorders by utilizing ECG-based models. Yasin et al. [22] reviewed previous works published from 2015 to 2020 focusing on analyzing mental disorders, Major Depressive Disorder (MDD) and Bipolar Disorder (BD) using EEG and artificial neural networks. They found various limitations in consideration of data availability, advancement of signal filtering techniques to process EEG data to remove noises compared to using notch and band-pass filtering, feature extraction capability compared to conventional machine learning algorithms, and utilization of DL. Although there were high demands using DL techniques, they emphasized the weakness of interpreting DL results precisely. Mulaffer et al. [12] studied sleep disorders by analyzing EEG utilizing ML and SVM. C3 and C4 EEG channels were used to detect insomnia. They compared EEG-based features and hypnogram-based features with SVM and found that EEG-based features showed a better performance in detecting insomnia. SVM has been emphasized as an important technique for analyzing EEG signals to detect dyslexia using EEG [14, 15]. Although numerous studies have been performed in identifying neurodevelopmental disorders, including ADHD, MDD, or BD, extracting important features to find the patterns of the disorders, performing either machine learning or deep learning approach, and interpreting the extracted features associated with the disorders are still needed for further studies. Therefore, this study focuses on finding important features utilizing wavelet transform utilizing neurophysiology EEG database [18], classifying the neurological disorders from normal subjects, and enhancing the understanding of the differences and similarities of the features utilizing visual analysis.

3 Methods

3.1 Data Description

In this study, we used TDBRAIN dataset (https://brainclinics.com/resources/) [18]. It contains a total of 1274 EEG data (with a sampling frequency of 500 Hz) for psychiatric patients (the ages of 5-89 years old) collected from 2001 to 2021. In this study, we analyzed 24 channels (C3, C4, CP3, CP4, CPz, F3, F4, F7, Cz, F8, P3, P4, FC3, FC4, FCz, Fp2, Fz, O1, O2, Oz, P8, T7, Mass, and T8). The dataset was collected based on two tasks - Eyes Open (EO) and Eyes Closed (EC). The dataset includes an attribute 'indication' to represent an unofficial diagnosis for EEG assessments by a general practitioner or psychologist/psychiatrist and a formal diagnosis. Many of the subjects were identified as 'Unknown' for their formal diagnosis. Thus, in this study, we only focused on the subjects that have matched information between indication (unofficial diagnosis as a referral-indication) and formal diagnosis (confirmed diagnosis by a licensed clinician) ($N = 305$) containing 'Burnout', 'Dyslexia', 'Chronic pain', 'MDD', 'OCD', 'ADHD', 'Parkinson', 'Insomnia', and 'Healthy.' Due to the high disorder types in the dataset, we analyzed all subjects' data classified into two groups 'Normal (i.e., Healthy)' and 'Abnormal' (i.e., Unhealthy).'

3.2 Proposed Approach

Our proposed approach is designed to have four steps: pre-processing, feature extraction and selection, channel-based predictive model generation, and visual analysis. Our primary consideration for proposing this new approach is to support finding important features presenting the characteristics of neurological and/or neurodevelopmental disorders, classifying them to differentiate the groups (Normal and Abnormal), and understanding distinctive patterns through visual analysis.

3.2.1 Pre-processing

As pre-processing steps, filtering and normalization are applied. First, a sliding moving average window was used to smooth EEG data. It is a commonly used technique that smooths data to capture trends from the data. Second, 50 Hz frequency was removed with a notch filter, and the band-pass filter between $0.5 \sim 100$ Hz frequencies was applied. Last, data normalization is applied.

3.2.2 Feature Extraction and Selection

Feature extraction is an essential step that influences the overall classification performance because it enhances the capability of understanding data associated with neurological developmental disorders. Fourier transform is a good analysis approach because it determines frequency information. Although it helps find frequency information from data, any local behaviors may not be detected. Thus, it is suitable for analyzing stationary data. Alternatively, wavelet transform is suitable for analyzing non-stationary data due to its ability to extract both frequency (scales) and time information. We used wavelet transform (WT) to examine the capability of identifying the disorders. Wavelets represent small waves that have limited duration, and zero average values [3]. WT is a suitable technique for analyzing data at a specific time and frequency or finding information with different scales [16]. It helps present local information with different frequencies to show trends, discontinuities, and repeated patterns underlying data.

In our study, discrete wavelet transform (DWT) is used to extract features and examine their capability in presenting the disorder conditions. It decomposes data until a pre-defined level. The input data are split each level into two sub-bands containing different frequency ranges (i.e., low and high frequencies). The high frequency represents detail coefficients, and the low frequency indicates approximate coefficients. Since the detail coefficients can detect rapid changes, they are broadly used to identify discontinuities or sudden changes. A non-overlapping sliding window was used to extract features. The sliding window size was defined as the sampling rate (i.e., 500) to extract the following wavelet features.

$$\omega_1^j = \sum_1^n \frac{d_{i,j}}{n}, \quad \omega_2^j = \sqrt{\frac{\sum_{i=1}^n (d_{i,j} - \overline{d_{i,j}})^2}{n}}, \quad \omega_3^j = \frac{\mu_{d_{i,3}}}{\mu_a},$$

$$\omega_4^j = \frac{\sum_1^n |\delta_{i,j}|}{n}, \quad \omega_5 = \sum_1^n \frac{(|d_{i,j}| - \overline{d_{i,j}})}{m_{\delta i,j}}, \quad \omega_6 = \frac{\mu_{d_{i,4}}}{\mu_a}$$

where $d_{i,j} = \{d_{1,j}, d_{2,j}, \ldots, d_{i,j}\}$ represent wavelet coefficients at the jth level, where $j = 1, 2, \ldots, l$, and l is a pre-defined decomposition level, and $i = 1, 2, \ldots, n$ is the number of elements of coefficients. $\overline{d_{i,j}}$ represents mean of $d_{i,j}$, $\delta_{i,j}$ denotes the sequences of elements that are greater than $\overline{d_{i,j}}$, especially when $d_{i,j}$ is sorted in descending order, $m_{\delta i,j}$ indicates the number of elements for $\delta_{i,j}$, μ_a is average approximation coefficients, and n indicates the length of coefficients.

To select the significant features from wavelet features, two different approaches–a statistical approach (analysis of variance (ANOVA)) and principal component analysis (PCA) [10]–are used. With the statistical approach, statistically significant features are determined. With PCA, principal components are identified. These feature selection approaches are named, in short, DWT-ANOVA and DWT-PCA, respectively.

3.3 Channel-Based Predictive Model Generation

An extensive classification study is performed to investigate the effectiveness of the selected features in identifying abnormal conditions (i.e., Unhealthy) from each channel. With the validated wavelet features (DWT-ANOVA and DWT-PCA), three ML classification algorithms, SVM, KNN, and Naive Bayes, are applied to to generate predictive learning models and compare their performances. For evaluating their classification performances, multiple metrics, including precision, recall, F1 score, and the area of the receiver operating characteristic (AUC), are calculated with five-fold cross-validation (5CV).

3.4 *Visual Analysis*

Measuring classification performances with various metrics would benefit in determining the effectiveness of differentiating the groups (Normal and Abnormal). However, it has a limitation in showing the underlying patterns of the features. Thus, visualization is considered to represent the selected features by mapping them into visual glyphs. In EEG data analysis associated with neurodevelopmental disorder studies, visual representations can help understand the patterns and structures of the data. Since the data consist of numerous variables, dimension reduction is applied to show the data in a lower dimensional space (i.e., 2D space). Various dimension reduction techniques are broadly used when representing high-dimensional data, such as PCA, LDA (Linear Discriminant Analysis), MDS (Multi-dimensional scaling), t-SNE (t-distributed stochastic neighbor embedding), UMAP (Uniform Manifold Approximation and Projection), and more. Among them, t-SNE and UMAP have been broadly used recently in representing data because they showed a better separation of the classes than others. However, since they use approximation methods, they require more computation time. Thus, PCA is used in our study to represent the data. The first and second principal components are used to create a scatterplot representing all instances.

4 Results

Figure 1 presents average EEG band frequencies (α, β, γ, δ, θ) between the two tasks (EO and EC). The EEG band frequencies for the subjects' data with ADHD showed similar patterns in the tasks. But, the other subjects' EEG band frequencies showed slightly different patterns. For instance, healthy subjects' data presented high δ and θ values in the EO and EC tasks, respectively. We also found a similar result when analyzing the data with Burnout and MDD having higher α values in the EC task than the EO task. The Dyslexia subjects' data indicated the highest δ value among all other frequencies (see Fig. 1f). When comparing the δ value for the Dyslexia between the EO and EC tasks, it showed a high δ value in the EC task. This was an opposite pattern compared to δ in healthy subjects' data because the δ value was high in the EO task in the healthy subjects' data (see Fig. 1a). Interestingly, three waves (α, δ, and θ) showed similar high amplitude for the Parkinson subjects. When comparing Dyslexia, ADHD, and MDD, we found indistinguishable patterns showing similar amplitudes in the EO task. But, their amplitudes were different in the EC task.

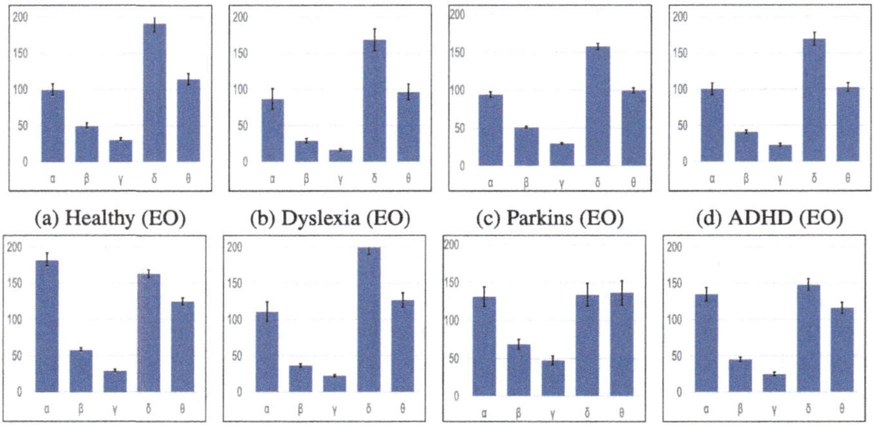

(a) Healthy (EO) (b) Dyslexia (EO) (c) Parkins (EO) (d) ADHD (EO)

(e) Healthy (EC) (f) Dyslexia (EC) (g) Parkinson (EC) (h) ADHD (EC)

Fig. 1 The presentation of average EEG wave bands ($\alpha, \beta, \gamma, \delta, \theta$) of the EEG channel (C4) between eyes open (EO) and eyes close (EC). The x-axis shows the EEG band frequencies and the y-axis represents the average of mean amplitude

By using level six decomposition with db4 wavelet, forty-two features were extracted. As mentioned above, we employed ANOVA and PCA to determine significant features from the wavelet-based features. Since each channel may have different variation distributions, threshold (T) is used to select PCA components instead of fixing the number of components. We defined $T = 0.95$ for principal components selection in this study. With the statistical test (i.e., ANOVA), only statistically significant features ($p < 0.05$) were determined. Different numbers of features were selected in each channel. About $17 \sim 35$ features were selected with ANOVA and $15 \sim 16$ for PCA with the EC task. For the EO task, $17 \sim 40$ features were determined with ANOVA. The same number of PCA features (i.e., principal components) were selected for the EC and EO tasks. To compare the features across all channels, we identified common features that appeared in all channels. There were nine common features for the EC task and six features for the EO task. The common features were the decomposition levels (4,5,6 and approximate coefficients).

Figure 2 shows a comparison of the common features with ANOVA for the channels ('C4' and 'Fp2'). It represents comparisons of the features by gender (male and female) and group (Normal and Abnormal). We found that the wavelet features could distinguish the difference between male and female subjects in either group. The feature ('w9' in the EC task) was not visible due to its small value. Interestingly, the Abnormal group showed similar feature values, while there were some differences for the Normal group. In particular, the Normal group between male and female subjects showed a minor distinction between the features.

Table 1 shows a comparison of performance results by applying different classification ML algorithms to all wavelet-based features and the selected features with

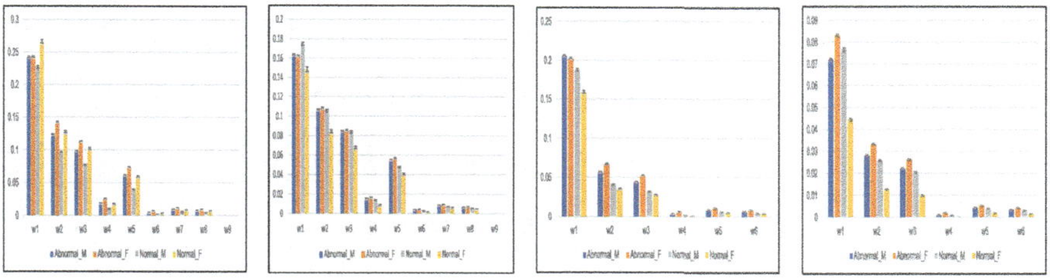

(a) C4 (b) Fp2 (c) C4 (d) Fp2

Fig. 2 The comparisons of the common features with ANOVA for the channels (C4 and Fp2) based on gender and class (i.e., Normal and Abnormal) for the EC task (**a** and **b**) and the EO task (**c** and **d**). Abnormal_M and Abnormal_F indicate male and female subjects with disorder conditions. The x-axis indicates the significant common features, and the y-axis represents the average value of the features. Each bar graph presents average ± SEM (standard error mean)

PCA and ANOVA. We found that KNN performed better than Naive Bayes for all channels except the AUC values. Improved classification performances were observed when using feature selection techniques. More specifically, the integration of using both wavelet features and PCA showed an improved performance result compared to using the features with ANOVA. We also found that Naive Bayes provided a better AUC performance except for the features with PCA. It is important to note that we excluded SVM classification performance from the table because it showed almost the same performance results (producing all 99%) for all channels.

Extensive visual analysis has been performed to evaluate all features for different channels to understand the patterns of all features. Figure 3 shows example visualizations of two channels (C4 and Fp2) with significant features. For other channels, we found similar results. Figure 3a and c represent the features collected when participants closed their eyes. Figure 3b and d show the representation of the features indicating the participants with eyes open. Different color attributes are used to indicate normal (blue) and abnormal (red) subjects. Although normal subjects appeared in a small region of the PCA space, there was no distinctive separation between them. However, numerous outliers were determined by showing whiskers and outliers using Tukey box-and-whisker plot (see Fig. 3c).

5 Conclusions and Future Works

In this paper, we introduced the integration of wavelet transform and visual analysis to analyze EEG to distinguish abnormal neurologically based disorders from normal neurological functioning. . We used wavelet transform to extract features by employing mathematical and statistical concepts to capture any changes in data. In addition, we compared two feature selection techniques, ANOVA and PCA, on the

Table 1 Performance evaluation results of the features from the EO and EC tasks with KNN and Naive Bayes classification algorithms. The results for the EO data are included in parentheses

	KNN												Naive Bays											
	ALL				ANOVA				PCA				ALL				ANOVA				PCA			
	P	R	F	A	P	R	F	A	P	R	F	A	P	R	F	A	P	R	F	A	P	R	F	A
C3	0.87 (0.88)	0.99 (0.91)	0.88 (0.89)	0.53 (0.53)	0.88 (0.91)	0.91 (0.91)	0.88 (0.88)	0.54 (0.54)	0.99 (0.99)	0.99 (0.99)	0.99 (0.99)	0.99 (0.99)	0.88 (0.87)	0.67 (0.62)	0.74 (0.70)	0.63 (0.63)	0.88 (0.89)	0.72 (0.55)	0.78 (0.65)	0.64 (0.66)	0.98 (0.98)	0.98 (0.98)	0.97 (0.98)	0.88 (0.93)
C4	0.86 (0.88)	0.90 (0.91)	0.88 (0.88)	0.52 (0.53)	0.87 (0.87)	0.91 (0.91)	0.89 (0.89)	0.53 (0.54)	0.99 (0.99)	0.99 (0.99)	0.99 (0.99)	0.99 (0.99)	0.87 (0.89)	0.68 (0.6)	0.75 (0.69)	0.62 (0.63)	0.88 (0.89)	0.73 (0.59)	0.78 (0.68)	0.63 (0.63)	0.98 (0.98)	0.98 (0.98)	0.97 (0.98)	0.85 (0.9)
CP3	0.73 (0.87)	0.91 (0.9)	0.88 (0.88)	0.53 (0.53)	0.88 (0.88)	0.91 (0.91)	0.89 (0.89)	0.54 (0.55)	0.99 (0.99)	0.99 (0.99)	0.99 (0.99)	0.99 (0.99)	0.87 (0.88)	0.87 (0.63)	0.64 (0.7)	0.72 (0.65)	0.88 (0.90)	0.72 (0.56)	0.78 (0.65)	0.64 (0.67)	0.98 (0.98)	0.98 (0.98)	0.98 (0.98)	0.88 (0.87)
CP4	0.86 (0.86)	0.90 (0.9)	0.88 (0.88)	0.53 (0.52)	0.88 (0.87)	0.91 (0.91)	0.89 (0.88)	0.54 (0.53)	0.99 (0.99)	0.99 (0.99)	0.99 (0.99)	0.99 (0.99)	0.87 (0.9)	0.87 (0.57)	0.64 (0.66)	0.6 (0.69)	0.88 (0.90)	0.55 (0.57)	0.65 (0.66)	0.64 (0.69)	0.98 (0.98)	0.98 (0.98)	0.97 (0.98)	0.86 (0.85)
CPz	0.87 (0.86)	0.91 (0.9)	0.88 (0.88)	0.52 (0.52)	0.88 (0.87)	0.91 (0.91)	0.89 (0.88)	0.53 (0.53)	0.99 (0.99)	0.99 (0.99)	0.99 (0.99)	0.99 (0.99)	0.88 (0.89)	0.64 (0.57)	0.72 (0.66)	0.62 (0.66)	0.87 (0.90)	0.71 (0.56)	0.77 (0.66)	0.61 (0.68)	0.98 (0.97)	0.98 (0.97)	0.98 (0.97)	0.89 (0.84)
F3	0.66 (0.87)	0.91 (0.9)	0.88 (0.88)	0.52 (0.53)	0.88 (0.88)	0.87 (0.91)	0.88 (0.89)	0.60 (0.55)	0.99 (0.99)	0.99 (0.99)	0.99 (0.99)	0.99 (0.99)	0.88 (0.88)	0.64 (0.61)	0.72 (0.7)	0.62 (0.63)	0.88 (0.89)	0.73 (0.62)	0.79 (0.70)	0.64 (0.64)	0.98 (0.98)	0.98 (0.98)	0.98 (0.98)	0.91 (0.90)
F4	0.86 (0.87)	0.90 (0.91)	0.88 (0.9)	0.52 (0.53)	0.88 (0.89)	0.87 (0.91)	0.88 (0.89)	0.60 (0.56)	0.99 (0.99)	0.99 (0.99)	0.99 (0.99)	0.99 (0.99)	0.87 (0.87)	0.70 (0.62)	0.77 (0.70)	0.61 (0.59)	0.88 (0.87)	0.73 (0.62)	0.79 (0.70)	0.62 (0.60)	0.98 (0.98)	0.98 (0.98)	0.98 (0.98)	0.90 (0.90)
F7	0.87 (0.87)	0.91 (0.71)	0.88 (0.88)	0.52 (0.52)	0.88 (0.88)	0.88 (0.92)	0.88 (0.89)	0.62 (0.55)	0.99 (0.99)	0.99 (0.99)	0.99 (0.99)	0.99 (0.99)	0.87 (0.88)	0.70 (0.58)	0.77 (0.88)	0.61 (0.63)	0.89 (0.89)	0.60 (0.57)	0.69 (0.66)	0.65 (0.64)	0.98 (0.98)	0.98 (0.98)	0.98 (0.98)	0.93 (0.91)
Cz	0.86 (0.86)	0.90 (0.91)	0.88 (0.88)	0.53 (0.52)	0.87 (0.87)	0.91 (0.91)	0.88 (0.88)	0.53 (0.53)	0.99 (0.99)	0.99 (0.99)	0.99 (0.99)	0.99 (0.99)	0.88 (0.89)	0.55 (0.58)	0.65 (0.67)	0.60 (0.66)	0.89 (0.90)	0.60 (0.57)	0.69 (0.66)	0.64 (0.67)	0.98 (0.97)	0.98 (0.97)	0.98 (0.97)	0.90 (0.84)
F8	0.86 (0.86)	0.90 (0.91)	0.88 (0.88)	0.51 (0.52)	0.87 (0.88)	0.91 (0.91)	0.88 (0.89)	0.53 (0.54)	0.99 (0.99)	0.99 (0.99)	0.99 (0.99)	0.99 (0.99)	0.88 (0.60)	0.55 (0.88)	0.65 (0.60)	0.60 (0.62)	0.88 (0.88)	0.56 (0.60)	0.65 (0.69)	0.61 (0.63)	0.98 (0.97)	0.98 (0.97)	0.98 (0.97)	0.90 (0.85)
P3	0.86 (0.86)	0.90 (0.90)	0.99 (0.90)	0.99 (0.53)	0.88 (0.87)	0.91 (0.91)	0.89 (0.88)	0.53 (0.54)	0.99 (0.99)	0.99 (0.99)	0.99 (0.99)	0.99 (0.99)	0.88 (0.88)	0.55 (0.62)	0.65 (0.69)	0.60 (0.64)	0.88 (0.90)	0.72 (0.55)	0.78 (0.64)	0.62 (0.67)	0.98 (0.97)	0.98 (0.97)	0.98 (0.97)	0.90 (0.86)
P4	0.87 (0.87)	0.90 (0.9)	0.88 (0.88)	0.52 (0.52)	0.88 (0.87)	0.91 (0.91)	0.89 (0.88)	0.54 (0.53)	0.99 (0.99)	0.99 (0.99)	0.99 (0.99)	0.99 (0.99)	0.87 (0.89)	0.65 (0.59)	0.73 (0.67)	0.60 (0.67)	0.88 (0.91)	0.70 (0.55)	0.77 (0.65)	0.61 (0.69)	0.98 (0.97)	0.97 (0.97)	0.97 (0.97)	0.86 (0.85)
FC3	0.87 (0.87)	0.90 (0.9)	0.88 (0.88)	0.53 (0.53)	0.87 (0.88)	0.91 (0.91)	0.89 (0.89)	0.54 (0.55)	0.99 (0.99)	0.99 (0.99)	0.99 (0.99)	0.99 (0.99)	0.88 (0.88)	0.63 (0.58)	0.71 (0.67)	0.60 (0.64)	0.89 (0.89)	0.66 (0.56)	0.73 (0.66)	0.67 (0.64)	0.97 (0.98)	0.97 (0.98)	0.97 (0.98)	0.86 (0.90)
FC4	0.87 (0.87)	0.90 (0.91)	0.88 (0.88)	0.53 (0.53)	0.88 (0.88)	0.91 (0.91)	0.89 (0.89)	0.55 (0.55)	0.99 (0.99)	0.99 (0.99)	0.99 (0.99)	0.99 (0.99)	0.89 (0.89)	0.61 (0.59)	0.70 (0.68)	0.65 (0.64)	0.90 (0.89)	0.56 (0.59)	0.66 (0.68)	0.66 (0.64)	0.98 (0.97)	0.98 (0.98)	0.98 (0.97)	0.91 (0.88)
FCz	0.86 (0.87)	0.90 (0.9)	0.88 (0.88)	0.51 (0.52)	0.87 (0.87)	0.91 (0.91)	0.88 (0.88)	0.52 (0.53)	0.99 (0.99)	0.99 (0.99)	0.99 (0.99)	0.99 (0.99)	0.87 (0.89)	0.69 (0.58)	0.76 (0.67)	0.59 (0.65)	0.87 (0.89)	0.73 (0.57)	0.79 (0.66)	0.60 (0.66)	0.98 (0.98)	0.98 (0.98)	0.98 (0.98)	0.89 (0.89)
Fp2	0.87 (0.87)	0.91 (0.91)	0.88 (0.88)	0.52 (0.52)	0.88 (0.88)	0.91 (0.91)	0.89 (0.89)	0.55 (0.56)	0.99 (0.99)	0.99 (0.99)	0.99 (0.99)	0.99 (0.99)	0.85 (0.87)	0.86 (0.6)	0.86 (0.69)	0.52 (0.57)	0.86 (0.87)	0.89 (0.59)	0.87 (0.68)	0.52 (0.57)	0.98 (0.97)	0.98 (0.97)	0.98 (0.97)	0.90 (0.88)
Fz	0.86 (0.87)	0.90 (0.91)	0.88 (0.88)	0.52 (0.53)	0.88 (0.88)	0.91 (0.91)	0.89 (0.89)	0.54 (0.55)	0.99 (0.99)	0.99 (0.99)	0.99 (0.99)	0.99 (0.99)	0.87 (0.89)	0.64 (0.62)	0.72 (0.71)	0.61 (0.67)	0.88 (0.89)	0.63 (0.63)	0.71 (0.71)	0.62 (0.67)	0.99 (0.98)	0.99 (0.98)	0.99 (0.98)	0.94 (0.93)

(continued)

Table 1 (continued)

	KNN												Naive Bays											
	ALL				ANOVA				PCA				ALL				ANOVA				PCA			
	P	R	F	A	P	R	F	A	P	R	F	A	P	R	F	A	P	R	F	A	P	R	F	A
Mass	0.87 (0.88)	0.91 (0.91)	0.88 (0.89)	0.52 (0.54)	0.88 (0.89)	0.91 (0.92)	0.89 (0.90)	0.55 (0.57)	0.99 (0.99)	0.99 (0.99)	0.99 (0.99)	0.99 (0.99)	0.87 (0.87)	0.59 (0.58)	0.68 (0.68)	0.58 (0.58)	0.87 (0.87)	0.58 (0.57)	0.67 (0.66)	0.59 (0.58)	0.97 (0.98)	0.97 (0.98)	0.97 (0.98)	0.88 (0.89)
O1	0.86 (0.87)	0.90 (0.91)	0.88 (0.88)	0.51 (0.53)	0.87 (0.88)	0.91 (0.91)	0.89 (0.89)	0.53 (0.54)	0.99 (0.99)	0.99 (0.99)	0.99 (0.99)	0.99 (0.99)	0.88 (0.89)	0.55 (0.6)	0.64 (0.69)	0.63 (0.66)	0.89 (0.90)	0.55 (0.55)	0.65 (0.64)	0.63 (0.66)	0.97 (0.98)	0.97 (0.98)	0.97 (0.98)	0.86 (0.92)
O2	0.86 (0.87)	0.90 (0.91)	0.88 (0.88)	0.52 (0.53)	0.88 (0.87)	0.91 (0.91)	0.89 (0.88)	0.54 (0.53)	0.99 (0.99)	0.99 (0.99)	0.99 (0.99)	0.99 (0.99)	0.89 (0.89)	0.56 (0.6)	0.66 (0.69)	0.64 (0.66)	0.90 (0.89)	0.54 (0.59)	0.64 (0.68)	0.65 (0.66)	0.98 (0.98)	0.98 (0.98)	0.98 (0.98)	0.91 (0.90)
Oz	0.86 (0.87)	0.91 (0.91)	0.88 (0.88)	0.52 (0.53)	0.87 (0.88)	0.91 (0.91)	0.88 (0.89)	0.53 (0.55)	0.99 (0.99)	0.99 (0.99)	0.99 (0.99)	0.99 (0.99)	0.89 (0.9)	0.58 (0.62)	0.68 (0.71)	0.63 (0.69)	0.89 (0.91)	0.54 (0.56)	0.63 (0.66)	0.64 (0.70)	0.98 (0.98)	0.98 (0.98)	0.98 (0.98)	0.91 (0.9)
P8	0.87 (0.87)	0.91 (0.9)	0.88 (0.88)	0.52 (0.53)	0.87 (0.87)	0.91 (0.91)	0.89 (0.89)	0.54 (0.54)	0.99 (0.99)	0.99 (0.99)	0.99 (0.99)	0.99 (0.99)	0.89 (0.9)	0.60 (0.58)	0.69 (0.67)	0.64 (0.68)	0.90 (0.91)	0.56 (0.55)	0.65 (0.65)	0.66 (0.69)	0.98 (0.98)	0.98 (0.98)	0.98 (0.98)	0.90 (0.92)
T7	0.88 (0.88)	0.87 (0.91)	0.87 (0.89)	0.61 (0.54)	0.88 (0.89)	0.91 (0.92)	0.89 (0.89)	0.54 (0.56)	0.99 (0.99)	0.99 (0.99)	0.99 (0.99)	0.99 (0.99)	0.88 (0.86)	0.58 (0.79)	0.68 (0.82)	0.62 (0.56)	0.88 (0.86)	0.64 (0.86)	0.72 (0.86)	0.63 (0.54)	0.97 (0.98)	0.97 (0.98)	0.97 (0.98)	0.86 (0.89)
T8	0.87 (0.87)	0.91 (0.91)	0.88 (0.88)	0.54 (0.53)	0.88 (0.88)	0.91 (0.91)	0.89 (0.89)	0.54 (0.54)	0.99 (0.99)	0.99 (0.99)	0.99 (0.99)	0.99 (0.99)	0.87 (0.87)	0.77 (0.69)	0.82 (0.76)	0.63 (0.59)	0.88 (0.88)	0.79 (0.61)	0.83 (0.70)	0.63 (0.54)	0.97 (0.98)	0.97 (0.98)	0.97 (0.98)	0.86 (0.9)

*P: Precision, R: Recall, F: F1-score, A: Accuracy

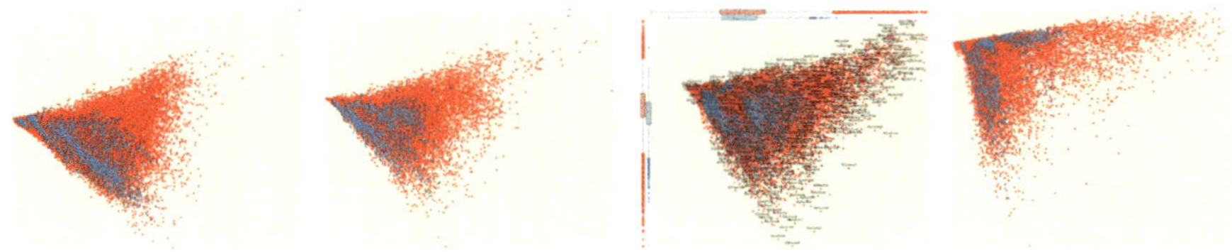

(a) EC - C4 (b) EO - C4 (c) EC - Fp2 (d) EO - Fp2

Fig. 3 Visualizations of the selected significant features in a PCA space. The x-axis indicates the 1st principal component and the y-axis denotes the 2nd principal component.

wavelet-based features. We found that DWT-PCA performed better in identifying abnormal conditions.

We performed two evaluations to assess the capability of the wavelet features. First, classification evaluation was performed with two machine learning techniques. KNN performed better than Naive Bays. However, Naive Bays with ANOVA provided better AUC than KNN. Since it is essential to examine the wavelet-based visually, considering the relationship with the class (i.e., normal and abnormal), visual analysis was also performed to enhance the understanding of the features.

For future works, we will utilize deep learning techniques to differentiate abnormal conditions by integrating all the channels within the dataset. Since appropriate wavelet selection is crucial in analyzing data through wavelet transform, different wavelet families will be tested to determine the best possible wavelet family to improve the overall ability to analyze EEG data. Our study has a limitation in finding dominant channel(s) associated with different neurological developmental disorders. Thus, an extensive analysis will be performed to determine an optimal approach for ADHD analysis. Also, we plan to investigate new features integrating different EEG band frequencies, the complexity of EEG, and power spectral density that can present different conditions of data. Lastly, we extend our study to test the proposed approach in different datasets.

Acknowledgements This material is based upon work supported by the National Science Foundation under Grant No. (2219634).

References

1. A. Abramovitch, B. McCormack, D. Brunner, M. Johnson, N. Wofford, The impact of symptom severity on cognitive function in obsessive-compulsive disorder: a meta-analysis. Clin. Psychol. Rev. **67**, 36–44 (2019)
2. A. Baddeley, Working memory. Science **255**(5044), 556–559 (1992)
3. I. Daubechies. *Ten Lectures on Wavelets* (SIAM, 1992)
4. H.C. Glass, Y. Li, M. Gardner, A.J. Barkovich, I. Novak, C.E. McCulloch, E.E. Rogers. Early identification of cerebral palsy using neonatal mri and general movements assessment in a cohort of high-risk term neonates. Pediatr. Neurol. **118**, 20–25 (2021)

5. J. Guevara, P. Lozano, T. Wickizer, L. Mell, H. Gephart, Utilization and cost of health care services for children with attention-deficit/hyperactivity disorder. Pediatrics **108**(1), 71–78 (2001)
6. A. Habib, L. Harris, F. Pollick, C. Melville, A meta-analysis of working memory in individuals with autism spectrum disorders. PLoS ONE **14**(4), e0216198 (2019)
7. S.K. Khare, V. Bajaj, U. Rajendra Acharya, Pdcnnet: an automatic framework for the detection of Parkinson's disease using eeg signals. IEEE Sens. J. **21**(15), 17017–17024 (2021)
8. J.E.W. Koh, C.P. Ooi, N.S.J. Lim-Ashworth, J. Vicnesh, H.T. Tor, O.S. Lih, R.-S. Tan, U Rajendra Acharya, D.S.S. Fung, Automated classification of attention deficit hyperactivity disorder and conduct disorder using entropy features with ecg signals. Comput. Biol. Med. **140**, 105120 (2022)
9. H.W. Loh, C.P. Ooi, P.D. Barua, E.E. Palmer, F. Molinari, U. Rajendra Acharya, Automated detection of adhd: current trends and future perspective. Comput. Biol. Med. 105525 (2022)
10. A. Maćkiewicz, W. Ratajczak, Principal components analysis (pca). Comput. Geosci. **19**(3), 303–342 (1993)
11. Z. Mohamed, M. El Halaby, T. Said, D. Shawky, A. Badawi, Characterizing focused attention and working memory using eeg. Sensors **18**(11), 3743 (2018)
12. L. Mulaffer, M. Shahin, M. Glos, T. Penzel, B. Ahmed, Comparing two insomnia detection models of clinical diagnosis techniques, in *2017 39th Annual International Conference of the IEEE Engineering in Medicine and Biology Society (EMBC)* (IEEE, 2017), pp. 3749–3752
13. E.D. Musser, J.T. Nigg, Emotion dysregulation across emotion systems in attention deficit/hyperactivity disorder. J. Clin. Child Adolesc. Psychol. **48**(1), 153–165 (2019)
14. H. Perera, M.F. Shiratuddin, K.W. Wong, A review of electroencephalogram-based analysis and classification frameworks for dyslexia, in *International Conference on Neural Information Processing* (Springer, 2016), pp. 626–635
15. P. Perera, H. Harshani, M.F. Shiratuddin, K.W. Wong, K. Fullarton. Eeg signal analysis of writing and typing between adults with dyslexia and normal controls (2018)
16. A. Samant, H. Adeli. Feature extraction for traffic incident detection using wavelet transform and linear discriminant analysis. Comput. Aided Civil Infrastruct. Eng. **15**(4), 241–250 (2000)
17. A. Singh, C.J. Yeh, N. Verma, A.K. Das, Overview of attention deficit hyperactivity disorder in young children. Health Psychol. Res. **3**(2), 2115 (2015)
18. H. van Dijk, G. van Wingen, D. Denys, S. Olbrich, R. van Ruth, M. Arns, The two decades brainclinics research archive for insights in neurophysiology (tdbrain) database. Sci. Data **9**(1), 1–10 (2022)
19. E.R. Watkins, H. Roberts, Reflecting on rumination: consequences, causes, mechanisms and treatment of rumination. Behav. Res. Therapy **127**, 103573 (2020)
20. D. Watts, R. Fernandes Pulice, J. Reilly, A.R. Brunoni, F. Kapczinski, I.C. Passos, Predicting treatment response using eeg in major depressive disorder: a machine-learning meta-analysis. Transl. Psychiatry **12**(1), 1–18 (2022)
21. M.L. Wolraich, J.F. Hagan, C. Allan, E. Chan, D. Davison, M. Earls, S.W. Evans, S.K. Flinn, T. Froehlich, J. Frost et al., Clinical practice guideline for the diagnosis, evaluation, and treatment of attention-deficit/hyperactivity disorder in children and adolescents. Pediatrics **144**(4) (2019)
22. S. Yasin, S.A. Hussain, S. Aslan, I. Raza, M. Muzammel, A. Othmani, Eeg based major depressive disorder and bipolar disorder detection using neural networks: a review. Comput. Methods Progr. Biomed. **202**:, 106007 (2021)

Auditing Algorithmic Fairness in Machine Learning for Health with Severity-Based LOGAN

Anaelia Ovalle, Sunipa Dev, Jieyu Zhao, Majid Sarrafzadeh, and Kai-Wei Chang

Abstract Auditing machine learning-based healthcare (ML4H) tools for bias is critical to preventing patient harm, especially in communities disproportionately facing health inequities. General frameworks are becoming increasingly available to measure ML fairness gaps between groups. However, ML4H auditing principles call for contextual, patient-centered approaches to model assessment. Therefore, ML auditing tools must be (1) better aligned with ML4H auditing principles and (2) able to illuminate and characterize communities vulnerable to the most harm. To address this gap, we propose supplementing ML4H auditing frameworks with SLOGAN (patient Severity-based LOcal Group biAs detectioN), an automatic tool for capturing local biases in clinical prediction tasks. SLOGAN adapts an existing tool, LOGAN (LOcal Group biAs detectioN), by contextualizing group bias detection in patient illness severity and past medical history. We investigate and compare SLOGAN's bias detection capabilities to LOGAN and other clustering techniques across patient subgroups in the MIMIC-III dataset. On average, SLOGAN identifies larger fairness disparities in over 75% of patient groups than LOGAN while maintaining clustering quality. Furthermore, in a diabetes case study, health disparity literature corroborates characterizations of the most biased clusters identified by SLOGAN. Our results contribute to the broader discussion of how machine learning biases may perpetuate existing healthcare disparities.

A. Ovalle (✉) · S. Dev · J. Zhao · M. Sarrafzadeh · K.-W. Chang
University of California, Los Angeles, USA
e-mail: anaelia@cs.ucla.edu

S. Dev
e-mail: sunipa@cs.ucla.edu

J. Zhao
e-mail: jieyuzhao@ucla.edu

M. Sarrafzadeh
e-mail: majid@cs.ucla.edu

K.-W. Chang
e-mail: kwchang@cs.ucla.edu

A. Shaban-Nejad et al. (eds.), *Artificial Intelligence for Personalized Medicine*,
Studies in Computational Intelligence 1106,
https://doi.org/10.1007/978-3-031-36938-4_10

1 Introduction

Fairness auditing frameworks are necessary for operationalizing machine learning algorithms in healthcare (ML4H). In particular, they must identify and characterize biases [1–3]. Ongoing directives to promote health equity must also translate to these spaces, with care placed on those historically vulnerable to the most harm, such as communities with chronic illnesses and racial and ethnic minorities [4, 5]. To do this, they must be prioritized when evaluating for fairness in ML4H [1–3, 6, 7].

Commercialized auditing tools are being increasingly leveraged for bias assessment in ML4H algorithms [4, 8]. However, we argue that applying out-of-the-box auditing tools without a clear patient-centric design is not enough. Existing auditing tools must align with health ethics principles that guide a framework's operationalization. In guiding ML4H auditing literature, this means the tool must be able to detect locally biased patient subgroups when monitoring the fairness of ML4H throughout its lifecycle [9]. To monitor disparities with health equity in mind, researchers must also engage critically with the broader sociotechnical context surrounding the use of ML auditing tools in healthcare [10].

This work addresses the gap by devising a patient-centric ML auditing tool called SLOGAN. SLOGAN adapts LOGAN [11], an unsupervised algorithm that uses contextual word embeddings [12] to cluster local groups of bias indicated by model performance differences. To better align auditing with measures of effective care planning and therapeutic intervention [13], SLOGAN identifies local group biases in clinical prediction tasks by leveraging patient risk stratification. Previous medical history is also commonly used for understanding health inequities through social, cultural, and structural barriers the patient experiences [14]. Therefore, SLOGAN characterizes these local biases using patients' electronic healthcare records (EHR) histories.

Experiments on in-hospital mortality prediction demonstrate how SLOGAN effectively identifies local group biases. We audit the model across 12 MIMIC-III patient subgroups. We then provide a case study to further examine fairness differences in patients with chronic illnesses such as Diabetes Mellitus. Results indicate that (1) SLOGAN, on average, captures more considerable biases than LOGAN, and (2) such identified biases align with existing health disparity literature.

2 Background and Related Work

2.1 Algorithmic Auditing in ML for Healthcare

[15] audit a commercialized ML4H algorithm by dissecting observed disparities between patient risk and overall health cost. The authors call for the continued probing of health inequity in these clinical systems. Likewise, [9, 10, 16, 17] create guidelines for operationalizing transparent assessments of ML4H models. Auditing frameworks

such as Aequitas[1] and AIFairness360[2] are operationalized for this purpose [4]. The tools provide reports relevant to protected groups and fairness metrics, indicating unfairness through preset disparity ranges.

2.2 Measuring Health Equity Barriers

Intersectional social identities are related to a patient's health outcomes [18, 19]. Therefore, measuring health equity in ML requires understanding a patient beyond their illness. In practice, this can include focusing on populations with histories of a significant illness burden or examining bias from the lens of social determinants of health (SDOH). Fairness literature has also dictated a need to measure biases from multidimensional perspectives [20]. Capturing social context beyond protected attributes is helpful for this cause. SDOH, such as unequal access to healthcare, language, stigma, racism, and social community, are underlying contributing factors to health inequities [14, 21, 22].

2.3 Fairness and Local Bias Detection

LOGAN [11], a method to detect local bias, adapts K-Means to cluster BERT embeddings while maximizing a bias metric within each cluster. LOGAN consists of a 2-part objective: a K-Means clustering objective (L_c) and an objective to maximize a bias metric (L_b, e.g. the performance gap between 2 groups) within each respective cluster.

$$\min_{C} \quad L_c + \lambda L_b \tag{1}$$

where $\lambda \leq 0$ is a tunable hyperparameter to control the tradeoff between the two objectives and indicates how strongly to cluster with respect to group performance differences. We define our bias metric as the model performance disparity between 2 groups, measured by accuracy. However, detecting biases by identifying similar contextual representations is not enough. The task must be adapted to the clinical domain to audit with health equity in mind. One way to do this is by incorporating domain-specific information. For example, severity scores stratify patients based on their immediate needs and help clinicians decide how to allocate resources effectively [23]. Therefore, we build off of LOGAN and create a tool that translates to the medical setting by mindfully using this information.

[1] http://aequitas.dssg.io/.

[2] https://aif360.mybluemix.net/.

3 Methodology

3.1 Clinical NLP Pretrained Embeddings

Several BERT models are publicly available for use in the clinical setting. These include various implementations of ClinicalBERT [24, 25]. We proceed with leveraging a variant of ClinicalBERT from [26] as this is an extension of ClinicalBERT with improvements such as whole-word masking.

3.2 Automatic Bias Detection

To create a patient-centric bias detection tool, we encourage SLOGAN to identify large bias gaps while accounting for similarity in patient severity. SLOGAN measures local biases in a model using patient-specific features and contextual embeddings of patient history for in-hospital mortality prediction. We do this via a patient similarity constraint. A variety of patient severity scores such as OASIS, SAPS II, and SOFA are available for use [27–29]. Following health literature and clinician advice, we select the SOFA acuity score. However, depending on clinician needs, a different constraint may be used (e.g., ICD-9 codes). Extending Eq. (1), this results in the following optimization problem:

$$\min_{C} \quad L_c + \lambda L_b + \gamma L_s \tag{2}$$

where L_s is added to encourage the model to group patients with similar acute severity. $\lambda \leq 0$ and $\gamma \geq 0$ are hyperparameters that control the tradeoff between the objectives of grouping patient similarity and clustering by local bias.

$$L_s = \sum_{j=1}^{k} \left| \sum_{x_i \in A} SOFA_{ij} - \sum_{x_i \in B} SOFA_{ij} \right|^2 \tag{3}$$

λ and γ are tuned via a grid search and we choose the combination that identifies the largest local group biases (Appendix Table 6).

We define the bias score as having at least a 10% difference in accuracy and at most a SOFA score difference of 0.8.[3] We compare SLOGAN to LOGAN and K-Means across three metrics. To measure the utility of the clusters found, we examine the ratio of biased clusters found (SCR) and the number of instances in those clusters (SIR). We use inertia to measure clustering quality, as it reflects how well the data clustered across respective centroids. Finally, we compare each algorithm's inertia to a baseline K-Means model normalized to 1.0.

[3] We choose the thresholds by splitting the data and creating bootstrap estimates 1000 times, then add three standard deviations.

4 Data and Setup

In order to maximize reproducibility, we perform experiments with the same patient cohorts defined in the benchmark dataset from the MIMIC-III clinical database [30, 31]. Following [32], to understand how BERT represents social determinants of health and captures possible stigmatizing language in the data, we extracted the history of present illness, past medical history, social history, and family history across physicians, nursing, and discharge summaries [33]. We employed MedSpacy [34] to extract any information related to a patient's social determinants of health. After preprocessing, this translated into a 70% train, 15% validation, and 15% test split of 1581, 393, and 309 patients, respectively. No patient appeared across the splits. Analyses were conducted across self-identified ethnicity, sex, insurance type, English speaking, presence of chronic illness, presence of diabetes (type I and II), social determinants of health, and negative patient descriptors to measure stigma. We also explored creating cross-sectional groups (Appendix Table 5).

We used SLOGAN to audit a fully connected neural network from [26] used to predict in-hospital mortality, a common MIMIC-3 benchmarking task [31].[4] Each patient note in the test set was encoded and concatenated with gender, OASIS, SAPS II, SOFA scores, and age. To provide a rich contextual representation of patient notes to SLOGAN, encodings consisted of the concatenated last four layers of Clinical-BERT [12]. The embeddings encoded 512 tokens, the maximum number of tokens for BERT. We followed the best hyperparameters of the model and chose the threshold that provides at least 80% accuracy on the validation set.

5 Results

5.1 Aggregate Analysis

We assessed SLOGAN's local bias clustering abilities and quality across 12 attributes in MIMIC-III, including demographic variables such as ethnicity and gender. The model was compared to K-Means and LOGAN using the SCR, SIR, |Bias|, and Inertia measurements defined in Sect. 3.2. We report these results in Table 1. In most attributes, SLOGAN was the best at identifying groups with fairness gaps. Identified groups contained more instances and larger biases, while maintaining clustering quality. In particular, SLOGAN identified the most and largest local group biases in at least 9/12 (75%) attributes, measured by SCR and |Bias|, respectively. When comparing LOGAN and K-Means, SLOGAN found the highest ratio of biased instances within biased clusters (SIR) in 7/12 (58%) MIMIC-3 attributes. We report audits across all attributes in Appendix Table 7.

[4] A patient who has passed within 48 hours of their ICU stay is assigned the label of 1, otherwise they are assigned the label 0.

Table 1 Average values for 12 MIMIC-III attributes across models and evaluation metrics. SCR, SIR, and |Bias| in %. |Bias| is the average absolute model performance difference in biased clusters. Bold is the best performance per row. Right-most column is number of MIMIC-III attributes where SLOGAN performs best. Arrows indicate desired direction of a number

	K-Means	LOGAN	SLOGAN	# of MIMIC-III attributes
Inertia (↓)	1.0	0.991	**0.981**	7/12 (58%)
SCR (↑)	15.3	22.9	**30.1**	12/12 (100%)
SIR (↑)	15.3	18.4	**23.4**	7/12 (58%)
|Bias| (↑)	12.5	21.5	**34.2**	9/12 (75%)

5.2 *Case Study: Diabetes Mellitus*

5.2.1 Cluster Analysis

Diabetes is one of the most common and costly chronic conditions worldwide, accompanied by serious comorbidities [35]. To further study this, we used SLOGAN to assess the local group biases on the **HAS DIABETES** attribute and identified fairness gaps in agreement with health literature (Fig. 1).

We report the accuracy and maximum absolute performance differences across identified biased clusters by K-Means, LOGAN, and SLOGAN in Table 2. The performance difference overall between patients that do and do not have diabetes was 9.1%. K-Means and LOGAN identified local groups with larger performance discrepancies (20% and 28.1%, respectively). Notably, SLOGAN performed the best at identifying a local region with the largest performance gap (37.1%). We also report the SCR, SIR, |Bias|, and Inertia in Table 3. Results indicate that SLOGAN found groups with a larger average bias magnitude than K-Means and LOGAN. While LOGAN and SLOGAN identified the same ratio of biased clusters (25.0%), SLOGAN identified the largest local bias region (28.6%) with a small tradeoff in inertia (Appendix Fig. 2).

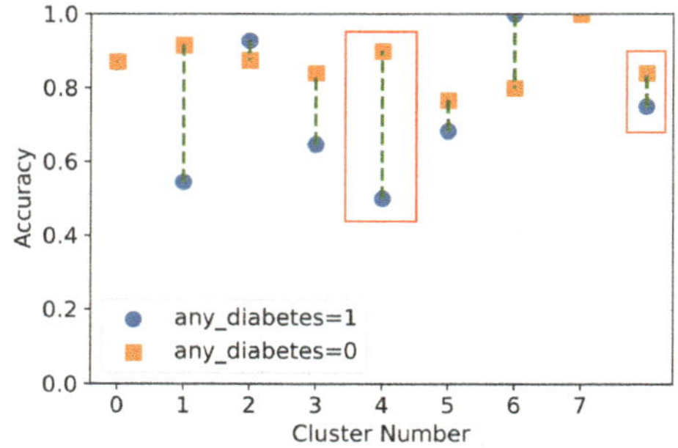

Fig. 1 Performance differences for **HAS DIABETES** attribute. Furthest right red box shows global bias, while SLOGAN finds a local area of much higher bias at cluster 4

Table 2 Bias detection (%) for in-hospital mortality task. Global indicates global bias. "Yes" indicates patient with diabetes. |Bias| is the max absolute model performance difference in biased clusters. SLOGAN identifies local biases greater than global bias observed in the data (bold)

	Method	Acc-Yes	Acc-No	|Bias|
Has diabetes	Global	75.0	84.1	9.1
	K-Means	55.0	75.0	20.0
	LOGAN	60.0	88.0	28.0
	SLOGAN	54.5	91.7	**37.1**

Table 3 Comparison under diabetes attribute. SCR and SIR are respectively the % of biased clusters and % of biased instances. |Bias|(%) is the average absolute bias score for the biased clusters. SLOGAN finds the largest bias (bold)

	Method	Inertia	SCR	SIR	|Bias|
Has diabetes	K-Means	1.00	33.3	27.1	14.2
	LOGAN	1.003	25.0	16.9	25.0
	SLOGAN	1.12	25.0	15.4	**28.6**

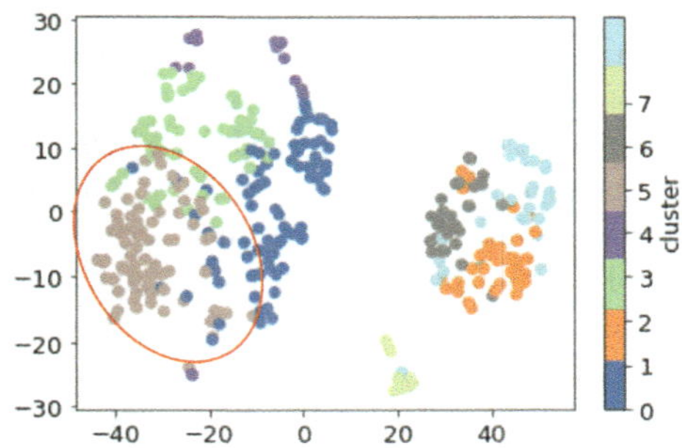

Fig. 2 t-SNE results with circled most biased cluster for **HAS DIABETES** attribute

To more carefully examine clusters formed by SLOGAN, we show respective performance deviations in Fig. 1. We found that SLOGAN identified fairness gaps documented in health literature. Two clusters exhibited a large local bias towards patients without diabetes, clusters 1 and 4. We analyzed differences in cluster characteristics between the most and least biased cluster. The most biased cluster, cluster 4, contained 38% more patients with chronic illnesses besides diabetes, with 33.3% suffering from chronic illnesses besides diabetes or hypertension. We then compared cluster 4 to all other clusters. Again, we found that it contained the largest percentage of (1) patients (62.5%) with chronic illnesses besides diabetes and (2) patients with chronic illnesses besides diabetes and hypertension (25%). Cluster 4 also had fewer patients with private insurance than the least biased cluster and the lowest percentage of English-speaking patients (4.6%) in the entire dataset (Appendix Table 8). Notably, these differences in disease burden, insurance, and language align with existing research indicating how populations with the largest health disparities

Table 4 Top 20 topic words in the most and least biased clusters using SLOGAN for **HAS DIABETES** attribute. Number is the bias score (%) of that cluster

Most biased (40.0)	Parent, given, recent, vanco, treat, fever, acinetobacter, ecg, negative, incubated, disorder, bottles, clozaril, complete, sputum, past, started, ed, found, admitted
Least biased (0.2)	Noted, past, recent, home, given, due, pain, two, offspring, mild, chest, initially, without, blood, vancomycin, children, shortness_breath, sibling, admitted, started

often suffer from a larger burden of disease and may experience significant structural language barriers [22, 36].

5.2.2 Bias Interpretation with Topic Modeling

Severe diabetes complications may result in various forms of deadly infections and respiratory issues [37–39]. Provided the in-mortality task, we asked if indications of severe diabetes complications were present when using SLOGAN. To do this, we ran Latent Dirichlet Allocation topic modeling [40] within identified SLOGAN clusters. We detail the preprocessing steps in the appendix. Table 4 lists the top 20 topic words for the most and least biased clusters. SLOGAN grouped patients with histories indicating deadly infections and respiratory issues in the most biased cluster. Terms included "sputum" (thick respiratory secretion), "Acinobacter" (bacteria that can live in respiratory secretions), and "Vanco" (used to treat infections).

Social determinants of health also correlate to effective self-management of diabetes [41, 42]. Therefore we also examined differences in social determinants of health between the least and most biased clusters. While LDA cannot determine the directionality of SDOH impact, the top 20 terms are among the most important when forming the cluster's topic distribution. In the least biased cluster, top words included terms around the community such as 'home', 'offspring', 'children', and 'sibling'. However, in the most biased cluster, just 1 of the 20 terms, 'parent', reflected possible existing social support.

6 Discussion

We developed SLOGAN as a framework to audit an ML4H task by identifying areas of patient severity-aware local biases. Results indicated that SLOGAN captures more and higher quality clusters across several subgroups than the baseline models, K-Means and LOGAN. To illustrate how to use SLOGAN in a clinical context, we conducted a case study that used SLOGAN to identify clusters of local bias in diabetic patients. We found that the biases observed aligned with existing health literature. In particular, the cluster with the *largest local bias* was also the cluster with the *largest disease burden*. This result demonstrates a need to further examine and repeat these experiments across patient cohorts and performance metrics. Interesting future works may include asking how models encode vulnerable communities in their representations and if health disparities consistently propagate into model biases.

In practice, SLOGAN can be used to determine biased clusters for review before model deployment in a healthcare setting. The tool may also track how biases shift due to changes in the data or across operationalization in different hospital networks. Furthermore, patient-centric local bias detection can supplement ML4H model auditing. With this information, ML researchers and clinicians can use auditing report cards to decide on the next steps for inclusive model development.

6.1 Ethical Statement and Limitations

Our analysis used MIMIC-III data, an open deidentified clinical dataset. Only credentialed researchers who fulfilled all training requirements and abided by the data use agreement accessed the data.[5] We review the data and clinical notes a second time to confirm the removal of any patient-related information, including location, age, name, date, or hospital.

In practice, further interdisciplinary discussion on how SLOGAN can best be integrated into the ML4H auditing pipeline is welcomed. While we do not analyze the factors influencing model fairness, we encourage this future work. Furthermore, it is important to note that the absence of flagged bias clusters is not an indicator of a total absence of risk for downstream unfair outcomes.

Appendix

LDA

LDA is run using the NLTK and gensim packages [43, 44]. Unigrams and bigrams are generated using gensim.phrases with min count=3 and threshold=5. The LDA is run on gensim with random state=100, updateevery=1, chunksize=100, and passes=100. To get achieve better topic modeling, words like child, son, daughter are tokenized as ‘offspring’. Words pertaining to father or mother are replaced with ‘parent’. Words such as hypertension and hypertensive are replaced with ‘hypert’. Similarly, words such as hypotension and hypotensive are replaced with ‘hypot’ (Table 10).
Negative Patient Descriptors

---

[5] https://physionet.org/content/mimiciii/1.4/#files.

Table 5 Percent of attribute in the MIMIC-3 data

Group	λ	γ
Has diabetes	–30	50
Has negative descriptor	–20	0
Has chronic illness	–30	50
Medicaid insurance	–70	30
Medicare insurance	–50	0
Private insurance	–70	40
Speaks English	–30	0
Assigned male at birth (AMAB)	–10	60
Assigned female at birth (AFAB)	0	70
Self-identifies white	–30	20
Self-identifies black	–20	60
AFAB + self-identifies black	–10	60

Table 6 Hyper parameter search for λ and γ after searching between combinations between –100–0 and 0–100, respectively

Group	Percent (%)
Has negative descriptor	8.86
Has diabetes	35.43
Has chronic illness	88.0
Medicaid insurance	7.71
Medicare insurance	60.86
Private insurance	28.0
Speaks English	86.57
Assigned male at birth (AMAB)	56.29
Assigned female at birth (AFAB)	43.71
Self-identifies white	75.14
Self-identifies black	13.43
AFAB + self-identifies black	8.86

We explored the SDOH dimension of stigma in clinical notes through the extraction of negative patient descriptors found in [32] and outline the results in the Appendix Table 9. However, further preprocessing beyond the usage of regexes is needed to reduce false positive rates.

Code

Please feel free to reach out to the authors for access to the code repository.

Table 7 Comparison between K-Means, LOGAN, and SLOGAN under each attribute type. SCR and SIR are respectively the ratio of biased clusters and ratio of biased instances. |Bias| is the averaged absolute bias score for these biased clusters. Results not shown in %

	Method	Inertia	SCR	SIR	|Bias|
Has diabetes	K-Means	1.00	0.33	0.27	0.14
	LOGAN	1.00	0.25	0.17	0.25
	SLOGAN	1.12	0.25	0.15	0.29
Has negative	Method	Inertia	SCR	SIR	|Bias|
	K-Means	1.00	0.00	0.00	0.00
	LOGAN	0.88	0.20	0.19	0.20
	SLOGAN	0.85	0.20	0.19	0.37
Has chronic illness	Method	Inertia	SCR	SIR	|Bias|
	K-Means	1.00	0.17	0.25	0.17
	LOGAN	1.15	0.40	0.32	0.40
	SLOGAN	0.89	0.50	0.47	0.23
Is medicaid insurance	Method	Inertia	SCR	SIR	|Bias|
	K-Means	1.00	0.40	0.46	0.23
	LOGAN	0.99	0.20	0.25	0.20
	SLOGAN	0.94	0.20	0.11	0.76
Is medicare insurance	Method	Inertia	SCR	SIR	|Bias|
	K-Means	1.00	0.13	0.13	0.21
	LOGAN	0.91	0.22	0.22	0.22
	SLOGAN	0.87	0.22	0.16	0.21
Is private insurance	Method	Inertia	SCR	SIR	|Bias|
	K-Means	1.00	0.22	0.20	0.12
	LOGAN	1.18	0.14	0.12	0.14
	SLOGAN	1.12	0.14	0.10	0.26
Is English speaker	Method	Inertia	SCR	SIR	|Bias|
	K-Means	1.00	0.00	0.00	0.00
	LOGAN	1.02	0.17	0.17	0.17
	SLOGAN	0.91	0.43	0.44	0.31
Assigned male at birth (AMAB)	Method	Inertia	SCR	SIR	|Bias|
	K-Means	1.00	0.00	0.00	0.00
	LOGAN	1.00	0.11	0.09	0.11
	SLOGAN	1.03	0.25	0.13	0.41
Assigned female at birth (AFAB)	Method	Inertia	SCR	SIR	|Bias|
	K-Means	1.00	0.00	0.00	0.00
	LOGAN	1.00	0.11	0.09	0.11
	SLOGAN	1.05	0.13	0.04	0.39
Self-identifies white	Method	Inertia	SCR	SIR	|Bias|
	K-Means	1.00	0.14	0.13	0.14
	LOGAN	0.86	0.38	0.37	0.38
	SLOGAN	0.98	0.40	0.41	0.26
Self-identifies black	Method	Inertia	SCR	SIR	|Bias|
	K-Means	1.00	0.40	0.28	0.20
	LOGAN	0.91	0.20	0.10	0.20
	SLOGAN	1.02	0.20	0.10	0.27
Self-identifies black + AFAB	Method	Inertia	SCR	SIR	|Bias|
	K-Means	1.00	0.20	0.13	0.28
	LOGAN	1.00	0.20	0.13	0.20
	SLOGAN	0.99	0.60	0.49	0.35

Table 8 Percentage differences (Δ, in %) in characteristics between most and least biased cluster for **HAS DIABETES** attribute. A positive number means the most biased cluster has more instances of this attribute versus the least biased cluster. N/A indicates division by zero

Group	Δ (%)
Private insurance	–100.0
Medicaid insurance	11.1
Medicaid insurance	51.5
Self-identifies white	36.5
Self-identifies black	N/A
Self-identifies hispanic	N/A
Self-identifies Asian	N/A
Self-identifies other	11.1
English speaker	–1.6
Assigned male at birth (AMAB)	–38.3
Has chronic illness, not diabetes	37.8
Has chronic illness, not diabetes or hypertension	33.3
Hypertensive	11.1
Has acute illness	27.8

Table 9 Most and Least Biased LDA top 20 words for **HAS NEGATIVE DESCRIPTOR** patient descriptor. Number in parentheses is the bias score (%) of that cluster

Most biased (38.7)	Denies, rehab, treat, pain, well, sputum, transferred, hx, valve, sent, course, cxr, chest pain, one, episodes, mild, cough, floor, worsening, disease, tobacco
Least biased (0.67)	Pain, given, denies, admit, home, time, last, well, hip, past, started, disease, found, noted, transferred, liver, developed, treat, symptoms, nausea, blood

Table 10 Top 20 topic words in the most and least biased cluster using SLOGAN under **IS ENGLISH SPEAKER**. Number in parentheses is the bias score (%) of that cluster

Most biased (32.7)	Disease, cardiac, lives, received, given, admit, denies, parent, family, cath, symptoms, cancer, positive, diabetes mellitus, type, past, time, alcohol, cad, recently, ct
Least biased (3.4)	Abdominal pain, denies, pain, started, chest pain, chronic, cough, disease, transferred, past, hyperlipidemia, patient, time, given, hypert, recent, cardiac, ros, shortness breath, complaints, found

References

1. M. Ghassemi, T. Naumann, P. Schulam, A. Beam, I. Chen, R. Ranganath, A review of challenges and opportunities in machine learning for health. AMIA Summits Transl. Sci. Proc. **2020**, 191 (2020)
2. V. Mhasawade, Y. Zhao, R. Chunara, Machine learning and algorithmic fairness in public and population health. Nat. Mach. Intell. **3**, 659–666 (2021)
3. I. Chen, E. Pierson, S. Rose, S. Joshi, K. Ferryman, M. Ghassemi, Ethical machine learning in healthcare. Ann. Rev. Biomed. Data Sci. **4**, 123–144 (2021)
4. L. Oala, J. Fehr, L. Gilli, P. Balachandran, A. Leite, S. Calderon-Ramirez, D. Li, G. Nobis, E. Alvarado, G. Jaramillo-Gutierrez, Others, Ml4h auditing: from paper to practice, in *Machine Learning For Health* (2020), pp. 280–317
5. L. Joszt, 5 vulnerable populations in healthcare, in *AJMC* (2022), https://www.ajmc.com/view/5-vulnerable-populations-in-healthcare
6. A. Rajkomar, M. Hardt, M. Howell, G. Corrado, M. Chin, Ensuring fairness in machine learning to advance health equity. Ann. Intern. Med. **169**, 866–872 (2018)
7. E. Röösli, S. Bozkurt, T. Hernandez-Boussard, Peeking into a black box, the fairness and generalizability of a MIMIC-III benchmarking model. Sci. Data **9**, 1–13 (2022)
8. A. Kumar, A. Ramachandran, A. De Unanue, C. Sung, J. Walsh, J. Schneider, J. Ridgway, S. Schuette, J. Lauritsen, R. Ghani, A machine learning system for retaining patients in HIV care. ArXiv Preprint ArXiv:2006.04944 (2020)
9. A. Hond, A. Leeuwenberg, L. Hooft, I. Kant, S. Nijman, H. Os, J. Aardoom, T. Debray, E. Schuit, M. Smeden, Others, Guidelines and quality criteria for artificial intelligence-based prediction models in healthcare: a scoping review. Npj Digit. Med **5**, 1–13 (2022)
10. S. Pfohl, A. Foryciarz, N. Shah, An empirical characterization of fair machine learning for clinical risk prediction. J. Biomed. Inform. **113**, 103621 (2021)
11. J. Zhao, K. Chang, LOGAN: local group bias detection by clustering, in *Proceedings Of The 2020 Conference On Empirical Methods In Natural Language Processing (EMNLP)* (2020), pp. 1968–1977
12. J. Devlin, M. Chang, K. Lee, K. Toutanova, Bert: pre-training of deep bidirectional transformers for language understanding. ArXiv Preprint ArXiv:1810.04805 (2018)
13. J. Katz, M. Minder, B. Olenchock, S. Price, M. Goldfarb, J. Washam, C. Barnett, L. Newby, S. Diepen, The genesis, maturation, and future of critical care cardiology. J. Am. Coll. Cardiol. **68**, 67–79 (2016)
14. L. Brennan Ramirez, E. Baker, M. Metzler, Promoting health equity; a resource to help communities address social determinants of health (2008)
15. Z. Obermeyer, B. Powers, C. Vogeli, S. Mullainathan, Dissecting racial bias in an algorithm used to manage the health of populations. Science **366**, 447–453 (2019)
16. T. Wiegand, R. Krishnamurthy, M. Kuglitsch, N. Lee, S. Pujari, M. Salathé, M. Wenzel, S. Xu, WHO and ITU establish benchmarking process for artificial intelligence in health. The Lancet. **394**, 9–11 (2019)
17. H. Siala, Y. Wang, SHIFTing artificial intelligence to be responsible in healthcare: a systematic review. Soc. Sci. Med. 114782 (2022)
18. J. McGinnis, P. Williams-Russo, J. Knickman, The case for more active policy attention to health promotion. Health Aff. **21**, 78–93 (2002)
19. A. Katz, D. Chateau, J. Enns, J. Valdivia, C. Taylor, R. Walld, S. McCulloch, Association of the social determinants of health with quality of primary care. Ann. Family Med. **16**, 217–224 (2018)
20. A. Hanna, E. Denton, A. Smart, J. Smith-Loud, Towards a critical race methodology in algorithmic fairness, in *Proceedings of the 2020 Conference on Fairness, Accountability, and Transparency* (2020), pp. 501–512
21. L. Aday, Health status of vulnerable populations. Annu. Rev. Public Health **15**, 487–509 (1994)
22. M. Peek, A. Cargill, E. Huang, Diabetes health disparities. Med. Care Res. Rev. **64**, 101S-156S (2007)

23. F. Ferreira, D. Bota, A. Bross, C. Mélot, J. Vincent, Serial evaluation of the SOFA score to predict outcome in critically ill patients. JAMA **286**, 1754–1758 (2001)
24. E. Alsentzer, J. Murphy, W. Boag, W. Weng, D. Jindi, T. Naumann, M. McDermott, Publicly available clinical BERT embeddings, in *Proceedings of the 2nd Clinical Natural Language Processing Workshop* (2019), pp. 72–78, https://aclanthology.org/W19-1909
25. K. Huang, J. Altosaar, R. Ranganath, Clinicalbert: modeling clinical notes and predicting hospital readmission. ArXiv Preprint ArXiv:1904.05342 (2019)
26. H. Zhang, A. Lu, M. Abdalla, M. McDermott, M. Ghassemi, Hurtful words: quantifying biases in clinical contextual word embeddings, in *Proceedings of the ACM Conference on Health, Inference, and Learning* (2020), pp. 110–120
27. J. Le Gall, S. Lemeshow, F. Saulnier, A new simplified acute physiology score (SAPS II) based on a European/North American multicenter study. JAMA **270**, 2957–2963 (1993)
28. A. Jones, S. Trzeciak, J. Kline, The sequential organ failure assessment score for predicting outcome in patients with severe sepsis and evidence of hypoperfusion at the time of emergency department presentation. Crit. Care Med. **37**, 1649 (2009)
29. A. Johnson, A. Kramer, G. Clifford, A new severity of illness scale using a subset of acute physiology and chronic health evaluation data elements shows comparable predictive accuracy. Crit. Care Med. **41**, 1711–1718 (2013)
30. A. Johnson, T. Pollard, L. Shen, L. Lehman, M. Feng, M. Ghassemi, B. Moody, P. Szolovits, L. Anthony Celi, R. Mark, MIMIC-III, a freely accessible critical care database. Sci. Data **3**, 1–9 (2016)
31. H. Harutyunyan, H. Khachatrian, D. Kale, G. Ver Steeg, A. Galstyan, Multitask learning and benchmarking with clinical time series data. Sci. Data **6**, 1–18 (2019)
32. M. Sun, T. Oliwa, M. Peek, E. Tung, Negative patient descriptors: documenting racial bias in the electronic health record: study examines racial bias in the patient descriptors used in the electronic health record. Health Affairs 10–1377 (2022)
33. M. Marmot, Social determinants of health inequalities. The Lancet. **365**, 1099–1104 (2005)
34. H. Eyre, A. Chapman, K. Peterson, J. Shi, P. Alba, M. Jones, T. Box, S. DuVall, O. Patterson, Launching into clinical space with medspaCy: a new clinical text processing toolkit in Python, in *AMIA Annual Symposium Proceedings* (2021), http://arxiv.org/abs/2106.07799
35. A. Ceriello, L. Barkai, J. Christiansen, L. Czupryniak, R. Gomis, K. Harno, B. Kulzer, J. Ludvigsson, Z. Némethyová, D. Owens, Others, Diabetes as a case study of chronic disease management with a personalized approach: the role of a structured feedback loop. Diabetes Res. Clin. Pract. **98**, 5–10 (2012)
36. G. Flores, The impact of medical interpreter services on the quality of health care: a systematic review. Med. Care Res. Rev. **62**, 255–299 (2005)
37. N. Joshi, G. Caputo, M. Weitekamp, A. Karchmer, Infections in patients with diabetes mellitus. N. Engl. J. Med. **341**, 1906–1912 (1999)
38. L. Muller, K. Gorter, E. Hak, W. Goudzwaard, F. Schellevis, A. Hoepelman, G. Rutten, Increased risk of common infections in patients with type 1 and type 2 diabetes mellitus. Clin. Infect. Dis. **41**, 281–288 (2005)
39. F. De Santi, G. Zoppini, F. Locatelli, E. Finocchio, V. Cappa, M. Dauriz, G. Verlato, Type 2 diabetes is associated with an increased prevalence of respiratory symptoms as compared to the general population. BMC Pulm. Med. **17**, 1–8 (2017)
40. D. Blei, A. Ng, M. Jordan, Latent dirichlet allocation. J. Mach. Learn. Res. **3**, 993–1022 (2003)
41. M. Clark, S. Utz, Social determinants of type 2 diabetes and health in the United States. World J. Diabetes **5**, 296 (2014)
42. M. Adu, U. Malabu, A. Malau-Aduli, B. Malau-Aduli, Enablers and barriers to effective diabetes self-management: a multi-national investigation. PLoS ONE **14**, e0217771 (2019)
43. Loper, E. & Bird, S. Nltk, The natural language toolkit. ArXiv Preprint arXiv:Cs/0205028 (2002)
44. R. Řehuřek, P. Sojka, Others, Gensim-statistical semantics in python, in *Retrieved From Genism. Org.* (2011)

Identification, Explanation and Clinical Evaluation of Hospital Patient Subtypes

Enrico Werner, Jeffrey N. Clark, Ranjeet S. Bhamber, Michael Ambler, Christopher P. Bourdeaux, Alexander Hepburn, Christopher J. McWilliams, and Raul Santos-Rodriguez

Abstract We present a pipeline in which unsupervised machine learning techniques are used to automatically identify subtypes of hospital patients admitted between 2017 and 2021 in a large UK teaching hospital. With the use of state-of-the-art explainability techniques, the identified subtypes are interpreted and assigned clinical meaning. In parallel, clinicians assessed intra-cluster similarities and inter-cluster differences of the identified patient subtypes within the context of their clinical knowledge. By confronting the outputs of both automatic and clinician-based explanations, we aim to highlight the mutual benefit of combining machine learning techniques with clinical expertise.

The first three authors are joined first authors

E. Werner (✉) · J. N. Clark · R. S. Bhamber · M. Ambler · A. Hepburn · C. J. McWilliams · R. Santos-Rodriguez
University of Bristol, Bristol, UK
e-mail: enrico.werner@bristol.ac.uk

J. N. Clark
e-mail: jeff.clark@bristol.ac.uk

R. S. Bhamber
e-mail: ranjeet.bhamber@bristol.ac.uk

M. Ambler
e-mail: mike.ambler@bristol.ac.uk

A. Hepburn
e-mail: alex.hepburn@bristol.ac.uk

C. J. McWilliams
e-mail: chris.mcwilliams@bristol.ac.uk

R. Santos-Rodriguez
e-mail: enrsr@bristol.ac.uk

M. Ambler · C. P. Bourdeaux · C. J. McWilliams
University Hospitals Bristol NHS Foundation Trust, Bristol, UK
e-mail: christopher.bourdeaux@uhbristol.nhs.uk

A. Shaban-Nejad et al. (eds.), *Artificial Intelligence for Personalized Medicine*, Studies in Computational Intelligence 1106,
https://doi.org/10.1007/978-3-031-36938-4_11

Keywords Clustering · Clinical evaluation · Explainability · Patient subtypes

1 Introduction

Patients admitted to hospital constitute a heterogeneous population with different levels of illness severity, morbidities, response to treatments and outcomes [9]. Therefore, predicting the right treatment is challenging even when patients are initially diagnosed with the same conditions. For diagnosis and determining treatment options, physicians rely on factors including the patient's medical history [6], their own clinical experience and their professional intuition [9].

Advances in computing technologies and the introduction of electrical health records (EHR) mean that more information is available to physicians than ever before. However, hospitals are still in the process of transitioning from paper records to EHR, which leads to challenges when analyzing the data and inferring high-level information [6]. As intensive care units (ICUs) are the most data-rich hospital department, machine learning approaches have mostly focused on these environments [3, 9, 19, 27]. Recent progress has also been made for general wards [8, 10, 15, 21].

Outcome prediction and risk scoring are of high clinical importance. Several risk scoring methods have been developed and deployed, e.g. Rothman index [23], MEWS [26], APACHE IV [5], and SOFA [16]. The National Early Warning Score 2 (NEWS) is increasingly used in UK hospitals [1] and has good predictive ability in patients with infections and sepsis [2]. However, for respiratory diseases like COVID-19, the results are conflicting [4, 8, 17]. This raises the question: are early warning scores such as NEWS equally effective for all patient subtypes?

We argue that the predictive ability of scores could be further improved by incorporating insightful patient subtyping. Historically, patients were grouped based on their level of sickness, i.e., the creation of ICUs. The reorganization presented an innovation, as expertise in caring for the critically ill could be focused on a single location [11]. Patient subtyping could be the next innovative step, namely, instead of focusing solely on the severity of their sickness, patients could be further grouped based on their clinical needs [27]. In a pilot study, non-ICU patients were physically grouped in based on similar patient characteristics rather than diagnoses, leading to a reduced admittance of low-risk patients to ICU from 42% to 22%. Additionally, the average ICU length of stay was reduced from 4.6 to 4.1 days [13].

Automatic patient subtyping aims to assign patients to clinically meaningful groups using factors such as their disease progression, medical history, EHR, and ultimately paves the path to precision or personalised medicine by tailoring diagnostic and therapeutic strategies to the patient's needs [6, 20]. Subtyping can be framed as an unsupervised machine learning task, using clustering methods to identify distinct high-density regions separated by sparse regions within a dataset [12]. These clusters represent patients that are in some sense similar according to the data. Clustering algorithms such as k-means and hierarchical clustering have recently been applied to identify clusters in a general ICU population [27], cardiovascular clusters in sepsis patients [14], and corticosteroid response in patients with severe asthma [28].

However, clustering alone is insufficient to provide practical support to determine treatment options. The resulting clusters also must be interpreted such that clinicians can validate and learn from cluster assignments. While previous studies manually assigned clinical meaning to the clusters after their creation, before these models can be widely deployed in hospitals, the final users must 'trust' the models. This requires an in-depth understanding of the models' behaviour and confidence in individual predictions [22]. Model-agnostic explainability approaches such as LIME and variants [22, 24] can be used for explaining the predictions of clustered data [29]. This paper presents a pipeline in which we:

- Propose the use of unsupervised machine learning techniques to identify patient subtypes on admission for a new dataset of hospital patients from a large UK teaching hospital.
- Implement a combination of explainability techniques and statistical properties of the clusters to evaluate and assign clinical meaning to the identified subtypes.
- In parallel and independently, hospital clinicians derive the main clinical properties of the identified subtypes using additional records, a key and necessary step in developing human-in-the-loop machine learning systems in medical settings.

2 Method

2.1 Data Source and Conditions

Subjects are patients admitted to the Bristol Royal Infirmary, a large teaching hospital covering most medical and surgical specialties. The clinical characteristics of this historical data source is summarised in Table 1. Only patients were considered for which the NEWS and all corresponding vitals were available i.e. temperature, systolic blood pressure, heart rate, hemoglobin saturation with oxygen (SATS), respiratory rate, and level of consciousness. Only vitals taken within the first 24 h after hospital admission were considered. Patient visits lasting less than 2 h were considered as routine appointments and omitted. Some patients were admitted several times and each admission was considered as an independent event. Patients with restricted or limited level of consciousness are described as 'unconscious'.

2.2 Dimensionality Reduction and Clustering

To improve interpretability, dimensionality reduction was performed using Uniform Manifold Approximation and Projection (UMAP) [18] based on the six vitals: temperature, systolic blood pressure, heart rate, SATS, respiratory rate and level of

Table 1 Clinical characterization of the full dataset. NEWS = National early warning score, SATS = Hemoglobin saturation with oxygen. Value format is mean (standard deviation)

Number of patients	60,731
Number of admissions	95,825
Gender (% female)	51.1
Age (years)	58.96 (±21.13)
Length of stay (hours)	47.84 (± 107.26)
Mortality (%)	2.84
NEWS	1.53 (±1.79)
Temperature (°C)	36.81 (±0.54)
Systolic blood pressure (mmHg)	123.97 (±21.16)
Heart rate (bpm)	79.51 (±15.96)
SATS (%)	96.03 (±2.62)
Respiratory rate (bpm)	17.43 (±2.77)
Limited level of consciousness (%)	0.84

consciousness. The first three vitals were scaled, the latter three transformed with the logit function. After dimensionality reduction, HDBScan [7] was applied to the embedding to identify clusters subsequently used for clinical validation.

2.3 *Clustering Explanations*

Understanding how and why patients were clustered was a fundamental requirement to establish clinical trust and ultimately reduce the risk of unintended harm. Each vital sign's contribution for the assignment of patients into each cluster was determined by generating and mutating 25,000 surrogate samples per cluster using the TabularBlimeyTree decision tree explainer [24] within `FAT Forensics` (v0.1.1): an open source toolbox [25].[1] in order to identify decision boundaries and determine each vital's contribution. Input features were the scaled vital signs. The probabilistic argument was set to False. Default arguments and settings were otherwise used. All generated samples were visualised in the embedding space to ensure reasonableness.

2.4 *Clinical Evaluation*

Clinical validation was conducted, independently of the previously described analysis, by providing two clinicians with ten sample patients for each cluster. These

[1] https://github.com/fat-forensics/fat-forensics.

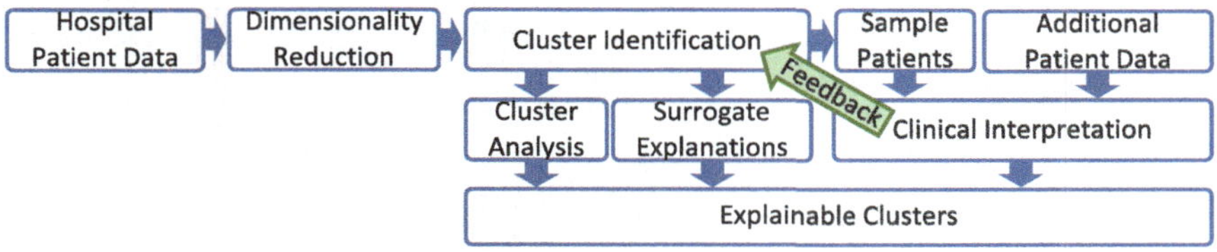

Fig. 1 Pipeline overview, from dataset import to output of explainable clusters

patients were selected uniformly at random and visually checked to ensure consistency with the underlying representation for each cluster. The clinicians then pulled additional information of these patients from the hospital databases and analysed their clinical records to assess and evaluate intra-cluster similarities and inter-cluster differences according to both the data and their clinical knowledge (Fig. 1).

3 Results

3.1 Cluster Characterization

The data extracted between Nov 2017 to March 2021 comprised 116,004 cases (70,452 patients). Of these, 95,825 cases (60,731 patients) had all vitals taken within the first 24 h of their $\geq$2 h hospital stay and were included in the study.

Dimensionality reduction and clustering revealed five clusters (Fig. 2A, summarised in Table 2). Most vitals plus gender and age reveal a gradient in values across or within clusters. Globally, the NEWS (Fig. 2D) is highest for cluster 0 but also shows high values in some areas in cluster 2. Temperature (Fig. 2E) reveals a gradient for all clusters, most noticeable in cluster 2. Systolic blood pressure (BP, Fig. 2F) behaves similarly. However, the increase in BP seems to be directed towards the unoccupied region between cluster 2, 3, and 4. Heart rate (Fig. 2G) reveals a gradient within all clusters. In contrast, SATS (Fig. 2H) has a more global gradient with values increasing from cluster 2 to cluster 4 to cluster 3 and 1. The intra-cluster SATS values are homogeneous within clusters 1, 3, and 4. Cluster 2 has a moderate gradient, and cluster 0 has a strong gradient.

Respiratory rate (Fig. 2I) is almost homogeneous with only moderate variations within all clusters. The level of consciousness is an exception (Fig. 2J). Cluster 0, the smallest cluster, contains almost all patients with limited level of consciousness and no fully "conscious" patients (Table 2). Only a small number of patients limited level of consciousness can be found distributed in the other clusters. Gender (Fig. 2B) and age (Fig. 2C) appear to show similarly distributed gradients, with female patients tending to be admitted at a lower age.

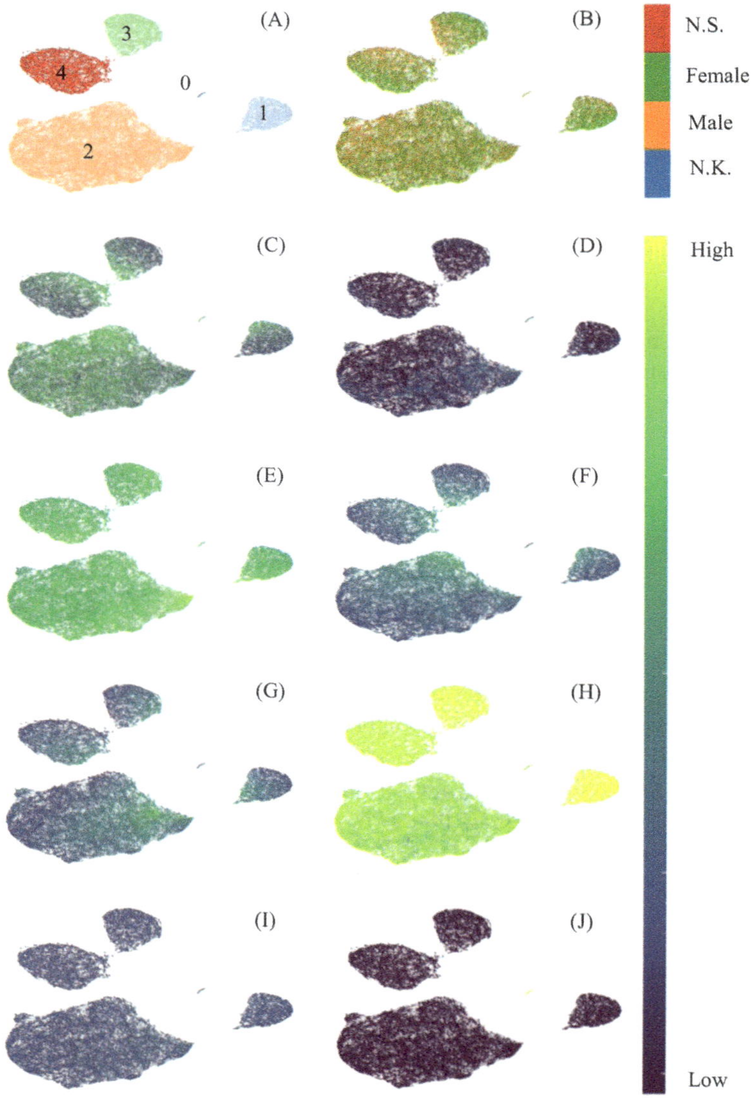

Fig. 2 Parameters mapped onto the embedding space after dimensionality reduction. Cluster assignment (**A**), gender (**B**) where N.K. = not known, N.S. = not specified, age (**C**), NEWS (**D**), temperature (**E**), systolic blood pressure (**F**), heart rate (**G**), hemoglobin saturation with oxygen (**H**), respiratory rate (**I**), level of consciousness where high = limited level of consciousness (**J**)

Table 2 Clinical characterization of all individual clusters. SATS = Hemoglobin saturation with oxygen. Value format is mean (standard deviation). ICD10 (10th revision of International Classification of Diseases) specifies codes for diseases and diagnoses, where for each cluster the most frequent code corresponds to the following; Cluster 0 sepsis, Cluster 1 abdominal and pelvic pain, Cluster 2, 3 & 4 chronic ischemic heart disease

	Cluster 0	Cluster 1	Cluster 2	Cluster 3	Cluster 4
Number of patients	434	7583	**41641**	9050	13752
Number of admissions	453	8713	**61022**	10080	15557
Gender (% female)	47.7	**63.6**	48.3	57.3	51.1
Age (years)	**69.9** (± 18.0)	49.6 (± 21.6)	64.1 (± 18.9)	52.3 (± 21.6)	56.2 (± 21.0)
Length of stay (hours)	**74.5** (± 162.8)	40.5 (± 102.5)	52.2 (± 112.6)	38.1 (± 90.6)	40.2 (± 94.6)
Mortality (%)	**21**	2	4	1	1
ICD10 code (most frequent)	A41.9	R10.3	I251	I251	I251
NEWS	**5.78** (± 2.78)	0.99 (±1.31)	1.65 (±1.88)	0.78 (±1.12)	0.69 (±1.03)
Temperature (°C)	36.68 (±0.68)	36.79 (±0.56)	**36.85** (±0.62)	36.75 (±0.46)	36.73 (±0.44)
Systolic blood pressure (mmHg)	121 (±26.33)	126 (±21.09)	**130** (±23.13)	128 (±21.19)	129 (±20.54)
Heart rate (bpm)	79.18 (±15.86)	78.10 (±16.23)	**81.62** (±17.53)	77.04 (±14.91)	76.54 (±13.82)
SATS (%)	95.72 (±3.13)	**100.00** (±0.02)	95.53 (±2.02)	99.00 (0)	98.00 (0.09)
Respiratory rate (bpm)	**18.14** (±4.60)	16.72 (±2.47)	17.53 (±2.95)	16.63 (±2.11)	16.63 (±2.06)
Limited level of consciousness (%)	**100**	0.18	0.12	0.06	0.03

3.2 *Cluster Visualisation and Vitals*

A goal of analysing the individual clusters is to identify their unique characteristics relative to each other and the overall population. Figure 3 shows the percentage difference in measurements for each cluster relative to the overall population.

Cluster 0 shows the clearest difference compared to all other clusters with higher than average NEWS, respiratory rate, consciousness, and longest hospital stay, and a decreased temperature and systolic blood pressure. The high consciousness score indicates that cluster 0 contains almost all patients with limited level of consciousness. Additionally, cluster 0's confidence interval covers the largest range, indicating higher variability. For clusters 1–4, the confidence intervals all have similar magnitudes. Features of cluster 2 appear distinct as they are directionally opposite to clusters 1, 3, and 4.

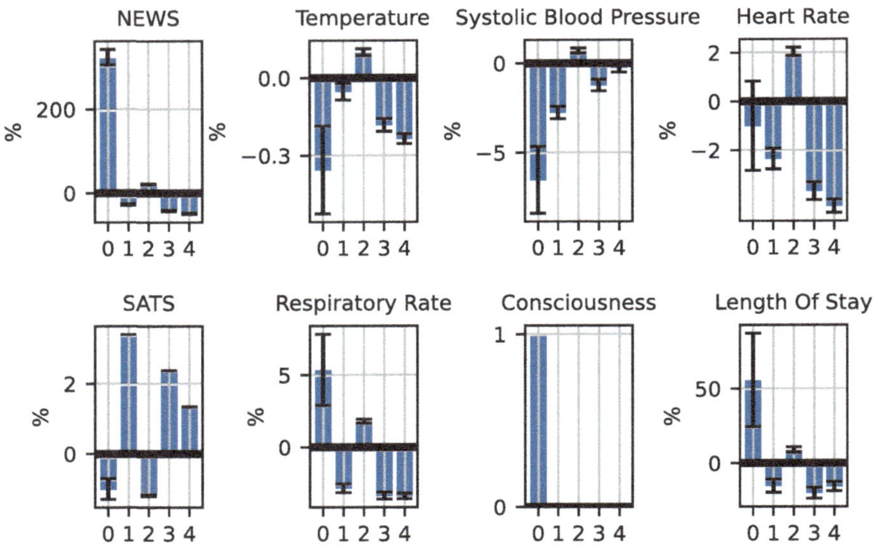

Fig. 3 Vitals, NEWS and length of stay for individual clusters. The mean value of each cluster is compared to the mean value (black line) of the whole population. Error bars represent 95% confidence intervals

3.3 *ICD10 Codes*

Figure 4 shows the frequency of identified primary 10th revision of International Classification of Diseases (ICD10) code groupings per cluster. The highest incidence ICD10 code group is 'Circulatory system (I00-I99)' with 20.9% and 21.0% in clusters 2 and 4, respectively. Diseases of circulatory system are overall the most prevalent group. Diseases of the respiratory system are common in cluster 0 (14.8%) and 2 (14.2%), with approximately three times higher occurrence than in the other clusters. 'Infectious/parasitic diseases (A00-B99)', 'Neuropsychiatric disease (F01-F99)' and 'Nervous system (G00-G99)' appear predominantly in cluster 0, whereas cases of 'Neoplasms (C00-D49)' and 'Digestive system (K00-K95)' are more common in the other clusters, ranging from 9.1% to 12.8% of all cases. Pregnancy-related incidences are most common in cluster 1 with 11.4%.

3.4 *Surrogate Explanations*

Surrogate explanations suggest that the most important feature in determining cluster 0 assignment is consciousness (0.63 feature importance), followed by SATS (Fig. 5). For all other clusters, SATS is the most dominant factor, with a feature importance factor approaching 1.0. Temperature had a marginal influence for Clusters 0 and 2.

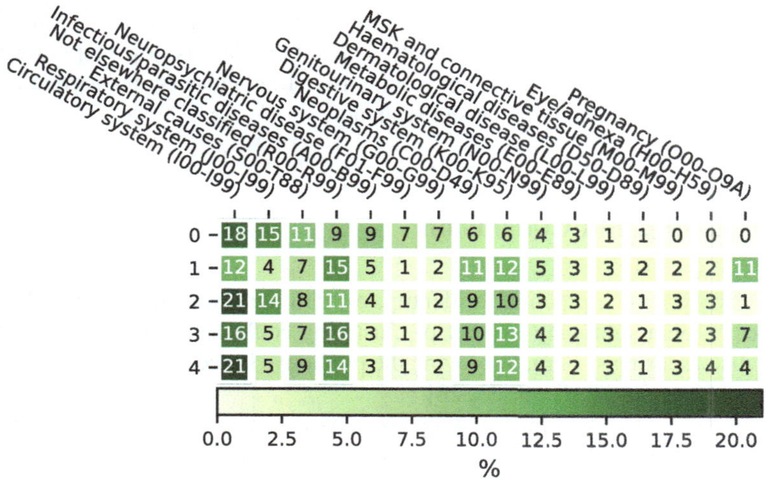

Fig. 4 Heatmap of primary ICD10 codes as recorded by clinicians at the time of patient admission and collated by top-level grouping. For display purposes only ICD10 codes with $\geq 2\%$ incidence for at least one cluster are displayed. Since only a subset of ICD10 codes are visualised, each row does not add up to 100. MSK = Musculoskeletal

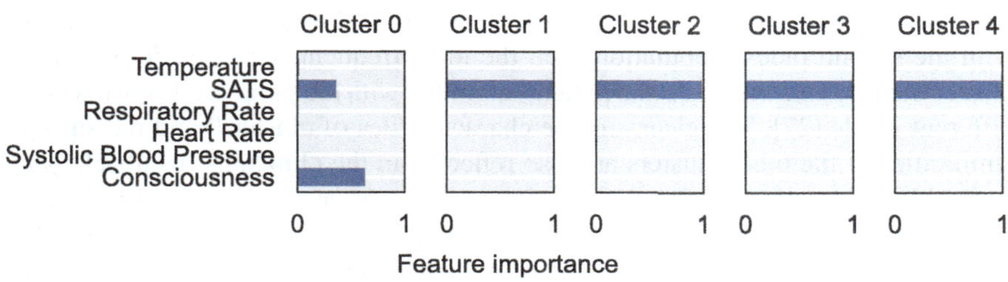

Fig. 5 Surrogate explanations for the contribution of each vital in determining the assignment of patients into each cluster. SATS = Hemoglobin saturation with oxygen

3.5 Clinical Interpretation of Clusters

The clinicians were able to detect inter-cluster differences and intra-cluster similarities. The limited level of consciousness and high NEWS in cluster 0, and different SATS levels for clusters 1, 2, 3, and 4 were identified as the main features. This mirrors findings of the vitals importance gained from the surrogate explanations.

Cluster 0 was found to have the highest NEWS. Yet, only one patient in the sampling group died. Cluster 1 was identified as a heterogeneous group with SATS of 100% and a low respiratory rate. Unsurprisingly, none of the analysed patients were admitted due to respiratory disease. That is in contrast to cluster 2 which showed high level of infection and/or respiratory disease. Based on the provided sample patients, this cluster appeared to have the highest mortality rate. SATS of 99% and a low heart rate was the most characteristic feature of cluster 3. And patients in cluster 4 tended to have minimal background with medical conditions, low respiratory rate, low NEWS and SATS of 98%. Apart from cluster 1, all clusters appeared to contain patients that did not match the cluster tendency.

4 Discussion and Conclusion

This study presents a pipeline to identify, explain and evaluate clusters of patient subtypes. Patient subtyping by way of clustering could be the first step towards a personalised scoring system, improving the predictive success of currently deployed risk scoring metrics [1, 5, 16, 23, 26].

The clusters identified in this study were based on just six vitals from the first set of readings taken during a hospital stay. Using only six vitals and including hospital departments beyond the ICU are in contrast to previous studies. Forte et al. [9] and Vranas et al. [27] included 76 and 23 clinical features, respectively, resulting in the identification of six subtypes of ICU patients. Here, cluster 0 appeared to predominantly include high-risk patients with a high NEWS and high level of limited consciousness, which was reflected in their elevated length of stay. The inter-cluster differences of the other clusters are harder to identify. Cluster 2 is the largest cluster with the second oldest population. Yet, the most frequent ICD10 code is equal in cluster 0 and 4. This indicates that patients admitted with the same condition may have different needs [27]. The clear, unique characteristics of cluster 0 and the stronger similarities of the other clusters are also reflected in the clinicians' feedback. They were able to identify SATS and level of consciousness as the key features of inter-cluster differences and successfully linked the cluster characteristics to the prevalent ICD10 codes. However, they also pointed out patients that did not match the cluster tendencies. This could be the result of the small sample size relative to the cluster size, distorting the clinicians' perception of the cluster characteristics. Additionally, after reading the first draft, the clinicians pointed out that the selected sample patients had a significant impact on which elements they considered as cluster characteristics.

For further insights, the identified clusters were characterised and analysed regarding the occurrence of the most frequent ICD10 codes. Vranas et al. [27] found 'Sepsis' as the most common diagnosis in ICU patients in five out of six clusters, whereas Castela Forte et al. [9] determined a different leading cause for each cluster. Here, circulatory diseases have the highest impact on hospital admissions, followed by 'Not elsewhere classified' cases. Diseases of the Respiratory system appear to distinguish clusters 0 and 2 from the other clusters. Castela Forte et al. [9] also identified two clusters with high prevalence of respiratory failure.

Surrogate explainers were generated to improve cluster explainability. While consciousness is a key criterion for separating cluster 0 from the other clusters, SATS is the most dominant factor for distinguishing clusters 1–4. The level of consciousness has previously been identified as the key feature in predicting discharge from ICU [19]. The integration of surrogate explainers and clinicians helped validate and verify the presented results. Future studies and the deployment in hospital settings should consider this approach to increase fairness, accountability, and transparency. This also aids in building trust between the clinicians and machine learning systems. However, the identified patient subtypes should be treated with care as the whole analysis is based on a dataset from one hospital. Adding data from other hospitals as well as adding other features may reveal other or alter the identified patient subtypes. Furthermore, two clinicians were part of the team in order to co-design the process. Future studies will increase the number of clinicians for additional feedback, increasing the acceptance of and trust in the identified patient subtypes.

Acknowledgements This work was funded by Health Data Research UK via the Better Care Partnership Southwest (HDR CF0129). JC, AH and RSR are funded by the UKRI Turing AI Fellowship EP/V024817/1.

References

1. T.E. Abbott, N. Cron, N. Vaid, D. Ip, H.D. Torrance, J. Emmanuel, Pre-hospital National Early Warning Score (NEWS) is associated with in-hospital mortality and critical care unit admission: a cohort study. Ann. Med. Surg. **27**, 17–21 (2018)
2. N. Alam, I. Vegting, E. Houben, B. van Berkel, L. Vaughan, M. Kramer, P. Nanayakkara, Exploring the performance of the National Early Warning Score (NEWS) in a European emergency department. Resuscitation **90**, 111–115 (2015)
3. R.S. Anand, P. Stey, S. Jain, D.R. Biron, H. Bhatt, K. Monteiro, E. Feller, M.L. Ranney, I.N. Sarkar, E.S. Chen, Predicting mortality in diabetic ICU patients using machine learning and severity indices. AMIA joint summits on translational science proceedings. AMIA Joint Summits Transl. Sci. **2017**, 310–319 (2018)
4. K.F. Baker, A.T. Hanrath, I. Schim van der Loeff, L.J. Kay, J. Back, C.J. Duncan, National Early Warning Score 2 (NEWS2) to identify inpatient COVID-19 deterioration: a retrospective analysis. Clin. Med. **21**(2), 84–89 (2021)
5. B. Balkan, P. Essay, V. Subbian, Evaluating ICU clinical severity scoring systems and machine learning applications: APACHE IV/IVa case study. In: Annual International Conference IEEE Engineering Medical Biological Society, pp. 4073–4076. IEEE, Honolulu (2018)
6. I.M. Baytas, C. Xiao, X. Zhang, F. Wang, A.K. Jain, J. Zhou, Patient subtyping via time-aware LSTM networks. In: Proceedings of the 23rd ACM SIGKDD International Conference on Knowledge Discovery and Data Mining, KDD '17, pp. 65–74. ACM, Halifax (2017)
7. R.J.G.B. Campello, D. Moulavi, J. Sander, Density-based clustering based on hierarchical density estimates. In: Advances in Knowledge Discovery and Data Mining, vol. 7819, pp. 160–172. Springer, Berlin (2013). https://doi.org/10.1007/978-3-642-37456-2_14
8. E. Carr, R. Bendayan, D. Bean, M. Stammers, W. Wang, H. Zhang, T. Searle, Z. Kraljevic, A. Shek, H.T.T. Phan, W. Muruet, R.K. Gupta, A.J. Shinton, M. Wyatt, T. Shi, X. Zhang, A. Pickles, D. Stahl, R. Zakeri, M. Noursadeghi, K. O'Gallagher, M. Rogers, A. Folarin, A. Karwath, K.E. Wickstrøm, A. Köhn-Luque, L. Slater, V.R. Cardoso, C. Bourdeaux, A.R. Holten, S. Ball, C. McWilliams, L. Roguski, F. Borca, J. Batchelor, E.K. Amundsen, X. Wu, G.V. Gkoutos, J. Sun, A. Pinto, B. Guthrie, C. Breen, A. Douiri, H. Wu, V. Curcin, J.T. Teo, A.M. Shah, R.J.B. Dobson, Evaluation and improvement of the National Early Warning Score (NEWS2) for COVID-19: a multi-hospital study. BMC Med. **19**(1), 23 (2021)
9. J. Castela Forte, G. Yeshmagambetova, M.L. van der Grinten, B. Hiemstra, T. Kaufmann, R.J. Eck, F. Keus, A.H. Epema, M.A. Wiering, I.C.C. van der Horst, Identifying and characterizing

high-risk clusters in a heterogeneous ICU population with deep embedded clustering. Sci. Rep. **11**(1), 12109 (2021)
10. F.Y. Cheng, H. Joshi, P. Tandon, R. Freeman, D.L. Reich, M. Mazumdar, R. Kohli-Seth, M.A. Levin, P. Timsina, A. Kia, Using machine learning to predict ICU transfer in hospitalized COVID-19 patients. JCM **9**(6), 1668 (2020)
11. D.K. Costa, J.M. Kahn, Organizing critical care for the 21st century. JAMA **315**(8), 751 (2016)
12. D.L. Davies, D.W. Bouldin, A cluster separation measure. IEEE Trans. Pattern Anal. Mach. Intell. **PAMI-1**(2), 224–227 (1979)
13. Y.D. Dlugacz, L. Stier, D. Lustbader, M.C. Jacobs, E. Hussain, A. Greenwood, Expanding a performance improvement initiative in critical care from hospital to system. Jt. Comm. J. Qual. Patient Saf. **28**(8), 419–434 (2002)
14. G. Geri, P. Vignon, A. Aubry, A.L. Fedou, C. Charron, S. Silva, X. Repessé, A. Vieillard-Baron, Cardiovascular clusters in septic shock combining clinical and echocardiographic parameters: a post hoc analysis. Intensive Care Med. **45**(5), 657–667 (2019)
15. H.M. Giannini, J.C. Ginestra, C. Chivers, M. Draugelis, A. Hanish, W.D. Schweickert, B.D. Fuchs, L. Meadows, M. Lynch, P.J. Donnelly, K. Pavan, N.O. Fishman, C.W. Hanson, C.A. Umscheid, A machine learning algorithm to predict severe sepsis and septic shock: development, implementation, and impact on clinical practice*. Crit. Care Med. **47**(11), 1485–1492 (2019)
16. B. Khwannimit, A comparison of three organ dysfunction scores: MODS, SOFA and LOD for predicting ICU mortality in critically ill patients. J. Med. Assoc. Thai. **90**(6), 1074–1081 (2007)
17. I. Kostakis, G.B. Smith, D. Prytherch, P. Meredith, C. Price, A. Chauhan, A. Chauhan, P. Meredith, A. Mortlock, P. Schmidt, C. Spice, L. Fox, D. Fleming, L. Pilbeam, M. Rowley, H. Poole, J. Briggs, D. Prytherch, I. Kostakis, C. Price, P. Scott, G.B. Smith, The performance of the National Early Warning Score and National Early Warning Score 2 in hospitalised patients infected by the severe acute respiratory syndrome coronavirus 2 (SARS-CoV-2). Resuscitation **159**, 150–157 (2021)
18. L. McInnes, J. Healy, N. Saul, L. Großberger, UMAP: uniform manifold approximation and projection. JOSS **3**(29), 861 (2018)
19. C.J. McWilliams, D.J. Lawson, R. Santos-Rodriguez, I.D. Gilchrist, A. Champneys, T.H. Gould, M.J. Thomas, C.P. Bourdeaux, Towards a decision support tool for intensive care discharge: machine learning algorithm development using electronic healthcare data from MIMIC-III and Bristol. UK. BMJ Open **9**(3), e025925 (2019)
20. R. Mirnezami, J. Nicholson, A. Darzi, Preparing for precision medicine. N. Engl. J. Med. **366**(6), 489–491 (2012)
21. S.P. Oei, R.J. van Sloun, M. van der Ven, H.H. Korsten, M. Mischi, Towards early sepsis detection from measurements at the general ward through deep learning. Intell.-Based Med. **5**, 100042 (2021)
22. M.T. Ribeiro, S. Singh, C. Guestrin, "Why Should I Trust You?": explaining the predictions of any classifier, in *Proceedings of the 22nd ACM SIGKDD International Conference on Knowledge Discovery and Data Mining* (ACM, San Francisco, 2016), pp. 1135–1144
23. M.J. Rothman, S.I. Rothman, J. Beals, Development and validation of a continuous measure of patient condition using the Electron. Med. Record. JBI **46**(5), 837–848 (2013)
24. K. Sokol, A. Hepburn, R. Santos-Rodriguez, P. Flach, bLIMEy: surrogate prediction explanations beyond LIME, in *2019 Workshop on Human-Centric Machine Learning (HCML 2019) at the 33rd Conference on Neural Information Processing Systems (NeurIPS 2019)* (Vancouver, Canada, 2019), https://arxiv.org/abs/1910.13016. ArXiv preprint arXiv:1910.13016
25. K. Sokol, R. Santos-Rodriguez, P. Flach, FAT forensics: a python toolbox for algorithmic fairness, accountability and transparency. Softw. Impacts **14**, 100406 (2022)
26. C. Subbe, Validation of a modified Early Warning Score in medical admissions. QJM **94**(10), 521–526 (2001)
27. K.C. Vranas, J.K. Jopling, T.E. Sweeney, M.C. Ramsey, A.S. Milstein, C.G. Slatore, G.J. Escobar, V.X. Liu, Identifying distinct subgroups of intensive care unit patients: a machine learning approach. Crit. Care Med. **45**(10), 1607–1615 (2017)

28. W. Wu, S. Bang, E.R. Bleecker, M. Castro, L. Denlinger, S.C. Erzurum, J.V. Fahy, A.M. Fitzpatrick, B.M. Gaston, A.T. Hastie, E. Israel, N.N. Jarjour, B.D. Levy, D.T. Mauger, D.A. Meyers, W.C. Moore, M. Peters, B.R. Phillips, W. Phipatanakul, R.L. Sorkness, S.E. Wenzel, Multiview cluster analysis identifies variable corticosteroid response phenotypes in severe asthma. Am. J. Respir. Crit. Care Med. **199**(11), 1358–1367 (2019)
29. Z. Zhou, M. Sun, J. Chen, *A model-agnostic approach for explaining the predictions on clustered data, in ICDM* (Beijing, IEEE, 2019), pp.1528–1533

Automatically Extracting Information in Medical Dialogue: Expert System and Attention for Labelling

Xinshi Wang and Xunzhu Tang

Abstract Medical dialogue information extraction is becoming an increasingly significant problem in modern medical care. It is difficult to extract key information from electronic medical records (EMRs) due to their large numbers. Previously, researchers proposed attention-based models for retrieving features from EMRs, but their limitations were reflected in their inability to recognize different categories in medical dialogues. In this paper, we propose a novel model, Expert System and Attention for Labelling (ESAL). We use mixture of experts and pre-trained BERT to retrieve the semantics of different categories, enabling the model to fuse the differences between them. In our experiment, ESAL was applied to a public dataset and the experimental results indicated that ESAL significantly improved the performance of Medical Information Classification.

Keywords Natural language processing · Medical information extraction · Mixture of experts

1 Introduction

Increasingly, hospitals are prioritizing Medical Dialogue Information Extraction (MDIE) due to the adoption of Electronic Health Records (EHR). Using MDIE, detailed medical information can be extracted from doctor-patient conversations. MDIE can be viewed as a multi-label classification problem made up of different

https://github.com/Xinshi0726/Expert-System-and-Attention-for-Labelling.

X. Wang (✉)
Rensselaer Polytechnic Institute, 110 8th St., Troy, NY 12180, USA
e-mail: wangx47@rpi.edu

X. Tang
University of Luxembourg, 2 Av. de l'Universite, 4365 Esch-sur-Alzette, Esch-sur-Alzette, Luxembourg
e-mail: xunzhu.tang@uni.lu

A. Shaban-Nejad et al. (eds.), *Artificial Intelligence for Personalized Medicine*, Studies in Computational Intelligence 1106,
https://doi.org/10.1007/978-3-031-36938-4_12

classes and their status labels. Specifically, the dataset we used in this paper includes symptoms, surgeries, tests, and other information.

Medical dialogue information extraction has received an increasing amount of attention from scholars, and various approaches have been developed. Doctor-patient dialogues were firstly converted to electronic medical records and the medical dialogue information extraction task was introduced, but no specific model to solve the task was proposed [11]. As a result, 186 symptom codes and their corresponding statuses were defined in a new dataset and proposed as a new task. By proposing two novel models, this problem was solved [4]. The first model was a span-attribute tagging model, and the second was a sequence-to-sequence model. Even though a wide range of symptoms was covered in the dataset, other critical medical information wasn't considered. To incorporate more medical information, a novel dataset that includes four main categories, namely symptoms, surgeries, tests, and other information, was introduced [3]. Furthermore, several specific items with corresponding statuses were predefined. In addition, a novel method of annotation was proposed, the sliding window technique, so that the dialogues included within the document could contain the proper amount of information. Meanwhile, a Medical Information Extractor (MIE) for multi-turn dialogues was developed [3]. A matching mechanism was used to match dialogues between predefined category-item representations and status representations. The utterance's category-item information was exploited to match its most suitable status in a window to aggregate its category-item and corresponding status information.

With the help of mixture of experts [7–9], we propose a model called Expert System and Attention for Labeling (ESAL) that extracts various representations of dialogue to address the different categories within the dialogue. To get category-specific representations, we first use BERT [2] to extract contextual representations of the dialogue and feed them to the category-specific BiLSTM [10] expert. After that, we calculate the attention value between the encoded candidate representation and the encoded dialogue representation in order to obtain the candidates. In a similar manner, we calculate the status using the same attention mechanism.

To summarize, this paper makes the following contributions:

- This paper proposes an expert system attention for labelling model for extracting medical dialogue information. Each specified category can be captured in terms of the utterance representation.
- In this study, we introduce an expert system that effectively strengthens the understanding of doctor-patient dialogue. To facilitate understanding, we also introduce a learnable embedding layer.
- On a widely used medical dialogue dataset, we perform extensive experiments. On window-level evaluation, our model achieves an F1 score of 70.00, while on dialogue-level evaluation, it scores 72.17. On the benchmark dataset, it outperforms the state-of-the-art approaches by a significant difference, demonstrating its effectiveness.

2 Related Work

2.1 *Medical Dialogue Information Extraction*

Medical Dialogue Information Extraction has attracted increasing scholar attention due to the growing priority of building Electronic Health Records in hospitals [19, 20]. To address this problem, a dataset with 4 predefined categories: i.e, symptom, test, surgeries, and other information, as well as their corresponding status was proposed [3]. The dataset can be viewed as a multi-label classification problem: there is a multi-label binary representation of the predefined category with its corresponding status for each doctor-patient dialogue window. The task takes a doctor-patient dialogue window as input and expects a multi-label binary representation of the category status pair as output. Each multi-label binary representation should have length equal to 355 as it is the number of elements in the Cartesian product of the items in the 4 categories with their corresponding status. To perform classification on the entire dialogue, results from each window will be merged to form a new set.

2.2 *Mixture of Experts*

Mixture of Experts is composed of many separate networks, each of which learns to handle a subset of the complete set of training cases [7]. The ensemble of individual experts has proven to be able to improve performance [14, 15]. Then, mixtures of experts system were converted into a basic building block [16, 17]. Mixture of Experts has been applied to various fields, such as multi-domain fake news detection [1] and recommendation systems [8].

3 Approach

In this section, we will elaborate the architecture of ESAL. The architecture is shown in Fig. 1. ESAL is composed of 4 different stages: (1). Embedding layer (2). Expert information extraction Layer (3). Self-Attention labelling Layer (4). Output Layer.

3.1 *Embedding layer*

For each doctor-patient dialogue, we first tokenize its content with Bert Tokenizer [2]. We then add special tokens for classification (i.e,[cls]) as well as separation (i.e,[sep]) to obtain a list of tokens $X = [[cls], token_1, token_2, ..., token_n, [sep]]$. We then feed the list of tokens into BERT to obtain word embedding $V = BERT(X)$.

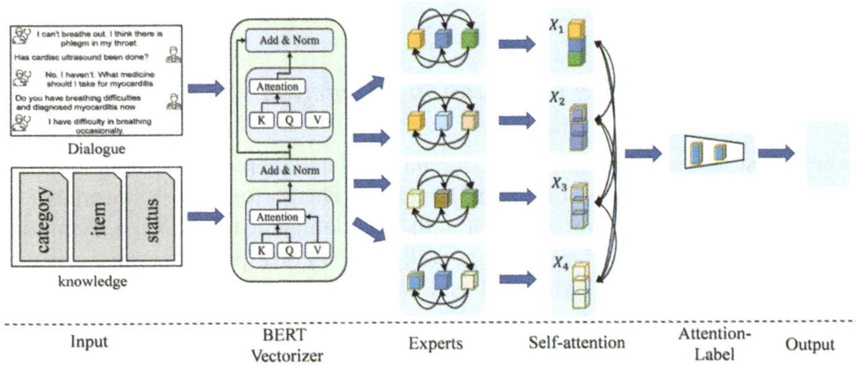

Fig. 1 Model architecture

Similarly, we perform the same operation on the candidates for matching to obtain the embedding $U = BERT(Q)$ for query Q.

3.2 *Expert information extraction layer*

With the advantage of Mixture-of-Experts, we employ multiple experts (i.e., network) to extract category-specific and status-specific representations of the utterance. We select the bidirectional long short-term memory network (BiLSTM) [10] with attention mechanism [13] as our individual network. BiLSTM has been widely used to extract contextual text features.

The equation below denotes the process for encoding each dialogue, where $H_C[i]$ consists of the contextual representation of embedding V specific to category i and H_S consists the status representation of embedding V.

$$H_C[i], H_S = BiLSTM(V), BiLSTM(U) \tag{1}$$

For candidates in the form of $\{Category : Item - Status\}$, we denote the Cartesian product between item and status given the category as Q_C. We then feed Q_C to the corresponding category expert to obtain the embedding and apply self-attention to the embedding to obtain a single vector C_C that compresses the information of the entire sequence in a weighted way. The procedures above can be described with the following equation, where $\sigma = \frac{exp(i)}{\sum_{i=1}^{n} \exp(i)}$ denotes the softmax operation.

$$\begin{aligned} U_C[i], U_S &= BiLSTM(Q_C), BiLSTM(Q_S) \\ A_C[i], A_S &= WU_C[i] + b, WU_S + b \\ P_C[i], P_S &= \sigma(A_C[i]), \sigma(A_S) \\ C_C, C_S &= \sum_i^n (P_C[i]U_C[i]), \sum_i^n (P_S U_S) \end{aligned} \tag{2}$$

3.2.1 Domain Gate

To incorporate information from all domains, we propose a domain gate with category representations from all domains as input. The output of the domain gate is the vector H_C indicating the weight ratio of each expert. Let $Gate(\cdot)$ denote the gate operation, we can describe the domain gate as the following equation:

$$H_C = Gate(\sum_{i=1}^{4} H_C[i]) \tag{3}$$

where the $Gate(\cdot)$ operation is a feed-forward network.

3.3 *Self-Attention Labeling Layer*

We employ self-attention to capture the most relevant candidate features from the utterance representation, where the candidate representation is treated as a query to calculate the attention value Q_C towards the category-specific utterance representation. Similarly, the candidate status representation is treated as another query to calculate the attention value toward the original utterances to obtain the most relevant status features from utterance representation. The process can be described with following equation:

$$\begin{aligned} P_C[i], P_S[i] &= \sigma(C_C[i]H_C), \sigma(C_S[i]H_S) \\ Q_C[i], Q_S[i] &= \sum_j^n (P_C[i,j]H_C[j]), \sum_j^n (P_S[i,j]H_S[j]) \end{aligned} \tag{4}$$

To assign the correct candidates to each dialogue window, we need to match every $Q_C[i]$ with the corresponding $Q_S[i]$. The category-item pair information and the status information does not necessarily appear in the same dialogue window, so we need to take the interactions between utterances among multiple dialogue windows into consideration. The process can be described with following equation, where *concat* denotes the concatenate operation:

$$C[i] = \sigma(Q_C[i]WQ_S[i])$$
$$\hat{Q}_S[i] = \sum_{j=1}^{n}(C[i,j]Q_S[i]) \tag{5}$$
$$F[i] = concat(C[i], \hat{Q}_S[i])$$

The output of the equation above gives the candidate information assigned to the doctor patient dialogue U.

3.4 Output Layer

We use the output from the Self-Attention Labeling Layer, (i.eF[i]) to generate the output for our model. Using a feedforward network, we can project the utterance's representation F[i] onto the 355 corresponding candidate positions, and then apply a softmax function to select the final prediction label. The process can be described with the following equations, where f denotes the feedforward network and $h_\theta(x) = \frac{1}{1+e^{-\theta^T x}}$ denotes the sigmoid function:

$$\begin{aligned} s[i] &= f(F[i]) \\ y &= h_\theta(max(s[i])) \end{aligned} \tag{6}$$

3.5 Loss Function

We adopt the cross entropy loss as our loss function. The function is defined as the following equation:

$$L = \frac{1}{I \times J}\sum_i \sum_j -y_j^i \ln(\hat{y_j^i}) + (1 - y_j^i)\ln(1 - \hat{y_j^i}) \tag{7}$$

The y_j^i is a binary encoding that denotes label of jth candidate from the ith label.I denotes the number of samples and J denotes the number of candidates. $\hat{y_j^i}$ denotes the ground truth value of label y_j^i.

4 Experiments

In this section, we will conduct experiments on the MIE dataset [3]. We will firstly describe the dataset and evaluation metrics. Then we will present results with a case study of the experiment.

4.1 Dataset Description

We evaluate our model on a public dataset MIE [3]. An example of a dialogue window is illustrated in Table 1.

The annotation of the sliding window dialogue is composed of several labels in the form of $\{Category : Item - Status\}$. An example of the annotated label is given in Table 2.

$Category$ contains four main categories (Symptom, Surgery, Test, and Other Info). $Item$ stands for the frequent items with respect to each category. There are 45, 4, 16, and 6 items, respectively. The $status$ is defined as doctor-pos, doctor-neg, patient-pos, patient-neg, or unknown. There are in total 1,120 dialogues, resulting in 18,212 windows. The data is divided into train/develop/test sets of size 800/160/160 for dialogues and 12,931/2,587/2,694 for windows respectively. In total, there are 46,151 annotated labels, averaging 2.53 labels in each window, 41.21 labels in each dialogue.

Table 1 Dialogue window

Role	Dialogue
Patient	Doctor, is it premature beat?
Doctor	Yes, Do you feel short breath?
Patient	No. Should I do radio frequency ablation?
Doctor	You should. Any discomfort in chest?
Patient	I always have bouts of pain

Table 2 Dialogue annotation

Category	Item (status)
Symptom	Premature beat (doctor-pos)
Test	Electrocardiogram (patient-pos)
Symptom	Cardiopalmus (patient-neg)
Symptom	Dyspnea (patient-neg)
Surgery	Radiofrequency ablation (doctor-pos)
Symptom	Chest pain (patient-pos)

4.2 Evaluation Metrics

We use the precision, recall, and F1 score to evaluate our results. We also follow the evaluation metrics MIE [3] employed to further analyze the model behavior from easy to hard. Category performance considers the correctness of the category. Item performance considers the correctness of the category and the item. Finally, the Full category takes the category, item, and the status into consideration, meaning all of them have to be completely correct. We will report the results in both the window-level and the dialogue level to further examine our results (Fig. 2).

Window-level: The results of each segmented window are evaluated and reported by the micro-average of all windows in the test set. Category evaluation means a prediction is assumed correct if the category matches the ground truth value. Item means a prediction is assume correct if both the category and the item match the ground truth value. Full evaluation is assumed correct if the category, item, and status match the ground truth value at the same time.

Dialogue-level: We merge the results with the same category and item of all the windows in the same dialogue. For category-item pair with multiple status assigned, we replace the unknown status with any other status occurred and replace the negative status with positive status if occurred (Fig. 3).

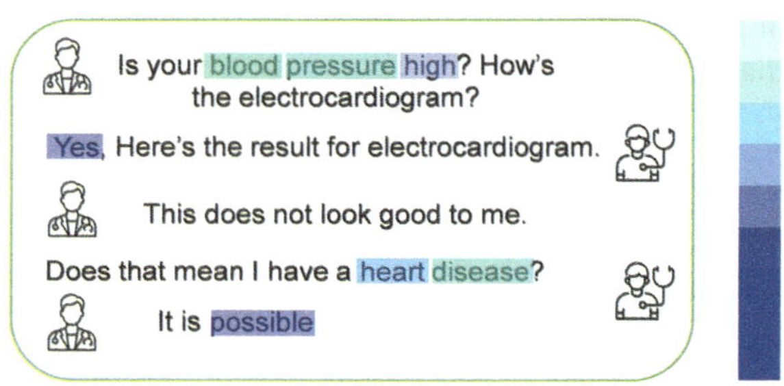

Fig. 2 Symptom expert attention heat map

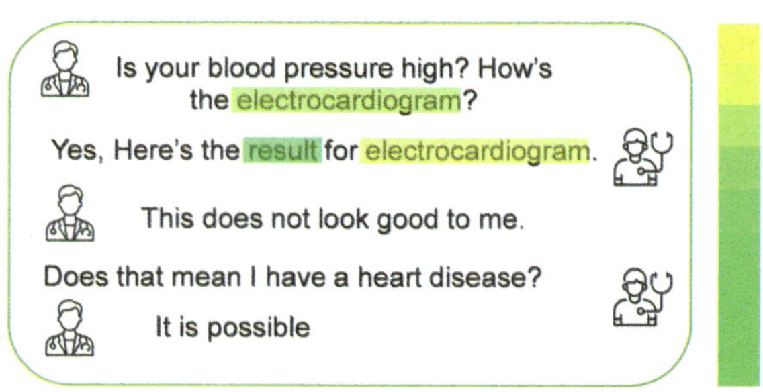

Fig. 3 Test expert attention heat map

Table 3 Window-level evaluation result, the results for MIE models are adopted from [3]

Models/levels	Window-level								
Model	Category			Item			Full		
	P	R	F1	P	R	F1	P	R	F1
Plain-classifier	67.21	63.78	64.92	60.89	49.20	53.81	53.13	49.46	50.69
MIE-classifier-single	80.51	76.39	77.53	76.58	64.63	68.30	68.20	61.60	62.87
MIE-classifier-multi	80.72	77.76	78.33	76.84	68.07	70.35	67.87	64.71	64.57
MIE-single	78.62	73.55	74.92	76.67	65.51	68.88	69.40	64.47	65.18
MIE-multi	80.42	76.23	77.77	77.21	66.04	69.75	70.24	64.96	66.40
ESAL	**92.42**	**89.66**	**90.26**	**89.46**	**83.38**	**84.85**	**72.08**	**70.93**	**70.00**

Table 4 Dialogue-level evaluation result, the results for MIE models are adopted from [3]

Models/levels	Dialogue-level								
Model	Category			Item			Full		
	P	R	F1	P	R	F1	P	R	F1
Plain-classifier	93.57	89.49	90.96	83.42	73.76	77.29	61.34	52.65	56.08
MIE-classifier-single	97.14	91.82	93.23	91.77	75.36	80.96	71.86	56.67	61.78
MIE-classifier-multi	96.61	92.86	93.45	90.68	82.41	84.65	68.86	62.50	63.99
MIE-single	96.93	90.16	92.01	94.27	79.81	84.72	75.37	63.17	67.27
MIE-multi	**98.86**	91.52	92.69	**95.31**	82.53	86.83	**76.83**	64.07	69.28
ESAL	96.51	**95.05**	**94.74**	92.52	**90.88**	**90.50**	73.68	**73.10**	**72.17**

4.3 Main Results

The experimental results are show in Table 1. From the table, we can make the following observations (Tables 3 and 4).

On both the window-level and dialogue level evaluation, Our model outperforms other models in most metrics. On window-level Full evaluation, our method has the performance improved by 5.4% compared to the MIE-multi in F1 score. On dialogue-level full evaluation, our method achieves an improvement of 4.17% in F1 score. These results demonstrate that the ESAL is performing better compared to the previous state-of-the-art model.

On Window-level evaluation, our model outperforms other models significantly in Category and Item evaluation. For Category evaluation, Our model has a performance improvement of 16.90% in F1 score. For Item evaluation, our model has a improvement of 21.65% in F1 score. Also, the improvement on Precision and Recall are significant. These results demonstrate that ESAL is able to extract a better domain-specific representation of the utterance.

4.4 Case Analysis

In this section, we perform an analysis on a specific case to verify the effectiveness of the mixture of experts. We did a data visualization on the attention value from Symptom expert and Test expert on the same utterance in graphs 2 and 3. Brighter Color suggests a higher attention value. The label for the utterance is {Symptom: high blood pressure- doctor-pos, Symptom: heart disease-unknown, Test: electrocardiogram-pos}. As we can see from graph 2, the highest attention value comes from "Yes", which suggests that our Symptom Expert captures the status information correctly. It also captures the status information for heart disease. Similarly, the test expert has captured the item and status. These two outputs gave category specific attention value on different items, thus proved the effectiveness of our model in capturing category-specific representations.

4.5 Conclusion

In this paper, we proposes an expert system attention for labelling model for extracting medical dialogue information, which utilizes two techniques: mixture of experts and an embedding layer. Experimental results on a public available dataset have shown that ESAL has the ability to capture category specific utterance representations and has better understanding of doctor-patient dialogue compared to previous models. For future work, We plan to investigate the interaction between doctor and patient to handle the pronoun ambiguity.

References

1. Q. Nan, J. Cao, Y. Zhu, Y. Wang, J. Li, MDFEND: multi-domain fake news detection, in *Proceedings of The 30th ACM International Conference on Information and Knowledge Management* (2021), pp. 3343–3347
2. J. Devlin, M. Chang, K. Lee, K. Toutanova, Bert: pre-training of deep bidirectional transformers for language understanding. ArXiv Preprint arXiv:1810.04805 (2018)
3. Y. Zhang, Z. Jiang, T. Zhang, S. Liu, J. Cao, K. Liu, S. Liu, J. Zhao, MIE: a medical information extractor towards medical dialogues, in *Proceedings of the 58th Annual Meeting of the Association for Computational Linguistics* (2020), pp. 6460–6469
4. N. Du, K. Chen, A. Kannan, L. Tran, Y. Chen, I. Shafran, Extracting symptoms and their status from clinical conversations. ArXiv Preprint arXiv:1906.02239 (2019)
5. M. Seo, A. Kembhavi, A. Farhadi, H. Hajishirzi, Bidirectional attention flow for machine comprehension. ArXiv Preprint arXiv:1611.01603 (2016)
6. S. Hochreiter, J. Schmidhuber, Long short-term memory. Neural Comput. **9**, 1735–1780 (1997)
7. R. Jacobs, M. Jordan, S. Nowlan, G. Hinton, Adaptive mixtures of local experts. Neural Comput. **3**, 79–87 (1991)
8. J. Ma, Z. Zhao, X. Yi, J. Chen, L. Hong, E. Chi, Modeling task relationships in multi-task learning with multi-gate mixture-of-experts, in *Proceedings of the 24th ACM SIGKDD International Conference on Knowledge Discovery and Data Mining* (2018), pp. 1930–1939

9. F. Zhuang, Z. Qi, K. Duan, D. Xi, Y. Zhu, H. Zhu, H. Xiong, Q. He, A comprehensive survey on transfer learning. Proc. IEEE **109**, 43–76 (2020)
10. M. Schuster, K. Paliwal, Bidirectional recurrent neural networks. IEEE Trans. Signal Process. **45**, 2673–2681 (1997)
11. G. Finley, E. Edwards, A. Robinson, M. Brenndoerfer, N. Sadoughi, J. Fone, N. Axtmann, M. Miller, D. Suendermann-Oeft, An automated medical scribe for documenting clinical encounters, in *Proceedings of the 2018 Conference of the North American Chapter of the Association for Computational Linguistics: Demonstrations* (2018), pp. 11–15
12. Ö. Uzuner, B. South, S. Shen, S. DuVall, 2010 i2b2/VA challenge on concepts, assertions, and relations in clinical text. J. Am. Med. Inform. Assoc. **18**, 552–556 (2011)
13. A. Vaswani, N. Shazeer, N. Parmar, J. Uszkoreit, L. Jones, A. Gomez, Ł. Kaiser, I. Polosukhin, Attention is all you need, in *Advances in Neural Information Processing Systems*, vol. 30 (2017)
14. R. Caruana, Multitask learning: a knowledge-based source of inductive bias, in *Proceedings of the Tenth International Conference on Machine Learning* (1993), pp. 41–48
15. G. Hinton, O. Vinyals, J. Dean, Others, Distilling the knowledge in a neural network. ArXiv Preprint arXiv:1503.02531 2 (2015)
16. D. Eigen, M. Ranzato, I. Sutskever, Learning factored representations in a deep mixture of experts. ArXiv Preprint arXiv:1312.4314 (2013)
17. N. Shazeer, A. Mirhoseini, K. Maziarz, A. Davis, Q. Le, G. Hinton, J. Dean, Outrageously large neural networks: The sparsely-gated mixture-of-experts layer. ArXiv Preprint arXiv:1701.06538 (2017)
18. J. Hopfield, Neural networks and physical systems with emergent collective computational abilities. Proc. Natl. Acad. Sci. **79**, 2554–2558 (1982)
19. R. Wachter, J. Goldsmith, *To combat physician burnout and improve care, fix the electronic health record* (Harvard Busin, Rev, 2018)
20. R. Xu, The burnout crisis in American medicine. The Atlantic (2018)
21. S. Hochreiter, J. Schmidhuber, Long short-term memory. Neural Comput. **9**(12), 1735–1780 (1997)

Transfer Learning and Class Decomposition for Detecting the Cognitive Decline of Alzheimer's Disease

Maha M. Alwuthaynani, Zahraa S. Abdallah, and Raul Santos-Rodriguez

Abstract Early diagnosis of Alzheimer's disease (AD) is essential in preventing the disease's progression. Therefore, detecting AD from neuroimaging data such as structural magnetic resonance imaging (sMRI) has been a topic of intense investigation in recent years. Deep learning has gained considerable attention in Alzheimer's detection. However, training a convolutional neural network from scratch is challenging since it demands more computational time and a significant amount of annotated data. By transferring knowledge learned from other image recognition tasks to medical image classification, transfer learning can provide a promising and effective solution. Irregularities in the dataset distribution present another difficulty. Class decomposition can tackle this issue by simplifying learning a dataset's class boundaries. Motivated by these approaches, this paper proposes a transfer learning method using class decomposition to detect Alzheimer's disease from sMRI images. We use two ImageNet-trained architectures: VGG19 and ResNet50, and an entropy-based technique to determine the most informative images. The proposed model achieved state-of-the-art performance in the Alzheimer's disease (AD) vs mild cognitive impairment (MCI) vs cognitively normal (CN) classification task with a 3% increase in accuracy from what is reported in the literature.

M. M. Alwuthaynani (✉) · Z. S. Abdallah · R. Santos-Rodriguez
University of Bristol, Bristol, UK
e-mail: maha.alwuthaynani@bristol.ac.uk

Z. S. Abdallah
e-mail: zahraa.abdallah@bristol.ac.uk

R. Santos-Rodriguez
e-mail: enrsr@bristol.ac.uk

M. M. Alwuthaynani
College of Computer Science and Information Systems, Najran University, Najran, Saudi Arabia

A. Shaban-Nejad et al. (eds.), *Artificial Intelligence for Personalized Medicine*,
Studies in Computational Intelligence 1106,
https://doi.org/10.1007/978-3-031-36938-4_13

1 Introduction

Dementia is a broad term for various mental pathologies that can cause memory troubles and brain changes. Alzheimer's disease (AD) is the cause of approximately 60–80% of dementia cases. People with Alzheimer's experience many symptoms that change over the years, reflecting the degree of damage to neurons in different parts of the brain. This disease starts years before its symptoms are present, and the speed with which symptoms progress from mild to moderate to severe varies from one individual to another [4].

The use of neuroimaging modalities has been demonstrated to significantly aid in the diagnosis of Alzheimer's disease. Structural magnetic resonance imaging (sMRI) is the most widely used neuroimaging modality for AD detection and has shown increased performance in the literature. Moreover, sMRI is capable of capturing grey matter atrophy related to the loss of neurons and synapses in AD as well as white matter atrophy linked to the loss of integrity of the white matter tract. Therefore, atrophy measured by sMRI is considered a robust AD biomarker [6].

According to [16], machine-learning approaches are valuable for diagnosing Alzheimer's. In addition, the use of deep learning models has become widespread for dealing with medical images. Deep learning has gained considerable interest in Alzheimer's detection research since 2013, and the number of publications on this topic has risen drastically since 2017 [6]. However, there are some limitations when training the model from scratch. The main limitation is that training models demand a significant amount of labelled data. Another limitation of using deep learning on sMRI data is that model training requires a large number of computational resources. It is also challenging to deal with irregularities in the dataset distribution.

Transfer learning is an alternative for training the model from scratch [3]. Transfer learning is an important mechanism in machine learning for addressing the issue of insufficient training data. It attempts to transfer knowledge from the source domain to the target domain [15].

Class decomposition assists with the issue of irregularities in the dataset distribution by making learning the class boundaries of a dataset more uncomplicated. The class decomposition aims to divide each class in the image dataset into subclasses, with each subclass being treated independently, simplifying the dataset's local structure to deal with any irregularities in the data distribution [18].

In this paper, we examine how transfer learning and class decompaction can be applied for enhanced diagnosis of AD. The essential motivation behind utilising transfer learning is tackling the challenges of the lack of availability of a large annotated training set. The contributions of our method can be summarised as follows:

- We proposed an efficient transfer learning-based approach for diagnosing Alzheimer's from sMRI scans.
- Investigated the influence of transfer learning across two different domains ImageNet data and sMRI images.

- We employed the image entropy strategy to select the most informative information for training the model when even using a small training dataset to achieve better performance
- We utilised class decompaction to uncover the hidden patterns within Alzheimer's images by dividing classes into sub-clusters and to overcome irregularities in distribution.

The rest of this paper is organised as follows: Sect. 2 describes the related work, the methodology is introduced in Sects. 3, and 4 presents our experiments and findings. Finally, Sect. 5 concludes the work.

2 Related Works

In this section, we aim to present state-of-the-art on how neuroimaging is utilised to diagnose and monitor Alzheimer's progression using voxel-based and slice-based methods, provide an overview of transfer learning approach and its application in the detection of Alzheimer's disease and explore how the class decomposition method can be used to assist in enhancing models performance.

Many studies documented in the literature have assessed structural brain variances to highlight the atrophy of AD and prodromal AD spatially distributed over many brain regions. In the following sections, we explore how neuroimaging is utilised to diagnose and monitor AD progression using voxel-based and slice-based diagnostic techniques.

Various studies have proposed models that rely on the voxel-based method. These involve voxel-wise analyses of local brain tissue to determine the pathological modifications in discriminative regions for AD diagnosis. The voxel-based method utilises voxel intensity values from whole neuroimaging modalities or tissue components. Each image demand is standardised to a 3D space [6]. The authors of [6] stated that approximately 70% of investigations utilising this approach involve a full-brain analysis. The advantage of a full brain analysis is that the spatial data are completely integrated, which allows for obtaining 3D data from neuroimaging scans. The disadvantage is that it causes increasing amounts of data dimensionality and computational load [2]. Many studies have employed distinct strategies using the voxel-based approach [17, 19].

The slice-based approach is employed to extract two-dimensional (2D) slices from 3D brain scans. As actual brain tissue is represented in a 3D format (3D brain scans), utilising a slice-based approach might result in data loss because it reduces volumetric data to 2D representations [2, 6]. Many investigations have utilised distinctive approaches to extract 2D image slices from 3D brain scans, whereas others have employed standard projections of the axial, sagittal, and coronal planes. However, none of these studies achieved a full-brain analysis because the 3D brain scans could not be converted into 2D slices. Therefore, a whole-brain analysis is not achievable using the slice-based approach [6]. In [2], the authors stated that using a 2D slice strategy decreases network complexity and the parameters required to train the model. At the same time, it has the drawback of spatial dependency loss between nearby slices. Many studies have employed distinct strategies to extract two-dimensional slices from 3D brain scans [8–10, 12, 14].

Transfer learning is a key mechanism in machine learning for dealing with the issue of insufficient training data [15]. It effectively extends knowledge previously learned in one task to a new task [13]. Another issue that can be addressed using transfer learning is that many machine learning approaches perform sufficiently only under a standard assumption: the training and testing dataset have the same feature space and distribution. Thus, most statistical models need to be rebuilt from scratch when the distribution changes using newly collected training data which in some cases could be an expensive re-collect the required training data and reconstruct the models. Transfer learning between task domains would be desirable to address these issues and reduce the need and effort to re-collect the training data [13]. According to [3], transfer learning approaches have shown robust performance because it transfers knowledge across domains. Moreover, many research used two transfer learning techniques: (1) employing a pre-trained network as a feature extractor, which is not demanding to train the network at all and (2) fine-tuning a pre-trained network on the data under study [11].

Many studies utilised pre-trained networks on the ImageNet dataset as feature extractors of medical images to overcome the lack of large-scale annotated data [3]. In [12], the authors proposed a transfer-based learning network that predicts Alzheimer's disease using sMRI scans. Authors of [10] suggested a layer-wise transfer learning-based model investigating the relationship between training size and classification accuracy in the context of transfer learning with intelligent data selection. The authors determined the most useful two-dimensional slices taken from three-dimensional sMRI images using an entropy-based method based on calculating the image entropy using a histogram. The model is very similar to the VGG-19 design. The fully connected layers were adjusted in all four configurations to test the model. Each configuration involved periodically freezing some blocks and altering the amount of the training dataset.

The class decomposition method was proposed in 2003 by [18], which is based on using clustering for pre-processing images. It enables the reduction of the impact of noisy data, finds hidden patterns within each class, and improves classification accuracy. The clustering-based class decomposition approach works by applying clustering to samples in a class to split it into sub-classes and then re-labelling each cluster's instances with a new class label [18]. Many studies have utilised class decomposition. Authors of [1] suggested CNN architecture called DeTraC, (Decompose, Transfer, and Compose) for adapting the class decomposition to medical image classification tasks. The suggested model was validated using three distinct datasets: chest X-rays, digital mammograms, and histological sections of human colorectal cancer. To increase the number of samples, data augmentation techniques like flipping, translation (translating, scaling, and rotation at various angles), colour processing, and minor random noise perturbation were used. The authors used the three primary scenarios of shallow-tuning, fine-tuning, and deep-tuning to evaluate DeTraC with five different pre-trained CNN networks (AlexNet, VGG16, VGG19, GoogleNet, and ResNet).

3 Methodology

The architecture of the proposed model is inspired by the [1] work. Our proposed approach uses sMRI scans from the Alzheimer's Disease Neuroimaging Initiative (ADNI) database. Figure 1 illustrates the proposed network using sMRI.

Images for all subjects go through three phases: feature extraction, class decomposition, and classification. In the class decomposition stage, the data samples of each class are divided into clusters (sub-classes) to feed into the classification phase. All sub-classes classified as AD_1, AD_2, MCI_1, MCI_2, CN_1 and CN_2 are then assembled back to construct the actual classes (AD, MCI, and CN) before the class decomposition process to produce the prediction.

3.1 Selection of the Most Informative Training Dataset

Alzheimer's Disease Neuroimaging Initiative (ADNI) database provided the structural magnetic resonance imaging (sMRI) data used in this investigation. Three-dimensional (3D) Neuroimaging Informatics Technology Initiative (NIFTI) formatted sMRI data were used for this study. However, handling 3D images necessitates powerful computing capabilities and a big memory space. Consequently, employing two-dimensional sMRI slices as an alternative to three-dimensional images is one option to lessen processing. Creating 2D slices from 3D sMRI scans produces a vast number of images, some of which contain noisy data while others are rich in information. Selecting the most relevant data is essential to the method's success. Therefore, we employed the image entropy method to select the most informative 2D slices, as opposed to most current techniques, which randomly select the two-dimensional images for training and testing the model. After extracting all two-dimensional slices for each subject, we calculate each slice's image entropy using the grey-level cooccurrence matrix (GLCM) [7]. It is a statistical technique for

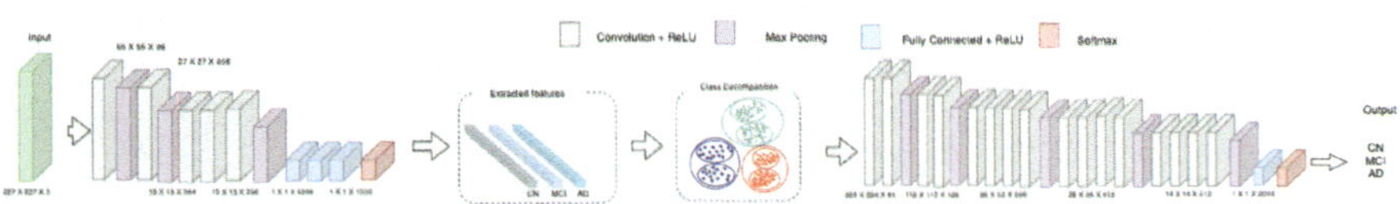

Fig. 1 The architecture of the proposed model consists of AlexNet for feature extraction, class decomposition for splitting classes into sub-classes, and VGG19 network for classification

analysing texture, investigates the spatial relationship between pixels and determines how frequently a combination of pixels appears in an image in a given direction and distance [5]. The two-dimensional images then sort in descending order based on the entropy of the images, and images with the greatest entropy values are the most informative. We pick only twenty slices with the highest entropy from each subject for training and testing the model. The following formula is applied for calculating the image entropy of a set of M symbols with probabilities $p_1, p_2, ..., p_M$:

$$H = -\sum_{i=1}^{M} p_i \log p_i. \tag{1}$$

3.2 Feature Extraction

For feature extraction, we use transfer learning that uses a pre-trained model on ImageNet to capture the general features from images by freezing some layers of the pre-trained network and adapting the top layer to be used for AD data. AlexNet network is used to extract features from two-dimensional images. The top layer of the pre-trained network is adopted for the three classes AD, MCI and CN. After extracting the features, we used principal component analysis (PCA) to reduce the dimensionality of the feature space, which assists in reducing memory requirements and enhancing the framework's efficiency. Then, the extracted features were passed to the cluster to perform the class decomposition.

3.3 Class Decomposition

Applying clustering as a pre-processing phase for each class is known as class decomposition. This approach was proposed by [18]. The idea of the clustering-based class decomposition approach is that clustering is applied to all data samples of each class to divide the class into clusters (sub-classes) and to re-label each cluster's instances with a new class label. This technique assists in decreasing the impact of noisy data, discovering the hidden patterns within each class and enhancing classification accuracy [18].

Suppose the feature space is illustrated by a two-dimensional matrix (A), where A is the image dataset, L is a set of class labels, and n, m, and k are the number of images, features and classes, respectively. A and L can be written as in (2).

$$A = \begin{bmatrix} a_{1_1} & a_{1_2} & ..... & a_{1_m} \\ a_{2_1} & a_{2_2} & ..... & a_{2_m} \\ . & . & . & . \\ a_{n_1} & a_{n_2} & ..... & a_{n_m} \end{bmatrix}, L = \{l_1, 1_2, ..., l_k\} \tag{2}$$

Class decomposition is applied to partition each class in a dataset (A) into k sub-classes, where each subclass is treated independently. The new class label will be a pair (c, k'), where c denotes the actual (class label) from label space Y, and k' represents the (cluster label) to which the sample belongs from the new cluster label space Y'. The class decomposition resulted in a new dataset (B) with new sub-classes. A and B datasets can be written as in (3).

$$A = \begin{bmatrix} a_{1_1} & a_{1_2} & \dots & a_{1_m} & l_1 \\ a_{2_1} & a_{2_2} & \dots & a_{2_m} & l_1 \\ . & . & . & . & . \\ a_{n_1} & a_{n_2} & \dots & a_{n_m} & l_2 \end{bmatrix}, B = \begin{bmatrix} b_{1_1} & b_{1_2} & \dots & b_{1_m} & l_{1_1} \\ b_{2_1} & b_{2_2} & \dots & b_{2_m} & l_{1_k} \\ . & . & . & . & . \\ b_{n_1} & b_{n_2} & \dots & b_{n_m} & l_{2_k} \end{bmatrix} \tag{3}$$

3.4 *Classification and Class Composition*

In the classification phase, VGG19 was employed for the classification task after the class decomposition, as shown in Fig. 1. We have also experimented with other pre-trained models to emphasise the effectiveness and robustness of our proposed method. The top layer of the VGG19 network is adopted for in Y'. The classifier was trained on the new dataset (B), which was produced after decomposing the classes. The classifier constructs a hypothesis h' and maps samples from class label space Y to the cluster label space Y'. The hypothesis $h'(x) = (a, b)$ produces a prediction consisting of a pair, a class label and a cluster label. The cluster label will be removed in the composition phase to obtain the actual prediction in the class label space Y. Then, the sub-classes will be reassembled to construct the predicted classes, depending on the dataset before decomposition.

4 Results and Discussion

This section provides the experimental results for the model. We executed our deep-learning approach using Keras with a TensorFlow backend. Our target is to differentiate AD subjects from MCI and CN subjects by analysing sMRI scans.

4.1 *Expierments Setup*

sMRI scans used in this study are from the Alzheimer's Disease Neuroimaging Initiative (ADNI) database (available at http://adni.loni.usc.edu). The dataset used in the experiments contains 134 three-dimensional T1-weighted magnetic resonance images, registered to MNI 152 template standard space. Table 1 shows the demographic characteristics of the subjects.

Table 1 Demographic characteristics of subjects for the ADNI sample

Characteristic	CN	MCI	AD
Subjects	44	45	45
Age range	61.1–89.7	60.4–87.4	62.7–89.7
Gender (M/F)	22/22	26/19	21/24
MMSE range	27–30	19–30	10–28

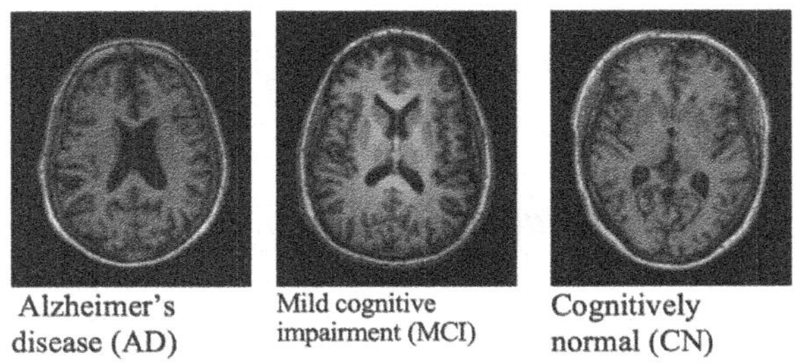

Fig. 2 Two-dimensional sMRI slices of Alzheimer's disease (AD), mild cognitive impairment (MCI), and cognitively normal (CN) subjects

We utilise NiBabel and OpenCV-Python libraries to process NIFTI files and obtain the 2D brain axial plane slices for training and testing the model. In addition, after extracting all two-dimensional slices, we used the scikit-image library to measure the entropy of the images and then picked only twenty slices with the highest entropy from each subject. The dataset is divided as follows: 80% of the subjects were randomly selected for training and validating the model, while the remaining 20% of the subjects were reserved for classifier testing. Figure 2 shows sample slices from the ADNI Dataset across the three classes.

We experimented with AlexNet to extract the features from 2D images. First, the top layers of the AlexNet network are adopted for the three classes. Then, the network is fine-tuned to extract the features. The selected features are scaled using a standard scaler and passed to the principal component analysis (PCA) for dimensionality reduction. The extracted features are finally passed to the next step for class decomposition.

For class decomposition, We use the elbow method to decide the optimal number of clusters for k-means clustering. After K-means was conducted with $k = 2$, the three class labels CN, MCI and AD are divided into six sub-classes. Table 2 shows the classes' distributions before and after class decomposition.

4.2 *Classification and Class Composition*

We aimed first to explore hyper-parameters for the model's training and determine which layers we should fine-tune. The performance of the ResNet50 and VGG19 networks was tested using the Adam algorithm to find the optimal number of learning

Table 2 Classes distribution before and after class decomposition

Class	Instances	Class	Instances
CN	704	CN_1	10
		CN_2	694
MCI	720	MCI_1	99
		MCI_2	621
AD	720	AD_1	26
		AD_2	694

Table 3 Hyper-parameters selection for classification phase

Network	Learning rate	Fine-tuned layers	Accuracy (%)	Specificity (%)	Sensitivity (%)
VGG19	0.01	block5 _conv4	92	95	91
		Dense	96	97	95
	0.001	Block5_conv1	94	96	93
		Block5_conv4	96	98	96
		Dense	98	99	98
ResNet50	0.01	conv5_block3_1_conv	86	93	86
	0.001	conv5_block3_1_conv	97	98	96
		dense	94	96	94

rates within (0.01 and 0.001) over 200 epochs and batch sizes of 64. As shown in Table 3, when fine-tuning the top two layers, the VGG19 network achieved the highest performance with an accuracy of 98% while ResNet50 achieved 94%. However, the ResNet50 performs well when fine-tuning the top thirteen layers with an accuracy of 97%, while the VGG19 network showed a drop in performance when fine-tuning more layers. As shown in Table 3, the VGG19 model achieved the highest accuracy of 98% when training the two top layers on ADNI data, while model accuracy decreased to 94 and 96% as more layers were trained. We notice that transferring knowledge from some layers outperforms strict training on the target task. For instance, when fine-tuning the top two layers, the VGG19 performance improved and achieved the most outstanding outcomes. With Resnet50, on the other hand, we had to fine-tune more layers in order to get the best results, which demonstrates that knowledge transferability varies between networks and even between layers. Figure 3 illustrates the learning curve accuracy and loss for model training and testing obtained using VGG19.

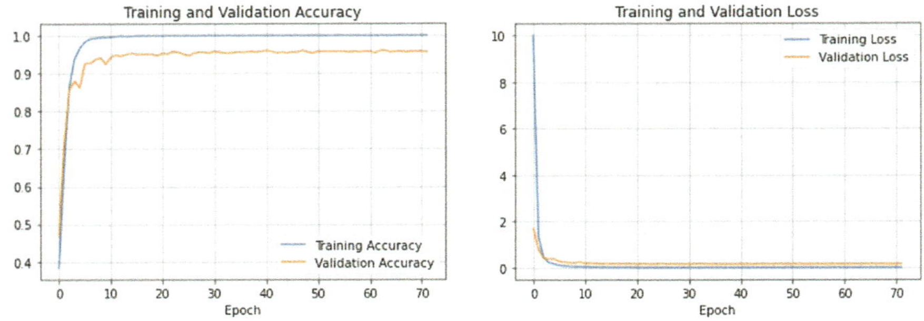

Fig. 3 The learning curve accuracy and loss error obtained by VGG19 pre-trained network

Table 4 Comparison of classification performance with state-of-art studies

No.	Training size	AD versus CN (%)	AD versus MCI (%)	MCI versus CN (%)	AD versus MCI versus CN (%)
[10]	2,560	99.4	99.2	99.0	95.2
[14]	1,731	95.4	82.2	90.1	85.5
[9]	–	90.4	77.2	72.4	
[12]	37,590	95.9 98.9	99.3 99.1	96.8 97.1	
Proposed model	1,715	99.0	99.0	98.0	98.0

4.3 Comparison with Existing Methods

Comparison with previous research findings has been challenging because studies differ in datasets, data preparation strategies and dimensional reduction methods and measurements. This section discusses the results in relation to other recent methods in terms of training size and accuracy. It is noteworthy that the methods' results are comparable, even though the studies may employ different experimental setups.

The training size was calculated for all the reported methods based on the sample used in training these models. For example, our work used 2,680 images, which were divided into 80% for training and validation and 20% for classifier testing, resulting in a training size of 1,715. Table 4 reports the results of the binary and ternary classifiers of the model. To further validate our proposed architecture, we adapted the same architecture for a three-way binary classification by changing the final classification layer.

We compared our model with four other state-of-the-art models; the results for these models are what is reported in their papers. The results in Table 4 shows that our model ternary classification outperforms the other approaches with an accuracy of 98%, a 3% improvement over the state-of-the-art performance, which archived

95.19, and 85.53. Furthermore, except for [10], our model's binary classification results outperformed the other approaches. Compared to this study, our model used a smaller training size (1,715 images) extracted from fewer subjects.

Our proposed model utilises transfer learning and class decomposition. Transfer learning deals with the challenge of the limited availability of annotated data while using class decomposition enhances model performance because it makes learning the class boundaries of a dataset uncomplicated and, as a result, can deal with any irregularities in data distribution. Using transfer learning with class decomposition leads to better accuracy than other state-of-the-art methods. Class decomposition makes learning the class boundaries of a dataset uncomplicated and deals with any irregularities in data distribution. It divided the Alzheimer's classes into six new subclasses, and this assisted in revealing the hidden patterns in each class and made more accurate predictions.

5 Conclusion

In this paper, we propose a model that integrates transfer learning with a class decomposition approach for diagnosing Alzheimer's from structural sMRI images. In addition, we use the entropy-based approach to select the training dataset that contains the most informative data. We use the VGG19 ImageNet-trained weights network to obtain highly accurate results. We compared our findings to those of four other cutting-edge procedures using the ADNI dataset. With an accuracy of 98.3%, our model ternary classification outperforms the other approaches, representing a 3% improvement over the state-of-the-art performance. In addition, except for [10], our model outperformed the others in the binary classification task.

For future work, we will conduct more analysis to investigate the impact of each component of our method (namely, class decomposition and the extracted features). Also, the two-dimensional slices have limitations in covering all of the brain's regions, therefore causing information loss. Other approaches for image segmentation can be considered in future work. Additionally, we aim to combine the model with other data modalities for the diagnosis of Alzheimer's disease, such as genomic data and Electronic Health Records. We seek to use the model to discover the hidden patterns in the mild cognitive impairment (MCI) category to reveal the conversion of mild cognitive impairment (MCI) patients to Alzheimer's disease by discriminating MCI levels based on cognitive decline. We also aim to extend the model architecture to be more scalable and also to include other factors that can assist in diagnosing Alzheimer's disease, such as the Mini-Mental State Examination (MMSE) score and age, which could increase model performance.

Acknowledgements This research was funded by the Saudi Ministry of Education and Najran University as part of a PhD scholarship. Also, we thank the Alzheimer's Disease Neuroimaging Initiative (ADNI) for providing the dataset.

References

1. A. Abbas, M.M. Abdelsamea, M.M. Gaber, Detrac: transfer learning of class decomposed medical images in convolutional neural networks. IEEE Access **8**, 74901–74913 (2020)
2. D. Agarwal, G. Marques, I. de la Torre-Díez, M.A. Franco Martin, B. García Zapiraín, and F. Martín Rodríguez, Transfer learning for alzheimer's disease through neuroimaging biomarkers: a systematic review. Sensors **21**(21), 7259 (2021)
3. Z. Ardalan, V. Subbian, Transfer learning approaches for neuroimaging analysis: a scoping review. Front.Artif. Intell. **5** (2022)
4. A. Association et al., 2018 alzheimer's disease facts and figures. Alzheimer's Dementia **14**(3), 367–429 (2018)
5. P. Bhagat, P. Choudhary, K.M. Singh, A comparative study for brain tumor detection in mri images using texture features, in *Sensors for Health Monitoring* (Elsevier, 2019), pp. 259–287
6. M.A. Ebrahimighahnavieh, S. Luo, R. Chiong, Deep learning to detect alzheimer's disease from neuroimaging: a systematic literature review. Comput. Methods Programs Biomed. **187**, 105242 (2020)
7. R.M. Haralick, K. Shanmugam, I. Dinstein, Textural features for image classification. IEEE Trans. Syst. Man, Cybern. **SMC-3**(6), 610–621 (1973)
8. M. Hon, N.M. Khan, Towards alzheimer's disease classification through transfer learning, in *2017 IEEE International Conference on Bioinformatics and biomedicine (BIBM)* (IEEE, 2017), pp. 1166–1169
9. W. Kang, L. Lin, B. Zhang, X. Shen, S. Wu, A.D.N. Initiative et al., Multi-model and multi-slice ensemble learning architecture based on 2d convolutional neural networks for alzheimer's disease diagnosis. Comput. Biol. Med. **136**, 104678 (2021)
10. N.M. Khan, N. Abraham, M. Hon, Transfer learning with intelligent training data selection for prediction of alzheimer's disease. IEEE Access **7**, 72726–72735 (2019)
11. G. Litjens, T. Kooi, B.E. Bejnordi, A.A.A. Setio, F. Ciompi, M. Ghafoorian, J.A. Van Der Laak, B. Van Ginneken, C.I. Sánchez, A survey on deep learning in medical image analysis. Med. Image Anal. **42**, 60–88 (2017)
12. S. Naz, A. Ashraf, A. Zaib, Transfer learning using freeze features for alzheimer neurological disorder detection using adni dataset. Multimedia Syst. **28**(1), 85–94 (2022)
13. S.J. Pan, Q. Yang, A survey on transfer learning. IEEE Trans. Knowl. Data Eng. **22**(10), 1345–1359 (2010)
14. A. Payan, G. Montana, Predicting alzheimer's disease: a neuroimaging study with 3d convolutional neural networks. arXiv preprint arXiv:1502.02506 (2015)
15. C. Tan, F. Sun, T. Kong, W. Zhang, C. Yang, C. Liu, A survey on deep transfer learning, in *International Conference on Artificial Neural Networks* (Springer, 2018), pp. 270–279
16. M. Tanveer, B. Richhariya, R.U. Khan, A.H. Rashid, P. Khanna, M. Prasad, C. Lin, Machine learning techniques for the diagnosis of alzheimer's disease: a review. ACM Trans. Multimedia Comput., Commun., Appl. (TOMM) **16**(1s), 1–35 (2020)
17. J. Venugopalan, L. Tong, H.R. Hassanzadeh, M.D. Wang, Multimodal deep learning models for early detection of alzheimer's disease stage. Sci. Rep. **11**(1), 1–13 (2021)
18. R. Vilalta, M.-K. Achari, C.F. Eick, Class decomposition via clustering: a new framework for low-variance classifiers, in *Third IEEE International Conference on Data Mining* (IEEE, 2003), pp. 673–676
19. E. Westman, A. Simmons, J.-S. Muehlboeck, P. Mecocci, B. Vellas, M. Tsolaki, I. Kłoszewska, H. Soininen, M.W. Weiner, S. Lovestone et al., Addneuromed and adni: similar patterns of alzheimer's atrophy and automated mri classification accuracy in europe and north america. Neuroimage **58**(3), 818–828 (2011)

Knowledge Augmentation for Early Depression Detection

Hrishikesh Kulkarni, Sean MacAvaney, Nazli Goharian, and Ophir Frieder

Abstract Individuals continue to share their mental health concerns on social media, providing an avenue to rapidly detect those potentially in need of assistance. While users of immediate need can be recognized with relative ease, early-stage disorder users in the boundary region pose a greater challenge to detect. The minimal posting histories of such users further complicate proceedings. However, these same boundary region users would benefit greatly from timely treatment; hence, detecting their mental health status is of utmost need. Additionally, pointers to identify the type of depression could be of great help. Augmenting knowledge for low posting users can help to solve this problem. We propose an NLP based method 'STBound' that intelligently determines the optimal region for knowledge augmentation. It answers three crucial questions: when?, for whom? and how much? to augment—to resolve this imbroglio. Our proposed selective knowledge augmentation method contributes to early depression detection performance improvement by an average of 11.9% in F1 score. Further, this approach shows promising performance enhancement of 12.1% in F1 score for the critical task of separating these boundary region users with bipolar depression. STBound identifies those depressed users in the boundary region who would otherwise go unidentified.

Keywords Depression detection · Social media · Mental health · Early risk detection

---

H. Kulkarni (✉)
Georgetown University, Washington DC, USA
e-mail: hpk8@georgetown.edu

S. MacAvaney
University of Glasgow, Glasgow, UK
e-mail: first.last@glasgow.ac.uk

N. Goharian
IR Lab, Georgetown University, Washington DC, USA
e-mail: first@ir.cs.georgetown.edu

O. Frieder
IR Lab, Georgetown University, Washington DC, USA
e-mail: first@ir.cs.georgetown.edu

A. Shaban-Nejad et al. (eds.), *Artificial Intelligence for Personalized Medicine*, Studies in Computational Intelligence 1106,
https://doi.org/10.1007/978-3-031-36938-4_14

1 Introduction

Depression, a mental state resulting in lack of hope and dejection along with persistent sadness [1], is a global health concern. Severe depression may result in self-harm, including suicide, and is also a major cause of disability worldwide. Around 5% [1] of the world population is estimated to be suffering from depression with 80% of those afflicted between 16 and 65 years of age, namely the working population. Thus, in addition to a social health issue, depression directly impacts the overall economy [2]. For at least the aforementioned reasons, early depression detection is of utmost importance. Unfortunately, the criticality to address depression only further increased recently due to reduced social interactions and lifestyle changes caused by the COVID-19 pandemic [3] resulting in a 25% increase in anxiety and depression [4].

A significant percentage i.e., around 16–20% of total depressed individuals qualify for bipolar disorder, and the median age of onset for bipolar disorder is 25 years [5]. Considering the high risk of unresolved morbidity associated with bipolar disorders, there is a pressing need to separate bipolar disorders in early stages.

Social media provides users venues to express their thoughts and feelings in the form of written posts. As mental health is a key factor leading to changes in textual patterns and articulations [6], noting these changes can contribute to identifying potential depression [7]. Unfortunately, existing depression detection systems generally require a substantial volume of data to predict [7, 8], resulting in depression being detected only at a later stage, potentially increasing disease severity. Complicating detection, many users have low posting activity, either due to their low posting frequency or simply being new to a platform. We refer to these users as low-resource users and focus our attention towards them, believing that early detection, irrespective of the number of posts, can help a larger population segment. We further empirically and mathematically define low-resource users with extensive analysis on Reddit Self-reported Depression Diagnosis (RSDD) dataset [8]. Usually, bipolar depression in case of low-resource users cannot be differentiated from major depression [9]. This results in sub-optimal treatment and poor outcome in this case [10]. The treatment for bipolar depression is significantly different. Hence, separating it at an early stage in such low-resource users could prove to be crucial.

Broadly speaking, our focus is on identifying low-resource users on the brink, namely those users in the boundary region of depression with a low posting frequency. We propose an NLP based method to intelligently identify low-resource social media users, potentially suffering from depression. The proposed method focuses on early depression detection and provides pointers to separate users with bipolar disorder. Depression detection is based on textual social media posts. These linguistic expressions by users are used to detect their mental state. Although, as described later, others have focused on similar detection, our primary attention is on those hard-to-detect users with low number of postings.

Our contributions are as follows:

- We identify low-resource users in need of help with detailed empirical and mathematical analysis and establish a lower bound on the δ parameter for correct re-evaluation of depressed users in the boundary region.
- We develop an approach to identify low-resource, at-risk, boundary users on the brink of depression.
- We demonstrate that our proposed intelligent and selective knowledge augmentation. significantly increases early depression detection accuracy.

2 Related Work

Increasing social media use has created additional venues for continuous mental health monitoring [8, 11–15]. A number of forums exit that help users with mental health problems via counselling by moderators. Identifying users with such immediate need is crucial in this process. Thus, triaging with high accuracy is necessary for prioritizing users to seek timely help [16–18]. Social media posts provide timely linguistic cues for mental health monitoring [6]. Researchers also worked on identification of linguistic cues for depression detection based on lexicons [19]. Depression is evident from social media behavior which comprises of use of language over time and sequence of posts [20]. Interestingly, language use itself is a prominent indicator of depression [21]. Typically, neural network based methods deliver better performance in identifying depressed users based on language usage [8].

RSDD dataset is an extensive self-reported depression diagnosis dataset constructed from Reddit [8]. Considering the relevance of timestamps of postings and resulting dynamic behavior, temporal cues were identified on the RSDD dataset [13]. While the RSDD dataset contains depression labels, Self-Reported Mental Health Diagnoses (SMHD) is a comprehensive dataset which provides self-reported diagnosis for bipolar and major depression along with other mental health conditions [15]. It contains Reddit posts of large set of control users, and a few thousand bipolar and depressed users. Dataset with diverse-mental health conditions can provide understanding into mental health related language which can be further leveraged to obtain crucial insights into mental health conditions [15]. Researchers also worked on extraction of medical concepts from large-scale datasets [22]. Datasets with mental health posts and corresponding human-written summaries have been constructed to facilitate mental health research [23]. RSDD and SMHD datasets have been widely used for performance evaluation of different methods.

Bipolar Disorder (BD) is the 10th most common cause of frailty in young population [24] and has triggered serious consequences. It affects life expectancy by 9–17 years [24]. It is a mental disorder with high prevalence, but can be misdiagnosed as a major depressive disorder [25]. 40% of patients with bipolar disorder are first misdiagnosed as major depressive disorder [24] and 17% of patients diagnosed as major depressive disorder were found to have undiagnosed bipolar disorder [26]. This makes it exceedingly important to separate users with bipolar disorder. Researchers worked on detecting bipolar disorders using neural network and radial basis function [27]. Different Machine Learning techniques like Decision Trees, Random Forest, SVM, Naïve Bayes, Logistic Regression and KNN were tried out to separate users with bipolar disorder [28].

On the other hand, it is also important to detect depression early to provide timely help before situation slips out of control. Apart from a major focus on a large number of posts, few researchers worked on early depression detection [17, 29, 30]. Researchers also proposed neural models to simplify medical text for consumption of general users using medical social media text [31]. Transformer based techniques were explored for applications in various fields like mental health [32]. Effectively adjusting sensitivity of classifiers can contribute to significant performance leap irrespective of classification methods [33].

Attempts thus far use large volumes of data. Limited exploration has been carried out on detection of depression as well as bipolar disorder in low-resource users clearly underlining the research gap.

3 Research Questions

Identifying users in the boundary region remains a challenge particularly with early depression detection in low-resource users. Prior attempts assumed a vast number of individual posts, failed to focus on detecting boundary region users, and seldom capitalized on disorder specifics, e.g., bipolar disorder. Specifically, we address the following research questions:

RQ 1: Which users can be termed as low-resource users?
RQ 2: Is it possible to leverage knowledge augmentation to improve early depression detection in case of these low-resource users?
RQ 3: Can boundary region re-evaluation help in deciding type of depression?

4 Method: Soft Thresholding for Boundary Region Users (STBound)

At-risk, boundary-region users are currently not classified as depressed and in need of immediate attention, but rather, are kept under watch. Their immediate ongoing actions are potentially indicative of their inclinations on the depression spectrum. Considering limited future activity of these users can help determine their mental health status.

We propose an intelligent, selective and timely augmentation for the boundary region users. This approach uses the Soft Thresholding based Boundary detection method (STBound) to identify low-resource users on the brink of depression. We clearly define low-resource users by conducting through empirical study and mathematical modelling on RSDD dataset as depicted in Fig. 1.

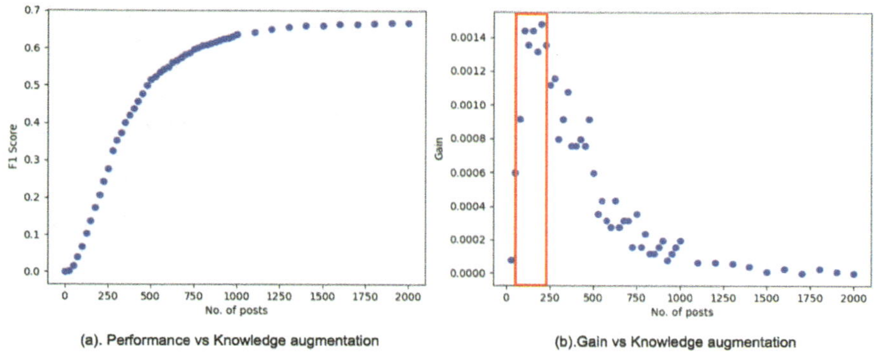

Fig. 1 Defining low-resource users

Figure 1a shows F1 score and (b) shows gain in F1 score per unit additional post for CNN. The red colored box in Fig. 1b is the low-resource user zone marked from number of posts where rate of change in gain in F1 is highest to number of posts where rate of change in gain in F1 is lowest addressing RQ 1. From this behavior depicted in Fig. 1 which is common across all machine learning methods on RSDD dataset we infer the following. Users with 50–200 posts can be termed as 'low-resource users' where intelligent use of information yields better results.

The above inference is for RSDD dataset and bounds might vary with a different dataset. To simulate low-resource users with variable number of posts we create a distribution identical to the original data distribution but bounded in established low resource user bounds. These bounds are defined considering the region of high gain per unit additional posts as per Fig. 1. For RSDD dataset we scale it by considering 20% of posts of each user and low-resource user bounds of 50 and 200. We decide the scaling percentage to be 20% as 20% of median number of posts lies perfectly between 50 and 200. As a result, we have successfully simulated a dynamic scenario of low-resource users with each user having number of posts between 50 and 200.

4.1 Hard Threshold Line

A hard threshold line is an empirically generated dynamic threshold separating users into depressed and not depressed categories. It is obtained by fitting linear regression to empirically determined threshold values on first 100, 200, 300 and 400 posts. The threshold values are obtained from validation set as per Eq. (1) where $F1[i]$ is the corresponding F1 score for threshold $th[i]$. Figure 2 depicts the distribution of users around respective hard threshold lines for methods SVM, LR, CNN and BERT, providing insights regarding users in the boundary region.

$$th_{ideal} = th[arg \max_{i}(F1[i])] \quad (1)$$

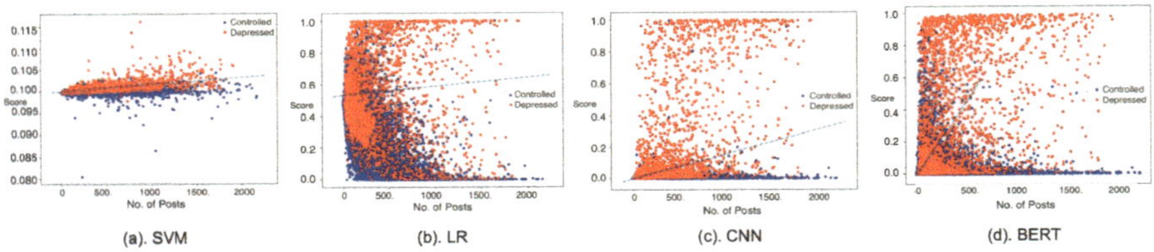

Fig. 2 Distribution of users on 20% data with respect to the established threshold line for respective methods

4.2 Soft Threshold Line

A soft threshold line, however is a variable line that determines the optimal region encompassing boundary users. It is obtained by subtracting β from the respective hard thresholds. Hard and soft threshold lines are specific to method under consideration. Hard threshold line helps us to get the threshold value dynamically—solely as a function of the number of posts a user has. All the users with score greater than their respective dynamic thresholds are classified as depressed users. In this case, the focus is on identifying the region where users are probably at risk but are not identified due to information inadequacy. Here soft threshold line plays the deciding role.

4.3 Boundary Region

Boundary region is defined as the region between hard and soft threshold lines. Users with scores lying in this boundary region are 'at-risk boundary region users'. These users are re-evaluated with intelligent knowledge augmentation. Algorithm 1 gives the method for classifying a user based on selective re-evaluation and controlled increase in user's posting data. It is carried out by increasing δ value step-wise until a termination criteria, i.e., δ_depth_count, is reached.

4.4 System Architecture

Figure 3 depicts an overview of the system architecture where user layer provides input to low resource user selection layer. The knowledge augmentation layer augments the input from previous layers and provide it to intelligent processing and learning layer. The output and decision layer performs early depression detection for boundary region users. We evaluated STBound with all four methods under consideration. STBound identifies low-resource users on the brink of depression successfully. Further, it improves early depression detection accuracy significantly with $\delta\%$ controlled increase in user posts.

Fig. 3 System architecture

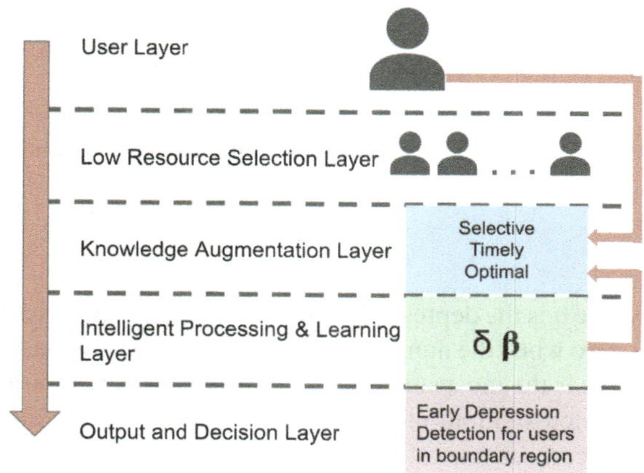

Algorithm 1 STBound algorithm

```
function EVALUATE(user, δ, β, i)
    i ← i + 1
    if i == δ_depth_count then
        RETURN(0)                                   ▷ User is not depressed
    else
        posts ← SIM_LOW_RES(user, δ)
        l ← LEN(posts)
        hard_thr ← GET_THR(method, l)
        soft_thr ← hard_thr − β
        score ← METHOD(posts)
        if score > hard_thr then
            RETURN(1)                               ▷ User is depressed
        else if score > soft_thr then
            EVAL(user, δ + inc, β, i)               ▷ Boundary user
        else
            RETURN(0)                               ▷ User is not depressed
        end if
    end if
end function
```

5 Analysis of Proposed Model

Let the initial number of posts of a low-resource user be p. We are adding $\delta\%$ extra posts for re-evaluation. Define:

$$\gamma = 1 + \delta \tag{2}$$

Hence, the total number of posts for re-evaluation are:

$$p + p\delta = p\gamma \tag{3}$$

We model the score of the ML method as a function of number of posts of the low-resource user by:

$$y = 1 - \frac{1}{e^{\frac{x}{a}}} \tag{4}$$

where a is the depressed user behavior parameter. Here, the output score of the model is zero when the number of posts is zero and y $\rightarrow$ 1 as the number of posts $\rightarrow \infty$. This curve is the most suited model for score as per empirical data. Intuitively, previously depicted sigmoid modeling for F1 score (Fig. 1a) also makes sense as even though the score increases with increase in posts for every user, its reflection in F1 score will occur late, i.e., after significant data are obtained.

Define the hard threshold line by:

$$y = t^{`}x + k \tag{5}$$

Proposition 1 *For a low-resource user who is depressed but is currently in the boundary region, lower bound on γ for extra posts needed for correct re-evaluation can be stated as:*

$$\gamma > \frac{1-k}{t^{`}p} + \frac{W_{-1}\left(\frac{-e^{-\frac{(1-k)}{at^{`}}}}{at^{`}}\right)}{\frac{p}{a}} \tag{6}$$

where W_{-1} is the lower branch of Lambert function and p is the number of posts of the low-resource user.

Proof For the low-resource user to be classified as depressed after $\delta\%$ increase in posts, score by the model for $p(1+\delta) = p\gamma$ posts should be greater than corresponding threshold. Hence:

$$1 - \frac{1}{e^{\frac{p\gamma}{a}}} > t^{`}p\gamma + k \tag{7}$$

Therefore,

$$\left(e^{\frac{-p}{a}}\right)^{\gamma} + (t^{`}p)\gamma + (k-1) < 0 \tag{8}$$

Equation $a^{x} + bx + c = 0$ can be expressed as:

$$ln(a)\left(-x - \frac{c}{b}\right)e^{ln(a)\left(-x-\frac{c}{b}\right)} = ln(a)\frac{a^{\frac{-c}{b}}}{b} \tag{9}$$

This is of form $ze^z = k$ and can be solved using Lambert's W function: $z = W(k)$. Hence we have,

$$ln(a)\left(-x - \frac{c}{b}\right) = W\left(ln(a)\frac{a^{\frac{-c}{b}}}{b}\right) \tag{10}$$

Hence,

$$x = \frac{-c}{b} - \frac{W\left(ln(a)\frac{a^{\frac{-c}{b}}}{b}\right)}{ln(a)} \tag{11}$$

The roots of the equation on the left in (8) can be found by substituting $a = e^{\frac{-p}{a}}$, $b = t^{`}p$ and $c = k - 1$ in (11):

$$root = \frac{1-k}{t^{`}p} + \frac{W\left(\frac{-e^{-\frac{(1-k)}{at^{`}}}}{at^{`}}\right)}{\frac{p}{a}} \tag{12}$$

Lower bound solution can be expressed using lower branch of Lambert's W function W_{-1} as:

$$\gamma > \frac{1-k}{t^{`}p} + \frac{W_{-1}\left(\frac{-e^{-\frac{(1-k)}{at^{`}}}}{at^{`}}\right)}{\frac{p}{a}} \tag{13}$$

$W_{-1}(.)$ can be evaluated using the Newton's method:

$$w_{j+1} = w_j - \frac{w_j e^{w_j} - z}{e^{w_j} + w_j e^{w_j}} \tag{14}$$

where w_0 for the lower branch can determined using Lajos Lóczi's formulation [34].

6 Experimentation

6.1 Datasets

The RSDD dataset [8] is a comprehensive dataset suitable for experimentation related to depression. It was created by annotating users from Reddit dataset[1] which is available publicly. RSDD is an extensive dataset spanning from Jan 2006 to Oct 2016. Using RSDD, we simulated low-resource platform users to facilitate experiments

[1] https://files.pushshift.io/reddit/.

related to early depression detection. RSDD has 107,274 control and 9210 diagnosed users. Each user has 969 posts on an average with a mean length of 148 tokens. The dataset has three components: training, validation and testing. The SMHD dataset consists of Reddit posts of users who have claimed to have been diagnosed with one or several of nine mental health conditions ('diagnosed users'), and matched control users. It is a large dataset that covers diverse mental health conditions. It has a total number of 385,476 users consisting of 6434 bipolar, 14,139 depressed and 335,952 control users. On an average, control users have 310 posts, depressed users have 162 posts and bipolar users have 158 posts. We used this dataset to evaluate STBound performance to separate bipolar disorder by selecting depressed and bipolar users.

6.2 Ethics and Privacy

Personalized healthcare data are sensitive. The data used are anonymized and due care is taken to minimize the risk while conducting experiments. The RSDD dataset contains the posts those are publicly available for academic use. All necessary care with reference to ethics and privacy was taken [8]. In this context, data were stored on secure servers, and no attempts were made to map, associate or re-identify users. Similar care was taken in the case of SMHD dataset also [15]. We refrain from making any details of these data publicly available. No attempt what so ever to link users to social media accounts was made.

6.3 Experimental Setup

We evaluate STBound on simulated low-resource users from RSDD dataset in combination with both traditional and connectionist machine learning methods. We considered lexical bag-of-words models Logistic Regression (LR) and SVM, conventional CNN and transformer based Bidirectional Encoder Representations from Transformers (BERT).

To assess the performance improvement of neural network based models over traditional machine learning models, we first evaluate the performance of STBound for LR and SVM. For these models, lexical bag-of-words (BoW) features were used as input. To minimize noise, we empirically determined a minimum document frequency of 12 for words to be considered. We fit the BoW tokenizer on the training set. For the CNN model, we had an embedding size of 50 per token. Additionally, the model had two Convolution1D layers with 25 filters, filter length of 3 and Rectified Linear Unit (ReLU) activation function. The model then had a 50 neuron dense layer and an output layer with Softmax activation function. A learning rate of 0.001 and 5 epochs led to the best results on the validation set.

RSDD has 969 posts per user on an average which results in memory constraints while running BERT. Considering these limitations, we considered 310x128 tokens per user and experimented with the following BERT [35] models: 'small bert/bert_en_uncased_L-4_H-512_A-8' and 'small_bert/bert_en_uncased_L-2_H-128_A-2'. The first model is more intricate and hence we could accommodate only 175x128 tokens. Acknowledging this trade-off, we infer that the latter gives best results. Note that here L denotes the number of layers, i.e., transformer blocks. Also note that H denotes the hidden size while A denotes the number of self-attention heads. A learning rate of $2e-5$ and 3 epochs led to the best results on the validation set.

6.4 Language Study

We conducted detailed language analysis to identify specific language characteristics which distinguish low-resource users from others. We observe that low-resource users have significantly higher usage of self-referencing phrases [36], i.e., 13% higher when calculated per unit sentence than others. Additionally we also notice significantly (i.e., 5%) lower occurrence of depression indicative phrases in low-resource users when compared with other users. Here we calculated depression indicative phrases [37] per unit sentence. Lower occurrence of depression indicative phrases in low-resource users justifies the need for thresholding and augmentation.

7 Results and Analysis

We have evaluated STBound on simulated low-resource users from RSDD dataset in combination with lexical bag-of-words, conventional CNN-based and transformer based ML methods. For comparison as a baseline, we obtain results on simulated low-resource users using dynamic thresholding. Here, users are clearly classified into depressed or not depressed using the obtained hard threshold. We use F1 score, Precision and Recall for comparing performance across different methods. We have focused on F1 score as Precision and Recall are equally important for depression detection. Table 1 shows the results of STBound with SVM, LR, CNN and BERT for different combinations of δ and β values. It adds δ% data to simulated low-resource users which satisfy the definition of boundary users. The boundary in this case is defined using the value of β. The variation in β values across models can be attributed to the differences in training algorithms and their respective parameters. Hence, optimal boundary region is determined separately for each method.

The highest F1 values are represented in bold in Table 1. We note that in Table 1, the first row with δ and β value equal to zero is our baseline and all performance improvements are represented with respect to it. We obtain an F1 of 0.359 for SVM for a δ value of 0.4 and β value of 0.002. Similarly we obtain an F1 of 0.421 for LR for δ value of 0.4 and β value of 0.3. In case of CNN, δ value of 0.4 and β value of 0.01 give the best F1 of 0.509. Further, in case of BERT, δ value of 0.4 and β

Table 1 STBound on low-resource RSDD users

	SVM				LR				CNN				BERT			
δ	β	F1	Pr	Rc	β	F1	Pr	Rc	β	F1	Pr	Rc	β	F1	Pr	Rc
0	0	0.321	0.327	0.311	0	0.388	0.419	0.361	0	0.446	0.431	0.462	0	0.422	0.425	0.419
0.1	0.0005	0.330	0.323	0.337	0.075	0.393	0.413	0.375	0.005	0.459	0.453	0.465	0.05	0.425	0.430	0.420
	0.0010	0.333	0.327	0.339	0.150	0.395	0.411	0.380	0.010	0.463	0.458	0.468	0.10	0.430	0.431	0.429
	0.0015	0.335	0.330	0.340	0.225	0.398	0.406	0.390	0.015	0.466	0.463	0.469	0.15	0.435	0.437	0.433
	0.0020	0.336	0.331	0.341	0.300	0.399	0.407	0.391	0.020	0.466	0.461	0.471	0.20	0.438	0.436	0.440
	0.0005	0.332	0.326	0.338	0.075	0.398	0.418	0.380	0.005	0.478	0.482	0.474	0.05	0.427	0.429	0.425
	0.0010	0.342	0.336	0.348	0.150	0.403	0.414	0.393	0.010	0.480	0.483	0.477	0.10	0.435	0.443	0.427
	0.0015	0.344	0.331	0.358	0.225	0.406	0.407	0.405	0.015	0.482	0.484	0.480	0.15	0.442	0.451	0.433
0.2	0.0020	0.346	0.328	0.366	0.300	0.408	0.404	0.412	0.020	0.482	0.484	0.480	0.20	0.452	0.460	0.444
0.3	0.0005	0.340	0.329	0.352	0.075	0.405	0.410	0.400	0.005	0.492	0.498	0.486	0.05	0.442	0.458	0.427
	0.0010	0.351	0.337	0.366	0.150	0.410	0.409	0.411	0.010	0.496	0.497	0.495	0.10	0.455	0.477	0.435
	0.0015	0.354	0.327	0.386	0.225	0.414	0.405	0.423	0.015	0.496	0.497	0.495	0.15	0.467	0.488	0.448
	0.0020	0.354	0.327	0.386	0.300	0.415	0.406	0.424	0.020	0.496	0.497	0.495	0.20	0.467	0.488	0.448
	0.0005	0.343	0.334	0.352	0.075	0.408	0.406	0.410	0.005	0.506	0.514	0.498	0.05	0.448	0.468	0.430
	0.0010	0.352	0.331	0.376	0.150	0.415	0.410	0.420	0.010	**<u>0.509</u>**	0.517	0.501	0.10	0.463	0.487	0.441
	0.0015	0.357	0.322	0.400	0.225	0.420	0.410	0.431	0.015	0.509	0.517	0.501	0.15	**<u>0.479</u>**	0.509	0.452
0.4	0.0020	**<u>0.359</u>**	0.315	0.418	0.300	**<u>0.421</u>**	0.411	0.432	0.020	0.509	0.517	0.501	0.20	0.479	0.509	0.452

Table 2 Efficacy

δ	SVM	LR	CNN	BERT
0.1	**0.467**	**0.284**	**0.448**	**0.379**
0.2	0.389	0.258	0.404	0.356
0.3	0.343	0.232	0.374	0.356
0.4	0.296	0.213	0.353	0.338

value of 0.15 give the best F1 of 0.479. Here an important point to note is that BERT model was trained on limited data of each user due to memory constraints.

To compare CNN and BERT, we ran experiments in identical setups. Here, we consider 310x128 tokens per user for training and testing both models. From this we infer that BERT model outperforms CNN model in the identical setup. Experimental findings and analysis illustrate significant improvement in the F1 values with selective incorporation of boundary region re-evaluation with knowledge augmentation addressing RQ 2. An increase in δ value leads to an increase in F1. Similar trends are observed with an increase in β values, until a saturation level is reached. For example, for CNN, as β approaches 0.020, the results start saturating. Similar trends were observed across all the methods for the 18 β values evaluated. This indicates a relative separation between 'at-risk boundary' users and users who are not depressed. Additionally, for higher δ values, an increase in β value leads to greater improvements. With an optimal selection of δ and β values, we can get up to 8.51% improvement for SVM, up to 11.84% improvement for LR, up to 14.13% improvement for CNN and up to 13.51% improvement for BERT.

$$Efficacy = \frac{F1(\beta_{saturation}) - F1(\beta_0)}{\delta * F1(\beta_0)} \tag{15}$$

The lower δ and β values with higher performance gain can help to identify the optimal boundary condition. We define efficacy as percentage improvement per unit δ. The obtained efficacy values can be found in Table 2. For SVM and LR the highest efficacy values are 0.467 and 0.284 respectively. These values are observed at δ value 0.1. In case of CNN and BERT the highest efficacy values are 0.448 and 0.379 respectively and are also observed at δ value 0.1. In a real life setting, if we encounter boundary users with slow posting rate then optimal δ values can be determined using an efficacy index based on the urgency. Highest efficacy points can help us to fine-tune boundary region parameters.

8 Bipolar Verses Depressed

We extended STBound to identify unipolar and bipolar depression. Misdiagnosis of bipolar depression is common. It is misdiagnosed as major depressive disorder (unipolar depression). Unipolar depression is treated by antidepressants while bipolar

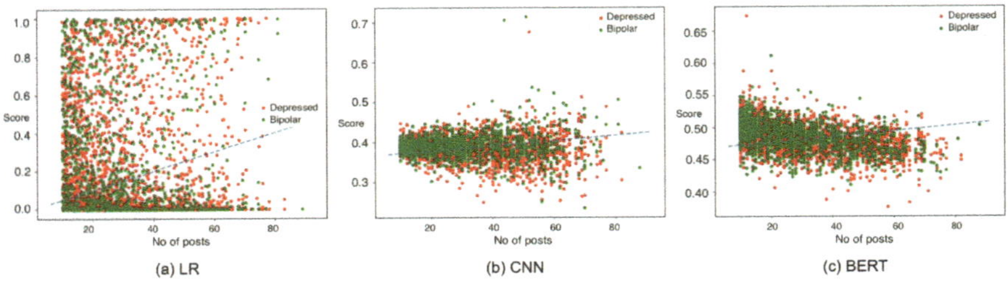

Fig. 4 STBound for separating Bipolar from Depressed users

Table 3 STBound on separating Bipolar from Depressed. † Hard threshold, ‡ STBound, on low-resource users

Method	F1	Precision	Recall	Method	F1	Precsion	Recall	Method	F1	Precsion	Recall
LR	0.316	0.353	0.286	CNN	0.049	0.518	0.026	BERT	0.321	0.394	0.270
LR†	0.349	0.349	0.349	CNN†	0.482	0.480	0.485	BERT†	0.495	0.389	0.680
LR‡	**0.414**	0.333	0.547	CNN‡	**0.524**	0.378	0.854	BERT‡	**0.539**	0.381	0.921

depression requires mood stabilizers [38]. It is important to identify bipolar depression because if missed, it will be treated like unipolar (with antidepressants) leading to increase in maniac episodes in patients—aggravating risks of self-harm and suicide multi-fold [39].

In early detection scenarios we have very limited data to detect bipolar depression. Higher percentage of bipolar cases lie in the boundary region defined by STBound. Re-evaluation of these users can help us to improve the classification of users in low-resource scenarios. Figure 4 depicts the distribution of users and use of STBound for separating bipolar and depressed users. Table 3 gives the performance of LR, CNN and BERT in separating bipolar disorder using hard threshold and using STBound. It is observed that performance of CNN gets a sharp leap using hard threshold. The performance further improves by 8.7% by use of STBound. In case of LR, after initial improvement obtained using hard threshold, STBound provides additional improvement of 18.6%. Similar trends are observed in case of BERT where after initial F1 score improvement, STBound provides 8.9% additional improvement. These results are indicative of possible use of STBound or it's enhancement to separate users with bipolar disorder addressing RQ 3. Though the improvements are realized at the cost of precision, the significant leap is definitely conclusive and promising for complex problem of separating users with bipolar depression.

9 Conclusion

We addressed RQ1 by defining low resource users empirically and mathematically. Depression detection for boundary region low-resource users is always an important and challenging task. The delay in depression detection for such users may result in delay in treatment and severe after effects. To address this issue, STBound performs selective intelligent knowledge augmentation and identifies boundary regions with higher precision. It further improves the accuracy of depression detection for the users in the boundary region by effective increase in δ value. The proposed method of selective and intelligent knowledge augmentation fetches improvement in overall F1 score on an average by 11.9% across all methods addressing RQ2. This substantial improvement helps in identifying the depressed boundary users on the brink of depression those otherwise would have gone unidentified. Early depression detection becomes even more crucial when it comes to separating bipolar disorder. Failure to separate bipolar disorder may result in delay in providing right treatment and further worsening the user's condition. STBound also improves the F1 score of separation of bipolar users by 12.1% addressing RQ3.

As a future work, other re-evaluation techniques can be combined with STBound. Fitting a curve with variable β values can optimally encompass the boundary region for further improvements. STBound with some contextual inputs can lead to promising techniques to separate bipolar disorder. Additionally, distribution of expressions over the time period can be used along with STBound to detect bipolar disorder in case of comorbidity.

References

1. WHO. World health organization depression. World Health Organization News (2021). Accessed: 2021-30-12
2. IHME. New global burden of disease analyses show depression and anxiety among the top causes of health loss worldwide, and a significant increase due to the covid-19 pandemic. IHME: Measuring what matters (2021). Accessed: 2022-01-05
3. M. Daly, E. Robinson, Lancet **399**(10324), 518 (2022)
4. CNN. More than 2,000 California mental health clinicians set to strike: CNN international. https://edition.cnn.com/2022/08/14/business/kaiser-mental-health-clinicians-strike/index.html (2022). Accessed 15-Aug-2022
5. S. Dattani, H. Ritchie, M. Roser, *Our World in Data* (2021). https://ourworldindata.org/mental-health
6. N. Vedula, S. Parthasarathy, in *Proceedings of the 2017 International Conference on Digital Health* (Association for Computing Machinery, New York, NY, USA, 2017), DH '17, pp. 127–136. https://doi.org/10.1145/3079452.3079465
7. A. Zafar, S. Chitnis, in *2020 10th International Conference on Cloud Computing, Data Science Engineering (Confluence)* (2020), pp. 88–93. https://doi.org/10.1109/Confluence47617.2020.9058189
8. A. Yates, A. Cohan, N. Goharian, in *Proceedings of the 2017 Conference on Empirical Methods in Natural Language Processing* (Association for Computational Linguistics, Copenhagen, Denmark, 2017), pp. 2968–2978. https://doi.org/10.18653/v1/D17-1322
9. Z. Jan, N. Al-Ansari, O. Mousa, A. Abd-alrazaq, A. Ahmed, T. Alam, M. Househ, J Med Internet Res **23**(11), e29749 (2021). https://doi.org/10.2196/29749
10. T.T. Erguzel, G.H. Sayar, N. Tarhan, Neural Comput. Appl. **27**, 1607 (2015)
11. S. Sidana, S. Mishra, S. Amer-Yahia, M. Clausel, M.R. Amini, in *Proceedings of the 39th International ACM SIGIR Conference on Research and Development in Information Retrieval* (Association for Computing Machinery, New York, NY, USA, 2016), SIGIR '16, pp. 849–852. https://doi.org/10.1145/2911451.2914697

12. G. Coppersmith, C. Hilland, O. Frieder, R. Leary, in *2017 IEEE EMBS International Conference on Biomedical Health Informatics (BHI)* (2017), pp. 393–396. https://doi.org/10.1109/BHI.2017.7897288
13. S. MacAvaney, B. Desmet, A. Cohan, L. Soldaini, A. Yates, A. Zirikly, N. Goharian, in *Proceedings of the Fifth Workshop on Computational Linguistics and Clinical Psychology: From Keyboard to Clinic* (Association for Computational Linguistics, New Orleans, LA, 2018), pp. 168–173. https://doi.org/10.18653/v1/W18-0618
14. G. Coppersmith, R. Leary, P. Crutchley, A. Fine, Biomed. Inf. Insights **10**, 1178222618792860 (2018). https://doi.org/10.1177/1178222618792860. PMID: 30158822
15. A. Cohan, B. Desmet, A. Yates, L. Soldaini, S. MacAvaney, N. Goharian, in *Proceedings of the 27th International Conference on Computational Linguistics (COLING)* (Association for Computational Linguistics, 2018), pp. 1485–1497. https://www.aclweb.org/anthology/C18-1126
16. A. Cohan, S. Young, N. Goharian, in *Proceedings of the Third Workshop on Computational Linguistics and Clinical Psychology* (Association for Computational Linguistics, San Diego, CA, USA, 2016), pp. 143–147. https://doi.org/10.18653/v1/W16-0316. https://aclanthology.org/W16-0316
17. A. Cohan, S. Young, A. Yates, N. Goharian, J. Assoc. Inf. Sci. Technol. (JASIST) (2018). https://doi.org/10.1002/asi.23865
18. M. Adrian, J. Coifman, M.D. Pullmann, J.B. Blossom, C. Chandler, G. Coppersmith, P. Thompson, A.R. Lyon, JMIR Ment Health **7**(7), e16338 (2020). https://doi.org/10.2196/16338
19. E. Villatoro-Tello, G. Ramírez-de-la Rosa, D. Gática-Pérez, M. Magimai.-Doss, H. Jiménez-Salazar, *Approximating the Mental Lexicon from Clinical Interviews as a Support Tool for Depression Detection* (Association for Computing Machinery, New York, NY, USA, 2021), pp. 557–566. https://doi.org/10.1145/3462244.3479896
20. G. Coppersmith, A. Fine, P. Crutchley, J. Carroll, in *Proceedings of the Seventh Workshop on Computational Linguistics and Clinical Psychology: Improving Access* (Association for Computational Linguistics, Online, 2021), pp. 25–31. https://doi.org/10.18653/v1/2021.clpsych-1.3
21. K. Kelly, A. Fine, G. Coppersmith, in *Proceedings of the Fourth Workshop on Natural Language Processing and Computational Social Science* (Association for Computational Linguistics, Online, 2020), pp. 184–192. https://doi.org/10.18653/v1/2020.nlpcss-1.20
22. L. Soldaini, N. Goharian, in *MedIR Workshop SIGIR* (2016)
23. S. Sotudeh, N. Goharian, Z. Young, in *Proceedings of the Thirteenth International Conference on Language Resources and Evaluation (LREC)* (European Language Resources Association (ELRA), 2022)
24. P.J.C. Suen, S. Goerigk, L.B. Razza, F. Padberg, I.C. Passos, A.R. Brunoni, Psychiatr. Res. **27**, 1607 (2021)
25. L. Shi, P. Thiebaud, J.S. McCombs, J. Affect. Disord. **82**(3), 373 (2004)
26. J. Daveney, M. Panagioti, W. Waheed, A. Esmail, Gen. Hosp. Psychiatr. **58**, 71 (2019)
27. M.n. Luján, A.M. Torres, A.L. Borja, J.L. Santos, J.M. Sotos, Electronics **11**(3) (2022). https://doi.org/10.3390/electronics11030343
28. N. Agnihotri, TechRxiv (2021). https://doi.org/10.36227/techrxiv.14346050.v1
29. A.M. Bucur, A. Cosma, L.P. Dinu. Early risk detection of pathological gambling, self-harm and depression using bert (2021)
30. S.G. Burdisso, M. Errecalde, M.M. Gómez, Expert Syst. Appl. **133**, 182 (2019). https://doi.org/10.1016/j.eswa.2019.05.023
31. N. Pattisapu, N. Prabhu, S. Bhati, V. Varma, *Leveraging Social Media for Medical Text Simplification* (Association for Computing Machinery, New York, NY, USA, 2020), pp. 851–860. https://doi.org/10.1145/3397271.3401105
32. F. Elsafoury, S. Katsigiannis, S.R. Wilson, N. Ramzan, *Does BERT Pay Attention to Cyberbullying?* (Association for Computing Machinery, New York, NY, USA, 2021), pp. 1900–1904. https://doi.org/10.1145/3404835.3463029

33. H. Kulkarni, S. MacAvaney, N. Goharian, O. Frieder, TBD3: A thresholding-based dynamic depression detection from social media for low-resource users. in *Proceedings of the Thirteenth Language Resources and Evaluation Conference*, pp. 2157–2165. European Language Resources Association, Marseille, France (2022). https://aclanthology.org/2022.lrec-1.232
34. L. Lóczi, Appl. Math. Comput. **433**, 127406 (2022). https://doi.org/10.1016/j.amc.2022.127406
35. J. Devlin, M.W. Chang, K. Lee, K. Toutanova, in *Proceedings of the 2019 Conference of the North American Chapter of the Association for Computational Linguistics: Human Language Technologies, Volume 1 (Long and Short Papers)* (ACL, Minneapolis, Minnesota, 2019), pp. 4171–4186. https://doi.org/10.18653/v1/N19-1423
36. A.M. Bucur, I. Podina, L. Dinu. A psychologically informed part-of-speech analysis of depression in social media (2021). https://doi.org/10.26615/978-954-452-072-4-024
37. A. Kumar, A. Sharma, A. Arora, Anxious depression prediction in real-time social data (2019). https://doi.org/10.48550/ARXIV.1903.10222
38. L.N. Yatham, S.H. Kennedy, S.V. Parikh, A. Schaffer, D.J. Bond, B.N. Frey, V. Sharma, B.I. Goldstein, S. Rej, S. Beaulieu, M. Alda, G. MacQueen, R.V. Milev, A. Ravindran, C. O'Donovan, D. McIntosh, R.W. Lam, G. Vazquez, F. Kapczinski, R.S. McIntyre, J. Kozicky, S. Kanba, B. Lafer, T. Suppes, J.R. Calabrese, E. Vieta, G. Malhi, R.M. Post, M. Berk, Bipolar Disord. **20**(2), 97 (2018)
39. M.J. Gitlin, Int. J. Bipolar Disord. **6**(1), 25 (2018) https://doi.org/10.1186/s40345-018-0133-9

Deep Annotation of Therapeutic Working Alliance in Psychotherapy

Baihan Lin, Guillermo Cecchi, and Djallel Bouneffouf

Abstract The therapeutic working alliance is an important predictor of the outcome of psychotherapy treatments. In practice, the working alliance is estimated from a set of scoring questionnaires in an inventory that both the patient and the therapist fill out. In this work, we propose an analytical framework of directly inferring the therapeutic working alliance from the natural language within the psychotherapy sessions in a turn-level resolution with deep embeddings such as the Doc2Vec and SentenceBERT models. The transcript of each psychotherapy session can be transcribed and generated in real-time from the session speech recordings, and these embedded dialogues are compared with the distributed representations of the statements in the working alliance inventory. We demonstrate, in a real-world dataset with over 950 sessions of psychotherapy treatments in anxiety, depression, schizophrenia and suicidal patients, the effectiveness of this method in mapping out trajectories of patient-therapist alignment and the interpretability that can offer insights in clinical psychiatry. We believe such a framework can be provide timely feedback to the therapist regarding the quality of the conversation in interview sessions.

1 Introduction

A fundamental concept in psychotherapy is the working alliance between the therapist and the patient or, more generally, the client seeking help [1]. The alliance involves several cognitive and emotional components of the relationship between these two agents, including the agreement on the goals to be achieved and the tasks

B. Lin (✉)
Columbia University, New York City, NY, USA
e-mail: baihan.lin@columbia.edu

G. Cecchi · D. Bouneffouf
IBM Thomas J Watson Research Center, Yorktown Heights, NY, USA
e-mail: gcecchi@us.ibm.com

D. Bouneffouf
e-mail: djallel.bouneffouf@ibm.com

A. Shaban-Nejad et al. (eds.), *Artificial Intelligence for Personalized Medicine*,
Studies in Computational Intelligence 1106,
https://doi.org/10.1007/978-3-031-36938-4_15

to be carried out, and the bond, trust and respect to be established over the course of the therapy. Qualitative methods to quantify therapy outcomes led to the conclusion that the strength of the alliance is one of the main factors that predict success [26]. Operational methods to quantify the alliance rely of evaluative reports by patients and therapists of whole sessions, typically limited to point-scales valuation [6]. This approach does not make use of the nuances afforded by natural language, is time-consuming and difficult to follow through systematically outside of research studies; even more so is the evaluation of individual dialogue turns over the course of each session.

Here we present an approach to quantify the degree of patient-therapist alliance by projecting each turn in a therapeutic session onto the representation of clinically established working alliance inventories, using language modeling to encode both turns and inventories. This allows us not only to quantify the overall degree of alliance but also to identify granular patterns its dynamics over shorter and longer time scales. We evaluate both qualitatively and quantitatively the effectiveness of this inference method in providing clinical insights for psychotherapy strategies in this work and improving the classification or diagnosis capability of deep learning models to predict psychiatric conditions from therapy transcripts in a later work [17]. Lastly, we discuss how our approach may be used as a companion tool to provide feedback to the therapist and to augment learning opportunities for training therapists (Fig. 1).

2 Problem Setting

2.1 Working Alliance Analysis

Algorithm 1 Working Alliance Analysis (WAA)

for i = 1,2,$\cdots$, T **do**
  Automatically transcribe dialogue turn pairs (S_i^p, S_i^t)
  for $(I_j^p, I_j^t) \in$ inventories (I^p, I^t) **do**
    Score $W_j^{p_i}$ = similarity$(Emb(I_j^p), Emb(S_i^p))$
    Score $W_j^{t_i}$ = similarity$(Emb(I_j^t), Emb(S_i^t))$
  end for
end for

The figure above is an outline of the analytic framework. We take the full records of a patient, or a cohort of patients belonging to the same condition. We either use it as is before the feature extraction, or we truncate them into segments based on timestamps or topic turns. As you can see, the original format is in pairs of dialogues. We can extract the features in three ways: first, we can use the full pairs of dialogues; second, we can only extract what the patient says; or the third option, we only extract what

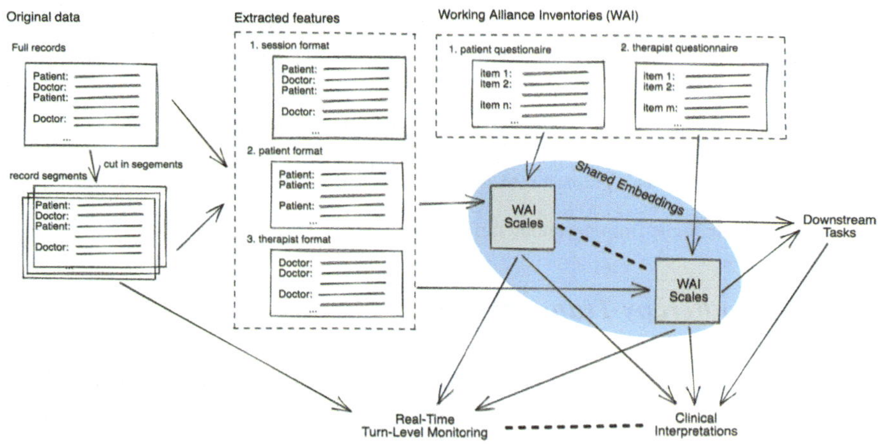

Fig. 1 Analytical pipeline of working alliance analysis

the doctor says. The three feature formats all have their pros and cons. The dialogue format contains all information, but the intents within the sentences come from two individuals, so they might mix together. The patient format contains the full narrative of the patients, which is usually more coherent, but it's only part of the story. The therapist format, which people in computational psychiatry also believes to be some kind of semantic labels of what the patient feels can be informative, but they can also be sometimes too simplistic.

When we have the features, we compare the working alliance inventories with the embeddings. Algorithm 1 outlines the process. During the session, the dialogue between the patient and therapist are transcribed into pairs of turns (such as the example in Fig. 2). We denote each patient response turn as S_i^p followed by a therapist response turn S_i^t. They are treated as a dialogue pair. The inventories of working alliance questionnaires also come in pairs: I^p for the patient (or client), and I^t for the therapist. They each consist of 36 statements. We embed both the dialogue turns and the inventories with deep sentence or paragraph embeddings, and then compute the cosine similarity between the embedding vectors of the turn and its corresponding inventory vectors. With that, for each turn (either by patient or by therapist), we obtain a 36-dimension working alliance score. We will describe in Sect. 3.1 the specific scales of our inferred working alliance scores which introduces interpretable information into our framework.

Here are a few downstream tasks and user scenarios that can plugged to our analytical frameworks. We can either use these extracted weighted topics to inform whether the therapy is going the right direction, whether the patient is going into certain bad mental state, or whether the therapist should adjust his or her treatment strategies. This can be built as an intelligent AI assistant to remind the therapist of such things.

```
PATIENT : I don't know. I was laughing at myself for thinking it.
THERAPIST : Yeah.
PATIENT : Um, like I don't know, was I being a martyr? I don't know.
THERAPIST : No it sort of sounds like you feel like you're doing me a favor or
something?
PATIENT : Um, I don't know.
THERAPIST : Like he's getting more out of this than I am. [00 : 03 : 07]
PATIENT : No, I don't think that. Naw, I'm just cranky.
THERAPIST : Yeah I'm hearing that, but I'm trying to explain what the, what
that's about. I mean you ought to be glad I'm here at all.
```

Fig. 2 Example dialogue from psychotherapy transcripts

2.2 *Psychotherapy Transcript Dataset*

The Alex Street *Counseling and Psychotherapy Transcripts* dataset[1] consists of transcribed recordings of over 950 therapy sessions between multiple anonymized therapists and patients. This multi-part collection includes speech-translated transcripts of the recordings from real therapy sessions, 40,000 pages of client narratives, and 25,000 pages of reference works. These sessions belong to four types of psychiatric conditions: anxiety, depression, schizophrenia and suicidal. Each patient response turn S_i^p followed by a therapist response turn S_i^t is treated as a dialogue pair. In total, these materials include over 200,000 turns together for the patient and therapist and provide access to the broadest range of clients for our linguistic analysis of the therapeutic process of psychotherapy.

3 Methods

3.1 *Working Alliance Inventories*

The Working Alliance Inventory (WAI) is a set of self-report measurement questionnaire that quantifies the therapeutic bond, task agreement, and goal agreement [6, 21, 25]. Since the original 12-item version [25], the inventory has used parallel versions for clients and therapist with good psychometric properties and helped establish the importance of therapeutic alliance in predicting treatment outcomes. The modern version of the inventory consists of 36 questions (Fig. 3), and the participant is asked to rate each item on a 7-point scale (1 = never, 7 = always)[21]. The WAI aims to (1) measure alliance factors across all types of therapy, (2) document the relationship between the alliance measure and the corresponding theoretical constructs underlying the measure, and (3) related the alliance measure to a unified theory of therapeutic change [7].

Operationally, the goal is to derive from these 36 items three alliance scales: the task scale, the bond scale and the goal scale. They measures the three major themes of psychotherapy outcomes: (1) the collaborative nature of the patient-therapist rela-

[1] https://alexanderstreet.com/products/counseling-and-psychotherapy-transcripts-series.

```
I felt uncomfortable with _.
_ and I agreed about the things I will need to do in therapy to help improve my situation.
I was worried about the outcome of the sessions.
What I was doing in therapy gave me new ways of looking at my problem.
_ and I understood each other.
_ perceived accurately what my goals were.
I find what I was doing in therapy confusing.
I believe _ liked me.
I wish _ and I could have clarified the purpose of our sessions.
I disagreed with _ about what I ought to get out of therapy.
I believe the time _ and I were spending together was not spent efficiently.
_ did not understand what I was trying to accomplish in therapy.
I was clear on what my responsibilities were in therapy.
The goals of the sessions were important for me.
I find what _ and I were doing in therapy was unrelated to my concerns.
I feel that the things I did in therapy helped me to accomplish the changes that I wanted.
```

Fig. 3 Example statements in working alliance inventory

TASK scale:	2,	4,	7,	11,	13,	15,	16,	18,	24,	31,	33,	35
Polarity	+	+	-	-	+	-	+	+	+	-	-	+
BOND scale:	1,	5,	8,	17,	19,	20,	21,	23,	26,	28,	29,	36
Polarity	-	+	+	+	+	-	+	+	+	+	-	+
GOAL scale:	3,	6,	9,	10,	12,	14,	22,	25,	27,	30,	32,	34
Polarity	-	+	-	-	-	+	+	+	-	+	+	-

Fig. 4 Keys to the three scales of working alliance inventory

tionship; (2) the affective bond between therapist and patient, and (3) the therapist's and patient's capabilities to agree on treatment-related short-term tasks and long-term goals. The score corresponding to the three scales comes from a key table (Fig. 4) which specifies the positivity or the sign weight to be applied on the questionnaire answer when summing in the end. The full scale is simply the sum of the scores of the three scales. The key table is like a weighting matrix that specifies the directionalities of the scales (Fig. 7).

3.2 Sentence Embeddings

In principle, any sentence or paragraph embeddings can help us characterize the dialogue turns and inventories. In this work, we used two deep embeddings. The Doc2Vec embedding [9] is a popular unsupervised learning model that learns vector representations of sentences and text documents. It improves upon the traditional bag-of-words representation by utilizing a distributed memory that remembers what is missing from the current context. The other embedding we evaluated is the SentenceBERT [24], which modifies a pretrained BERT network by using siamese and triplet network structures to infer semantically meaningful sentence embeddings. With these two deep embeddings, we embed the turn-level entries (either the dialogue turn in the transcripts, or the statement item in the working alliance inventories) into vectors of 300 or 384 dimensions. And then compute the cosine similarity between the vector at certain turn and an inventory entry. Given the space limit, the results for the Doc2Vec are shown in the main text, while the SentenceBERT results can be found in the supplementary materials.

4 Results

4.1 Insights from Analyzing Psychotherapy Transcripts

In this section, we present the findings of applying the working alliance analysis to the psychotherapy dataset.

Figure 5 is an example time series of an anxiety psychotherapy session. We see that the alliance scores varies across the scales. If we investigate the relationship among the scales, we observe that the task scale positively correlates with the bond scale in both versions, while the goal scale slightly negatively correlates with the task scale in the therapist version (Fig. 6).

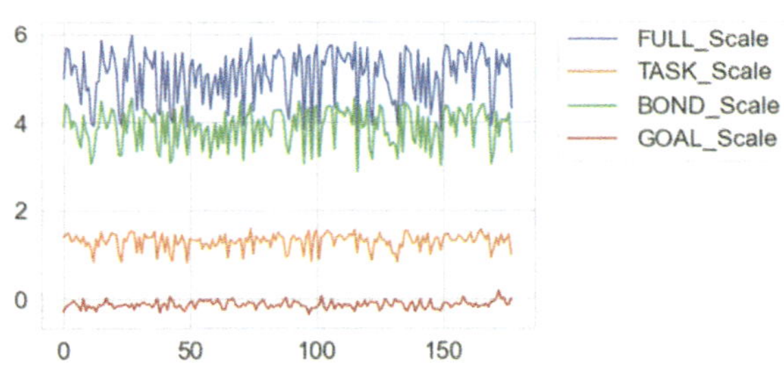

Fig. 5 Example trajectory of the working alliance scores

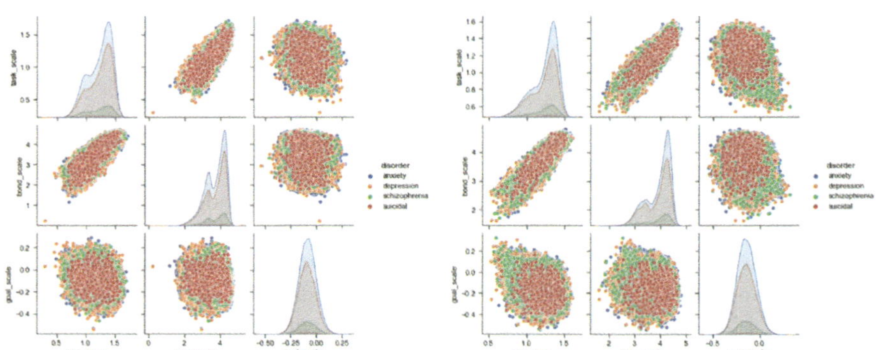

Fig. 6 Relational plots of the working alliance score scales (left: patient version; right: therapist version)

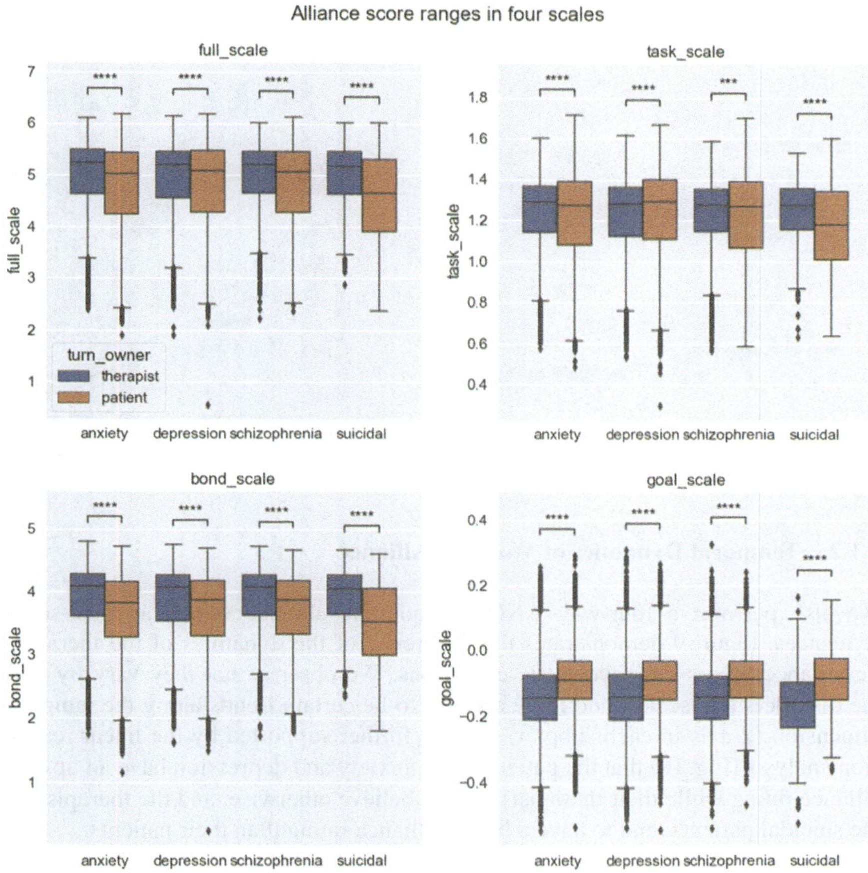

Fig. 7 Box plots of the working alliance scores

4.1.1 Patient-Therapist Consistency of Working Alliance

We investigate the consistency of the alliance estimation by the patient verses the therapist. Overall, comparing to the patient estimates, we observe that the therapist tends to overestimate the working alliance. More specifically, the therapists overestimates the task and bond scales, but underestimates the goal scale. These differences are all statistically significant ($p < 0.001$).

Between the disorders, the alliance scores between anxiety and depression, and between anxiety and schizophrenia, are all significantly different in both the therapist and patient versions ($p < 0.001$). As in Fig. 8, the suicidality can be significantly detected based on the working alliance scores of all four scales.

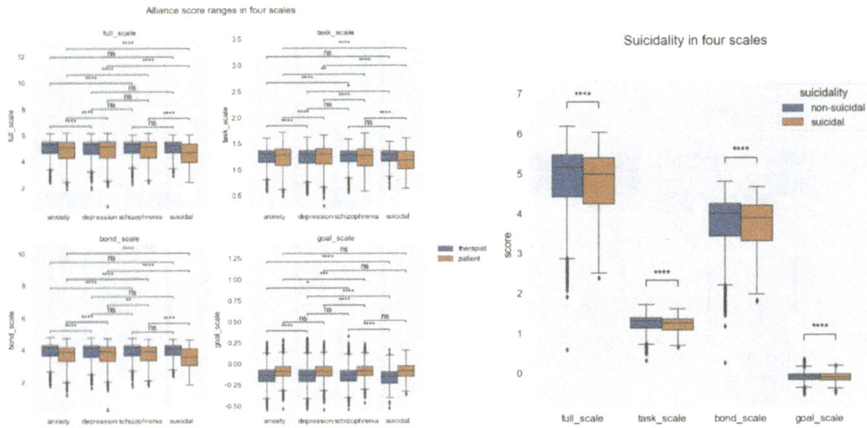

Fig. 8 Alliance scores across disorders

4.1.2 Temporal Dynamics of Working Alliance

We also perform a four-way ANOVA upon the alliance scores as time-series sequences. Figure 9 demonstrates the difference of the dynamics of the therapeutic alliance across the psychiatric conditions. We observe that they vary by both the disorders and scales, and there appears to be certain trends along the temporal dimension (x-axis in each subplot). This is further supported by the linear regression analysis (Fig. 10) that the patients with anxiety and depression have an upward alliance rating while their therapists tend to believe otherwise, and the therapists of the suicidal patients tend to have a higher alliance rating than their patients.

We can also map out their trajectories in the alliance space of the three major scales (task, bond and goal). As in Fig. 11, we plot the average trajectories of different psychiatric conditions and notice that the suicidal trajectories are much more spread out in the bond and task scales (which aligns with the findings in the ANOVA plots). Based on the directionality, the suicidality trace shows a significant divergence trend. This is the first step of a potential turn-level resolution temporal analysis of the working alliance. We can be generalize in a sense that with this approach one can go over your sessions (as a therapist) and analyze the dynamics afterwards.

Given these time-series, we can visualize them with dimension reduction techniques such as t-SNE. Because the psychotherapy sessions come in different lengths, we compute the dynamic time wrapping distances between the session trajectories of 36-dimension alliance scores, and then use this pairwise distance matrix to perform the t-SNE unsupervised learning. Figure 12 presents the difference between the manifolds of the therapist alliance trajectory space and the patient alliance space. We notice that the patient trajectories have two major clusters of alliance, while the therapist only has one. This is consistent with what we observed in the relational plots

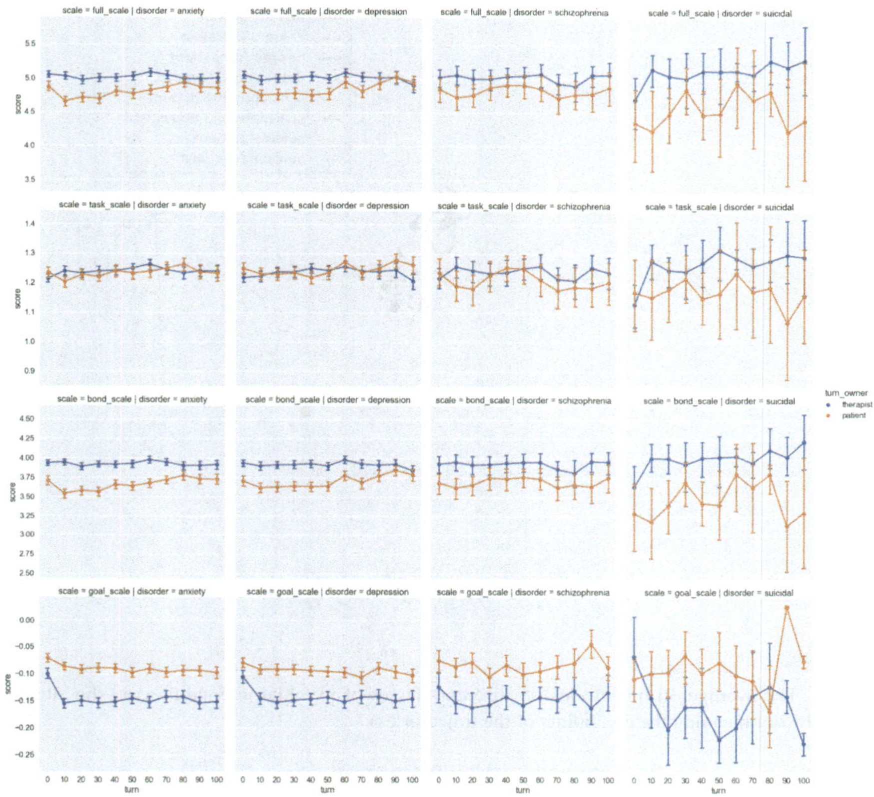

Fig. 9 Four-way ANOVA of the alliance dynamics

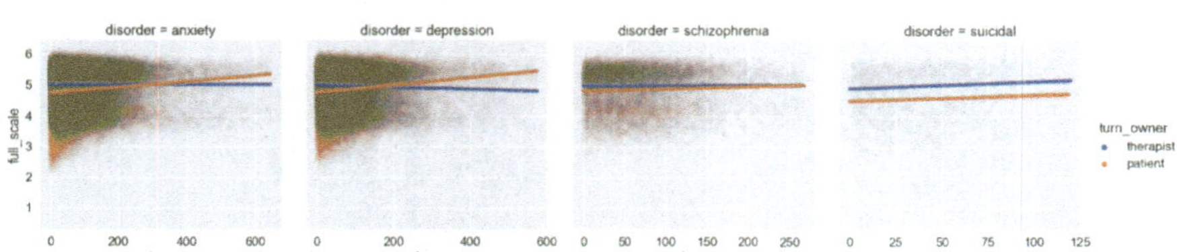

Fig. 10 Regression analysis of the temporal progression of the working alliance score

Fig. 6 that the patient alliance scores in the task and bond scales follow a bi-modal distribution.

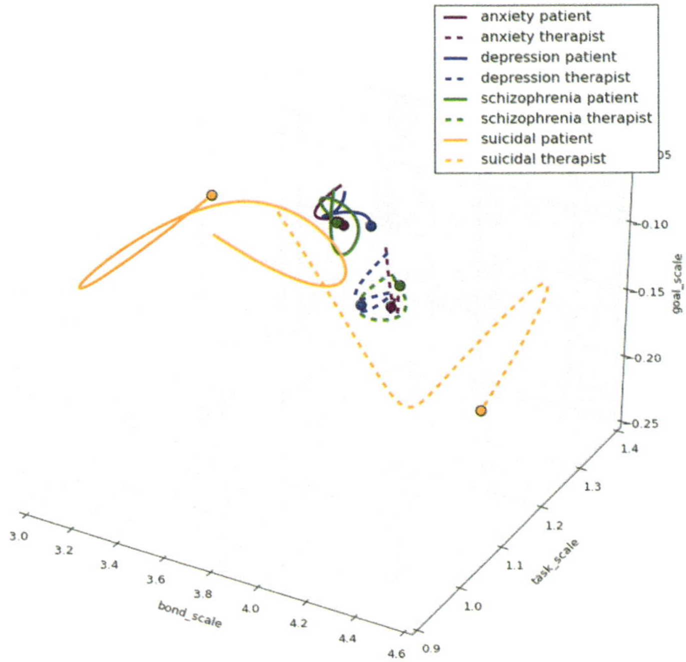

Fig. 11 The average 3d trajectories of different classes of psychiatric conditions in the alliance space (the dot meaning the end points of the trajectories)

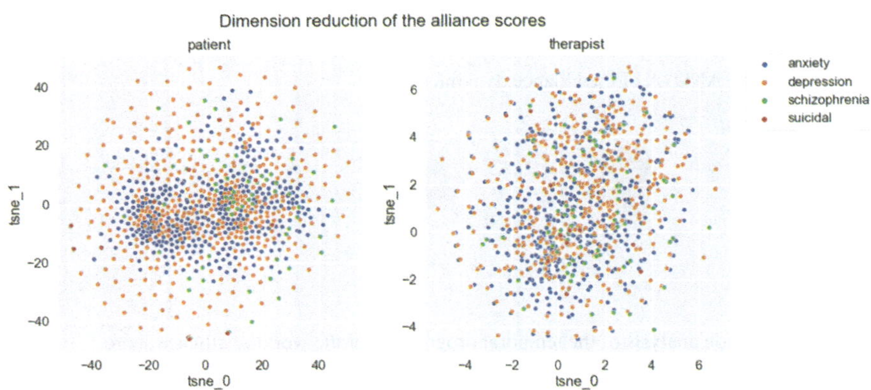

Fig. 12 Dimension reduction of the alliance trajectories

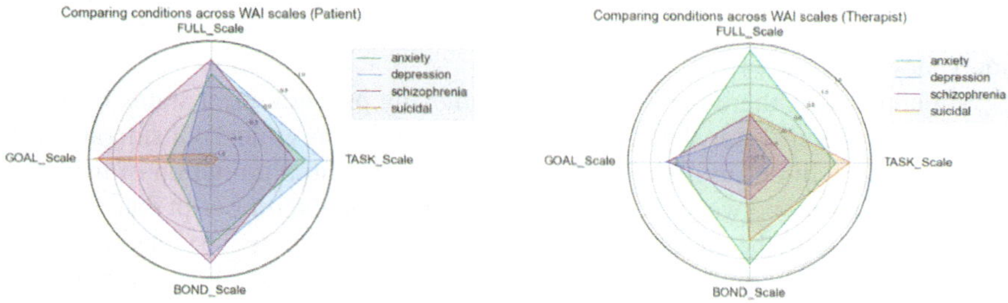

Fig. 13 Radar plots of the working alliance scores

4.1.3 Disorder-Specific Dialogue Prototypes of Working Alliance

We can further aggregate the alliance score by averaging all time points together all into the four scales. To plot the scale with respect to one another in a single plot, we normalize each scale to standard normal and present the radar plots of the scale-wise features of the patient and the therapist (Fig. 13). We observe that the suicidal patient are comparatively most imbalanced, large only in the goal scale and small in all others. While, on the therapist version, it is the opposite, which aligns with the observation made in the 3d trajectories.

5 Discussion

Our analytic approach reveals several insightful features of the therapeutic relationship. We observe systematic differences in the mean inferred alliance scores between patients and therapists, and also across disorders. However the in-session evolution of the inferred scores provide a much more interesting perspective. In particular, while all conditions show a systematic misalignment of scores between patients and therapists, this is significantly starker for suicidality, something that can be observed in the mean as well as in the time trace for full and sub-scales. In contrast, anxiety and depression display a clear trend for the full and the bond scales to *converge* as the sessions progress, something not present in the task and goal scales, nor in schizophrenia or suicidality. These features of the therapeutic dialogue can be mapped to what in psychiatry is usually called *alignment* and plays an important symptomatic and diagnostic role in several neuropsychiatric conditions, e.g., in relation to the hypothesis of Theory of Mind for schizophrenia [3]. By analyzing past sessions, and eventually sessions in real time, trained therapists may be able to identify key segments of the therapy leading to breakthroughs, compounding their expertise with further causal/predictive analytic modeling, while trainees may sharpen their intuition by reading or watching annotated versions of sessions conducted by experts. Needless to say, coupled with a generative language model and further statistical optimization, it may be possible to design limited chatbots to engage patients in triage and emergency response [4].

While effective, there are limitations in inferring psychological states from text data. In the session "Ethical Statements", we will cover the ethical considerations of this work. We will discuss other aspects here. One potential limitation of using the semantic similarity between the inventory statement and the transcript data is that, it measures how close the meaning or concept of a dialogue turn to the meaning or concept of the inventory item. But it doesn't necessarily fully capture the directionality of such similarity. For instance, which score is higher, a statement that is irrelevant to the inventory item, or a statement that is opposite to the inventory item. There are clear solutions to this problem. One alternative to the prototypical approach we use, is to use both an inventory item and its counter argument. For instance, if the inventory item x is "We share a mutual understanding of the expectation of this therapy", its counter argument $\neg x$ would be "We don't share a mutual understanding of the expectation of this therapy". We can then compute two similarity numbers v_x and $v_{\neg x}$. The score of this inventory item would then be $v_x - v_{\neg x}$. In practice, however, we observe no clear difference between this approach and ours, which can be found in the supplementary materials. This suggests that the sentence embedding we use already capture the concept of negation.

Another innate challenge of our line of research is the scarcity of clinical validation in the field. Working alliance, since its introduction in [1], is a proxy for the therapeutic alignment and outcome. As a result, existing behavioral and operational methods are also approximations to this psychological properties and not golden standards. We provide an alternative, but in much higher temporal resolution (turns are the timestamps). As far as we are aware, there are not available datasets in the field that have "ground truth" to validate in clinical setting. In future work, we aim to conduct clinical studies to further validate our results in field and intervention settings.

6 Ethical Statements

Till now, there are still a severe global shortage of workforce in mental health [23]. What we are proposing here, is not to replace existing work force of psychiatrists or therapists, but to assist them. In education setting, having an interpretable model that can inform the next-generation psychiatrists about the strategies adopted by experienced therapists. In this way, it can potentially alleviate this societal issue in both assisting and educating junior psychiatrists. As more and more successful applications of AI are deployed in clinical domains, there are many ethical considerations we practitioners of machine learning should be aware of and must take into considerations, as thoroughly pointed out in this review [10]. When dealing with patient data, the privacy and security is a top priority. Following the suggestion of best practices from [22], all examples in this paper as well as the dataset we analyzed are properly anonymized with pre- and post-processing techniques. In addition, the dataset itself was sourced with proper license from the Alexander Street platform. We remove all personally identifiable information (meta data, user name, identifiers, doctors' name) from the data.

As the clinical domain of this work is mental health and psychological well-being, there are additional ethical considerations. Emerging techniques in wearable devices, digital health, brain imaging measurements, smartphone applications and social media are gradually transforming the landscape of the monitoring and treatment of mental health illness. However, most of these attempts are proof of concept as identified by this review [5], and requires extensive caution to prevent from the pitfall of over-interpreting preliminary results. The limitations of these prior studies, including ours, reside in the difficulty of a systematic clinical validation and a uncertain future expectation of the technological readiness for patient care and therapeutic decision making approved by authorities. For instance, it was recently shown that despite the high predictability of statistical learning-based methods in analyzing large datasets in support of clinical decisions in psychiatry, existing machine learning solutions is highly susceptible to overfitting in realistic tasks which has usually a small sample sizes in the data, missing data points for some subjects, and highly correlated variables [8]. These properties in real-world applications limits the out-of-sample generalizability of the results.

Another ethical boundary to maintain is to make sure that the AI systems we use to diagnose, interpret and predict the mental health illness don't lead to increased risks to the patients. This requires both the practitioners and ML researchers to be fully aware of potentially bias and ethical challenges, such as gender bias, language-related ambiguity and ethnicity-related mental illness connections [2], in order to deploy the AI system in a responsible and safe way. Here we analyze a dataset with over 950 sessions of psychotherapy transcripts. Although it is the biggest dataset we find in this research domain, due to the anonymous nature of the dataset and a lack of details behind the collection process and demographics of the transcription, we cannot guarantee that the generalizability and representativeness to all populations. However, we believe that the insights we gain from these interpretable investigations is unlikely to increase the unforeseeable risks of the patients involved and can be potentially useful.

7 Conclusions

We have presented an approach that combines the state-of-the-art language modeling with the knowledge and practical expertise in psychotherapy, as captured in therapy-evaluation inventories, to provide a uniquely granular representation of the evolution of the interaction of patients and therapists. It is both insightful for post-session interpretations and useful for diagnosing the patients from linguistic features. While here we focus specifically on the Working Alliance Inventory, our method is generic and can be extended to the broader spectrum of assessment instruments. Finally, it would be possible to refine and further validate the language-based estimation of working alliance by providing punctuated rater evaluations as inference anchors. Next steps include predicting these inference anchors as states (like [13–15]) and training chatbots or dialogue topic recommendation system [16] as reinforcement learning agents given these states (like [12, 18–20]). Combining with other inference modules, we can eventually create AI knowledge management system with automatic annotation powered by NLP techniques [11] for the field of mental health.

8 Reproducibility Statement

The codes to reproduce all analytical and empirical results can be accessed and reproduced at the repository https://github.com/doerlbh/PsychiatryNLP.

References

1. E.S. Bordin, The generalizability of the psychoanalytic concept of the working alliance. Psychother.: Theory, Res. Pract. **16**(3), 252 (1979)
2. I.Y. Chen, P. Szolovits, M. Ghassemi, Can AI help reduce disparities in general medical and mental health care? AMA J. Ethics **21**(2), 167–179 (2019)
3. J.N. De Boer, S.G. Brederoo, A.E. Voppel, I.E. Sommer, Anomalies in language as a biomarker for schizophrenia. Curr. Opin. Psych. **33**(3), 212–218 (2020)
4. S. Garg, I. Rish, G. Cecchi, P. Goyal, S. Ghazarian, S. Gao, G. Ver Steeg, A. Galstyan, Modeling dialogues with hashcode representations: a nonparametric approach, in *Proceedings of the AAAI Conference on Artificial Intelligence*, vol. 34 (2020), pp. 3970–3979
5. S. Graham, C. Depp, E.E. Lee, C. Nebeker, X. Tu, H.C. Kim, D.V. Jeste, Artificial intelligence for mental health and mental illnesses: an overview. Curr. Psych. Rep. **21**(11), 1–18 (2019)
6. A.O. Horvath, An exploratory study of the working alliance: Its measurement and relationship to therapy outcome. Ph.D. thesis, University of British Columbia (1981)
7. A.O. Horvath, L.S. Greenberg, *The Working Alliance: Theory, Research, and Practice*, vol. 173 (Wiley, 1994)
8. R. Iniesta, D. Stahl, P. McGuffin, Machine learning, statistical learning and the future of biological research in psychiatry. Psychol. Med. **46**(12), 2455–2465 (2016)
9. Q. Le, T. Mikolov, Distributed representations of sentences and documents, in *International Conference on Machine Learning* (PMLR, 2014), pp. 1188–1196
10. B. Lin, Computational inference in cognitive science: Operational, societal and ethical considerations (2022). arXiv:2210.13526
11. B. Lin, Knowledge management system with NLP-assisted annotations: a brief survey and outlook, in *CIKM Workshops* (2022)
12. B. Lin, D. Bouneffouf, G. Cecchi, Split Q learning: reinforcement learning with two-stream rewards, in *Proceedings of the Twenty-Eighth International Joint Conference on Artificial Intelligence, IJCAI-19. International Joint Conferences on Artificial Intelligence Organization* (2019), pp. 6448–6449. https://doi.org/10.24963/ijcai.2019/913
13. B. Lin, D. Bouneffouf, G. Cecchi, Predicting human decision making in psychological tasks with recurrent neural networks. PloS one (2022)
14. B. Lin, D. Bouneffouf, G. Cecchi, Predicting human decision making with lstm, in *2022 International Joint Conference on Neural Networks (IJCNN)* (IEEE, 2022)
15. B. Lin, D. Bouneffouf, G. Cecchi, R. Tejwani, Neural topic modeling of psychotherapy sessions, in *International Workshop on Health Intelligence* (Springer, 2023)

16. B. Lin, G. Cecchi, D. Bouneffouf, Supervisorbot: Nlp-annotated real-time recommendations of psychotherapy treatment strategies with deep reinforcement learning (2022). arXiv:2208.13077
17. B. Lin, G. Cecchi, D. Bouneffouf, Working alliance transformer for psychotherapy dialogue classification (2022). arXiv:2210.15603
18. B. Lin, G. Cecchi, D. Bouneffouf, J. Reinen, I. Rish, A story of two streams: reinforcement learning models from human behavior and neuropsychiatry, in *Proceedings of the 19th International Conference on Autonomous Agents and MultiAgent Systems* (2020), pp. 744–752
19. B. Lin, G. Cecchi, D. Bouneffouf, J. Reinen, I. Rish, Unified models of human behavioral agents in bandits, contextual bandits and RL (2020). arXiv:2005.04544
20. B. Lin, G. Cecchi, D. Bouneffouf, J. Reinen, I. Rish, Models of human behavioral agents in bandits, contextual bandits and RL, in *International Workshop on Human Brain and Artificial Intelligence* (Springer, 2021), pp. 14–33
21. D.J. Martin, J.P. Garske, M.K. Davis, Relation of the therapeutic alliance with outcome and other variables: a meta-analytic review. J. Consult. Clin. Psychol. **68**(3), 438 (2000)
22. T. Matthews, K. O'Leary, A. Turner, M. Sleeper, J.P. Woelfer, M. Shelton, C. Manthorne, E.F. Churchill, S. Consolvo, Stories from survivors: privacy & security practices when coping with intimate partner abuse, in *Proceedings of the 2017 CHI Conference on Human Factors in Computing Systems* (2017), pp. 2189–2201
23. M. Olfson, Building the mental health workforce capacity needed to treat adults with serious mental illnesses. Health Affairs **35**(6), 983–990 (2016)
24. N. Reimers, I. Gurevych, Sentence-bert: sentence embeddings using siamese bert-networks (2019). arXiv:1908.10084
25. T.J. Tracey, A.M. Kokotovic, Factor structure of the working alliance inventory. Psychol. Assess.: J. Consult. Clin. Psychol. **1**(3), 207 (1989)
26. B.E. Wampold, How important are the common factors in psychotherapy? an update. World Psych. **14**(3), 270–277 (2015)

Neural Topic Modeling of Psychotherapy Sessions

Baihan Lin, Djallel Bouneffouf, Guillermo Cecchi, and Ravi Tejwani

Abstract In this work, we compare different neural topic modeling methods in learning the topical propensities of different psychiatric conditions from the psychotherapy session transcripts parsed from speech recordings. We also incorporate temporal modeling to put this additional interpretability to action by parsing out topic similarities as a time series in a turn-level resolution. We believe this topic modeling framework can offer interpretable insights for the therapist to optimally decide his or her strategy and improve psychotherapy effectiveness.

1 Introduction

Mental illness remains an issue in all countries and cultures across the globe. According to the National Institute of Mental Health (NIMH), nearly one in five U.S. adults live with a mental illness (52.9 million in 2020). One of the major causes of the mental illness is depression [4], followed by suicide which is the second cause of death among young people [22]. It is clear that there is a need for new innovative solutions in this domain. Psychotherapy is a term given for treating mental health problems by talking with a mental health provider such as a psychiatrist or psychologist [1]. To reduce the workload on mental health provider, natural language processing (NLP) is more and more adopted [26]. Noting that psychotherapy has

B. Lin (✉)
Columbia University, New York city, NY, USA
e-mail: baihan.lin@columbia.edu

D. Bouneffouf · G. Cecchi
IBM Thomas J Watson Research Center, Yorktown Height, NY, USA
e-mail: djallel.bouneffouf@ibm.com

G. Cecchi
e-mail: gcecchi@us.ibm.com

R. Tejwani
MIT Media Lab, Cambridge, MA, USA
e-mail: tejwanir@mit.edu

A. Shaban-Nejad et al. (eds.), *Artificial Intelligence for Personalized Medicine*,
Studies in Computational Intelligence 1106,
https://doi.org/10.1007/978-3-031-36938-4_16

been the first discipline using NLP. It started with a chat bot ELIZA [28] capable of mimicking a psychotherapist. Another chatbot, Parry [30], was able of simulating an individual with Schizophrenia. Natural language processing including topic modeling has shown interesting results on mental illness detection. In [25] the authors demonstrate that Latent Dirichlet Allocation (LDA) can uncover latent structure within depression-related language collected from Twitter. Authors [31] shows the add-value of using social media content to detect Post-Traumatic Stress Disorder.

Although previous works demonstrate the effectiveness of classical topic modeling, they are no longer the state-of-the-art. In recent years, deep learning progresses the fields and the Neural Topic Modeling shows up as the consistent better solution compared to the classical Topic modeling [19]. In this context, we propose in this work to use Neural Topic Modeling to learn the topical propensities of different psychiatric conditions from the psychotherapy session transcripts. We benchmark our findings on the Alex Street Counseling and Psychotherapy Transcripts dataset,[1] which consists of the transcribed recordings of over 950 therapy sessions between multiple anonymized therapists and patients with anxiety, depression, schizophrenia or suicidal intents. This multi-part collection includes speech-translated transcripts of the recordings from real therapy sessions, 40,000 pages of client narratives, and 25,000 pages of reference works. In total, these materials include over 200,000 turns together for the patient and therapist and provide access to the broadest range of clients for our linguistic analysis of the therapeutic process of psychotherapy.

The goal here is to evaluate the existing techniques on neural topic modeling and find the most adapted one to this domain. Second, we incorporate temporal modeling to put additional interpretability, where the goal of this framework is to offer interpretable insights for the therapist to optimally decide on psychotherapy strategy.

2 Related Work on Topic Modeling

In natural language processing and machine learning, a topic model is a type of statistical graphical model that help uncover the abstract "topics" that appear in a collection of documents. The topic modeling technique is frequently used in text-mining pipeline to unravel the hidden semantic structures of a text body. There are quite a few neural topic models evaluated in this work. The Neural Variational Document Model (NVDM) [19] is an unsupervised text modeling approach based on variational autoencoder. Reference [18] further shows that among NVDM variants, the Gaussian softmax construction (GSM) achieves the lowest perplexity in most cases, and thus recommended. We denote it as NVDM-GSM. Unlike traditional variational autoencoder based methods, Wasserstein-based Topic Model (WTM) uses the Wasserstein autoencoders (WAE) to directly enforce Dirichlet prior on the latent document-topic vectors [23]. Traditionally, it applies a suitable kernel in minimizing the Maximum

[1] https://alexanderstreet.com/products/counseling-and-psychotherapy-transcripts-series.

Mean Discrepancy (MMD) to perform distribution matching. We call this variant WTM-MMD. Similarly, we can replace the MMD priors with a Gaussian Mixture prior and apply Gaussian Softmax on top of it. We denote this method, WTM-GMM. In order to tackle the issue with large and heavy-tailed vocabularies, the Embedded Topic Model (ETM) [3] models each word with a matched categorical probablity distribution given the inner product between a word embedding and a vector embedding of its assigned topic. To avoid imposing improper priors, Bidirectional Adversarial Training Model (BATM) applies the bidirectional adversarial training into neural topic modeling by constructing a two-way projection between the document-word distribution and the document-topic distribution [29].

3 Therapy Topic Modeling Framework

Figure 1 is an outline of the analytic framework. During the session, the dialogue between the patient and therapist are transcribed into pairs of turns. We take the full records of a patient, or a cohort of patients belonging to the same condition. We either use it as is before the feature extraction, or we truncate them into segments based on timestamps or topic turns. When we have the features, we fit them into the topic models. The end results of the topic modeling would be a list of weighted topic words, that tells us what the text block is concerned with. These knowledges are usually very informative and interpretable, thus important in psychotherapy.

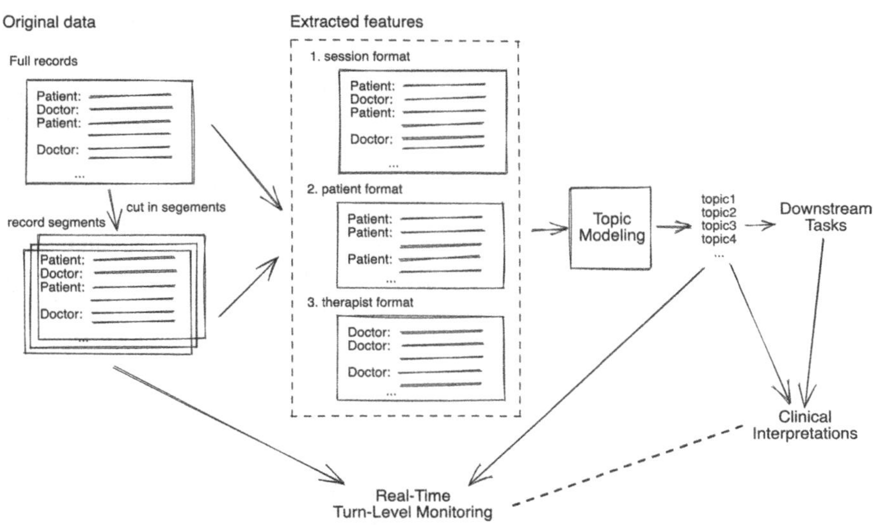

Fig. 1 Psychotherapy topic modeling framework with real time turn level monitoring)

Here are a few downstream tasks and user scenarios that can plugged to our analytical frameworks. We can either use these extracted weighted topics to inform whether the therapy is going the right direction, whether the patient is going into certain bad mental state, or whether the therapist should adjust his or her treatment strategies. This can be built as an intelligent AI assistant to remind the therapist of such things. Some topics can also be off-limit taboos, such as those in suicidal conversations, so if such terms arises from the topic modeling (say, a dynamic topic modeling), it can be flagged for the doctor to notice.

Algorithm 1 Temporal Topic Modeling (TTM)

```
Learned topics T as references
for i = 1,2,···, N do
  Automatically transcribe dialogue turn pairs (S_i^p, S_i^t)
  for T_j ∈ topics T do
    Topic score W_j^{p_i} = similarity(Emb(T_j), Emb(S_i^p))
    Topic score W_j^{t_i} = similarity(Emb(T_j), Emb(S_i^t))
  end for
end for
```

Given the learned topics, we can backtrack the transcript to get a turn-resolution topic scores. Algorithm 1 outlines the pipeline of our temporal topic modeling analysis (TMM). Say, if we have learned 10 topics, the topic score will be a vector of 10 dimensions, with each dimension corresponding to some notion of likelihood of this turn being in this topic. Because we want to characterize the directional property of each turn with a certain topic, we compute the cosine similarity of the embedded topic vector and the embedded turn vector, instead of directly inferring the probability as traditional topic assignment problem (which would be more suitable if we merely want to find the assignment of the most likely topic). In the result section, we will present the temporal modeling of the Embedded Topic Model (ETM), but this analytic pipeline can in principle be applied to any learned topic models. This Embedded Topic Model is special because, like our approach here, it also models each word with a categorical distribution whose natural parameter is the inner product between a word embedding and an embedding of its assigned topic. We use the same word embedding here (Word2Vec [20]) to embed our topic and turns.

4 Results

In this section, we compare five state-of-the-art neural topic modeling approaches introduced above, and analyze their learned topics. We separate transcript sessions into three categories based on the psychiatric conditions of the patients (anxiety, depression and schizophrenia), and train the topic models over each of them for over 100 epochs at a batch size of 16. As in the standard preprocessing of topic modeling training, we set the lower bound of count for words to keep in topic training to be 3, and the ratio of upper bound of count for words to keep in topic training to be 0.3. The evaluation procedure follows the same implementation for [29].[2]

4.1 Evaluation Metrics

Topic models are usually evaluated with the likelihood of held-out documents and topic coherence. However, it was shown that a higher likelihood of held-out documents does not necessarily correlate to the human judgment of topic coherence [2]. Therefore, we adopt a series of more validated measurements of topic coherence and diversity by following [27]. In the first evaluation, we compute four topic embedding coherence metrics (c_v, c_{w2v}, c_{uci}, c_{npmi}) to evaluate the topics generated by various models (as outlined in [27]). The higher these measurements, the better. In all experiments, each topic is represented by the top 10 words according to the topic-word probabilities, and the four metrics are calculated using Gensim library [24].[3] Other than these four topic embedding coherence evaluation provided by Gensim, we also included two other useful metrics. [21] proposed a robust and automated coherence evaluation metric for identifying such topics that does not rely on either additional human annotations or reference collections outside the training set. This method computes an asymmetrical confirmation measure between top word pairs (smoothed conditional probability). In addition, we compute the topic diversity by taking the ratio between the size of vocabulary in the topic words and the total number of words in the topics. Similarly, the higher these two measures are, the better the topic models.

4.2 Quantitative Evaluations of the Topic Models

Tables 1 and 2 summarize quantitative evaluations. We first observe that the different measures of the coherence gives different rankings of the topic models, but there are a few models that perform relatively well across the metrics. WTM and ETM both yield relatively high topic coherence and diversity.

4.3 Temporal Dynamics of the Topic Models

To ensure that the topics can be mapped from one clinical condition to another condition, we compute a universal topic model on the text corpus of the entire Alex Street psychotherapy database. And then, given the learned topics from this universal topic

[2] https://github.com/zll17/Neural_Topic_Models.

[3] https://github.com/RaRe-Technologies/gensim.

Table 1 Coherence embedding evaluations of the neural topic models (following [27])

	Anxiety				Depression				Schizophrenia			
	c_v	c_{w2v}	c_{uci}	c_{npmi}	c_v	c_{w2v}	c_{uci}	c_{npmi}	c_v	c_{w2v}	c_{uci}	c_{npmi}
NVDM-GSM	0.410	0.484	−0.844	−0.019	0.495	0.531	−3.522	−0.109	0.642	–	−1.954	−0.065
WTM-MMD	0.340	0.428	−2.827	−0.099	0.290	0.462	−3.797	−0.124	0.576	0.751	−0.997	−0.036
WTM-GMM	0.353	0.413	−3.259	−0.116	0.678	0.535	−0.126	−0.006	0.572	0.774	−1.587	−0.050
ETM	0.413	–	−2.903	−0.093	0.403	–	−2.399	−0.05	0.379	0.864	−7.232	−0.199
BATM	0.352	0.387	−5.056	−0.190	0.404	0.423	−4.238	−0.160	0.507	0.816	−9.655	−0.343

Table 2 Topic evaluations of the neural topic models (following [21])

	Anxiety		Depression		Schizophrenia	
	Topic coherence	Topic diversity	Topic coherence	Topic diversity	Topic coherence	Topic diversity
NVDM-GSM	0.653	**−380.933**	0.487	−316.439	0.527	−431.393
WTM-MMD	**0.927**	−453.929	0.907	−359.964	0.447	−403.694
WTM-GMM	0.907	−425.515	0.340	**−236.815**	0.467	**−204.930**
ETM	0.893	−449.000	**0.933**	−367.069	**0.973**	−310.211
BATM	0.720	−441.049	0.773	−443.394	0.500	−337.825

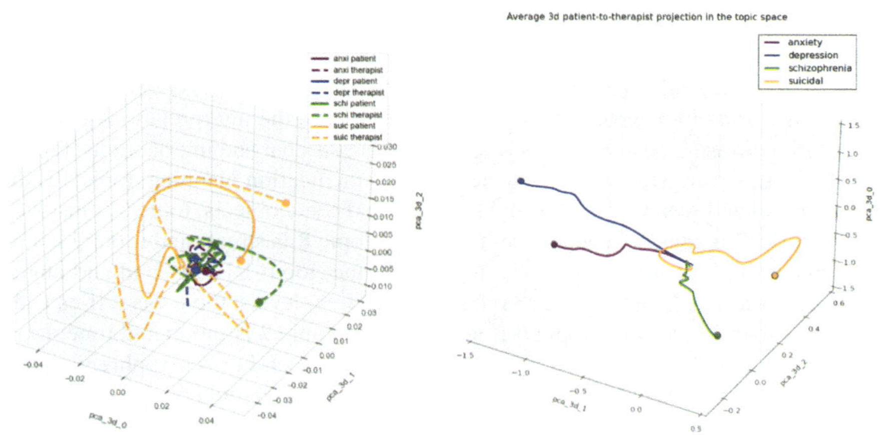

Fig. 2 The average 3d trajectories and the patient-to-therapist projections of different classes of psychiatric conditions in the principal topic space (dots are the trajectory end points).

models, we can compute a 10-dimensional topic score for each turn corresponding to the 10 topics. The higher the score is, the more positively correlated this turn is with this topic. Given this time-series matrix, we can potentially probe the dynamics of the these dialogues within the topic space. We can also provide more distinctive features for downstream tasks by performing a principal component analysis on the topic space. Figure 2 presents the average temporal trajectories of the patients and therapists, as well as the patient-to-therapist projection (i.e. the vector difference in the patient-therapist pair) in the principal topic spaces. We observe that the suicidal sessions cover a wider variety of topics (by having more spread-out trajectories), and have a more curved patient-therapist topic difference with multiple twists along the full session, while the other three clinical conditions have more consistent directions of such differences. This might suggest that a strategy by the therapist to divert from the sensitive topics. In schizophrenia sessions, the therapist appears to cover a bigger topical arc than the patient, suggesting a therapeutic strategy of visiting multiple topics to distract the patient from sensitive ones. The topical trajectories of the anxiety and depression sessions, comparatively, are more converged. This is a first step of identifying the prototypical therapeutic strategies in different psychiatric conditions and a potential turn-level resolution temporal analysis of topic modeling. With this approach one can go over the sessions (as a therapist) and analyze the dynamics afterwards.

4.4 Interpretable Insights from the Learned Topics

To provide interpretable insights, it is important to parse out the concepts behind these learned topics. To better understand what these topics are, we parse out the highest scoring turns in the transcripts that correspond to each topics.

First, we dive into the individual topic models trained on text corpus of each psychiatric condition separately. For instance, here are the interpretations from the top scoring turns in the anxiety sessions: topic 0 is chit-chat and interjections; topic 1 is low-energy exercises; topic 2 is fear; topic 3 is medication planning; topic 4 is the past, control and worry; topic 5 is other people and some objects; topic 6 is just well being; topic 7 is music, headache and emotion; topic 8 is stress; and topic 9 is fear and responsibilities. For depression, topic 0 is time; topic 1 is husband and anger; topic 2 is time and distance; topic 3 is energy and stress levels; topic 4 is self-esteem; topic 5 is money and time; topic 6 is age and time; topic 7 is mood and time; topic 8 is people and objects; topic 9 is holidays and chit-chats. For schizophrenia, topic 0 is family; topic 1 is extreme terms; topic 2 is energy level and positives; topic 3 is people and family; topic 4 is operational stuffs; topic 5 is calm things; topic 6 and 9 are critical topics.

For the universal topic models, the results are much more coherent. For instance, topic 0 is about figuring out, self-discovery and reminiscence. Topic 1 is about play. Topic 2 is about anger, scare and sadness. Topic 3 is about counts. Topic 4 is about tiredness and decision. Topic 5 is about sickness, self injuries and coping mechanisms. Topic 6 is about explicit ways to deal with stress, such as keep busying and reaching out for help. Topic 7 is about numbers. Topic 8 is about continuation and keep doing. Topic 9 is mostly chit-chat, interjections and transcribed prosody.

We notice that among all the clinical conditions we compare, the learned topics obtain a relatively poor mapping in the dialogue of suicidal cases. This might be due to the small sample size available in suicidal sessions, or the frequent hand annotations of behaviors (e.g. "patient crying for a few minutes" or "patient leaves the room") with time stamps, which doesn't conform to the annotation style of other sessions.

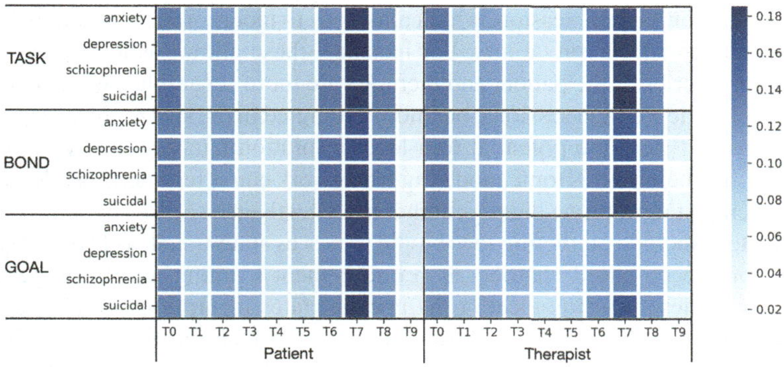

Fig. 3 Topic distributions of turns with top working alliance.

4.5 Ranked Topics Informed by Working Alliance

Although our approach can annotate the topics in each dialogue turns of the psychotherapy sessions, we don't know how informative they might be from the therapeutic point of view. In [14], we propose a computational technique to directly infer the therapeutic working alliance of a dialogue turn, which can be predictive of how effective the current therapy treatment is to the given patient at the given state. Combining this method with our topic modeling framework would enable us to highlight disorder-specific topics and dialogue segments that are potentially indicative of the therapeutic breakthroughs. For each disorder, we filter the turns to the top 100 working alliance scores, separately in three scales (task, bond and goal). Figure 3 is the heatmap of the averaged topic scores. We first observe no clear distinction among the working alliance scales, but notice a relatively uniform coverage of the topics when the patient and therapist are well aligned in the goal scale in all clinical conditions except for suicidal cases. Inspecting the top 10 turns with the highest topic scores, we notice that within the turns with high working alliance goal scale, suicidal patients tend to discuss sensitive terms like "alive", "stop" and "sexual".

Throughout the analytics, we follow the ethical guidelines pointed out in [5].

5 Conclusions

In this work, our first goal is to compare different neural topic modeling methods in learning the topical propensities of different psychiatric conditions. We first observe that different measures of the coherence gives different rankings of the topic models, but there are a few topic models that perform relatively well across metrics. For instance, Wasserstein Topic Models and Embedded Topic Models both yield relatively high topic coherence and diversity. Our second goal is to parse topics in different segments of the session, which allows us to incorporate temporal modeling and add additional interpretability. For instance, these allows us to notice that the session trajectories of the patient and therapist are more separable from one another in anxiety and depression sessions, but more entangled in the schizophrenia sessions. This is the first step of a potential turn-level resolution temporal analysis of topic modeling. We believe this topic modeling framework can offer interpretable insights for the therapist to improve the psychotherapy effectiveness.

Next steps include predicting these topic scores as states (such as [10, 11]), training text or speech-based chatbots as reinforcement learning agents (as reviewed in [8]) given these psychological or therapeutic states incorporating biological and cognitive priors (as in [9, 15–17]) and studying its factorial relations with other inference anchors (e.g. working alliance and personality [7, 13, 14]). The end goal is to construct a complete AI knowledge management system [6] or dialogue topic recommendation system [12] of mental health utilizing different NLP annotations in real time.

References

1. R. Ahmad, D. Siemon, U. Gnewuch, S. Robra-Bissantz, Designing personality-adaptive conversational agents for mental health care. Inf. Syst. Front. 1–21 (2022)
2. J. Chang, S. Gerrish, C. Wang, J. Boyd-Graber, D. Blei, Reading tea leaves: how humans interpret topic models. NIPS **22** (2009)
3. A.B. Dieng, F.J. Ruiz, D.M. Blei, Topic modeling in embedding spaces. Trans. Assoc. Comput. Linguist. **8**, 439–453 (2020)
4. J.T.S. Li, C.P. Lee, W.K. Tang, Changes in mental health among psychiatric patients during the covid-19 pandemic in Hong Kong-a cross-sectional study. Int. J. Environ. Res. Public Health **19**(3), 1181 (2022)
5. B. Lin, Computational inference in cognitive science: operational, societal and ethical considerations (2022). arXiv:2210.13526
6. B. Lin, Knowledge management system with NLP-assisted annotations: a brief survey and outlook, in *CIKM Workshops* (2022)
7. B. Lin, Personality effect on psychotherapy outcome: a predictive natural language processing framework (2022)
8. B. Lin, Reinforcement learning and bandits for speech and language processing: tutorial, review and outlook (2022). arXiv:2210.13623
9. B. Lin, D. Bouneffouf, G. Cecchi, Split Q Learning: Reinforcement Learning with Two-Stream Rewards, in *Proceedings of the Twenty-Eighth International Joint Conference on Artificial Intelligence, IJCAI-19. International Joint Conferences on Artificial Intelligence Organization* (AAAI Press, 2019). pp. 6448–6449. https://doi.org/10.24963/ijcai.2019/913
10. B. Lin, D. Bouneffouf, G. Cecchi, Predicting human decision making in psychological tasks with recurrent neural networks. PLoS ONE **17**(5), e0267907 (2022)
11. B. Lin, D. Bouneffouf, G. Cecchi, Predicting human decision making with lstm, in *2022 International Joint Conference on Neural Networks (IJCNN)* (IEEE, 2022)
12. B. Lin, G. Cecchi, D. Bouneffouf, Supervisorbot: Nlp-annotated real-time recommendations of psychotherapy treatment strategies with deep reinforcement learning (2022). arXiv:2208.13077
13. B. Lin, G. Cecchi, D. Bouneffouf, Working alliance transformer for psychotherapy dialogue classification (2022). arXiv:2210.15603

14. B. Lin, G. Cecchi, D. Bouneffouf, Deep annotation of therapeutic working alliance in psychotherapy, in International Workshop on Health Intelligence (Springer, 2023)
15. B. Lin, G. Cecchi, D. Bouneffouf, J. Reinen, I. Rish, A story of two streams: Reinforcement learning models from human behavior and neuropsychiatry, in *Proceedings of the 19th International Conference on Autonomous Agents and MultiAgent Systems* (2020), pp. 744–752
16. B. Lin, G. Cecchi, D. Bouneffouf, J. Reinen, I. Rish, Unified models of human behavioral agents in bandits, contextual bandits and RL (2020). arXiv:2005.04544
17. B. Lin, G. Cecchi, D. Bouneffouf, J. Reinen, I. Rish, Models of human behavioral agents in bandits, contextual bandits and RL, in *International Workshop on Human Brain and Artificial Intelligence* (Springer, 2021), pp. 14–33.
18. Y. Miao, E. Grefenstette, P. Blunsom, Discovering discrete latent topics with neural variational inference, in *International Conference on Machine Learning* (PMLR, 2017), pp. 2410–2419
19. Y. Miao, L. Yu, P. Blunsom, Neural variational inference for text processing, in *International Conference on Machine Learning* (PMLR, 2016), pp. 1727–1736.
20. T. Mikolov, I. Sutskever, K. Chen, G.S. Corrado, J. Dean, Distributed representations of words and phrases and their compositionality. Adv. Neural Inf. Process. Syst. **26** (2013)
21. D. Mimno, H. Wallach, E. Talley, M. Leenders, A. McCallum, Optimizing semantic coherence in topic models, in *Proceedings of the 2011 Conference on Empirical Methods in Natural Language Processing* (2011), pp. 262–272
22. A.M. Moe, E. Llamocca, H.M. Wastler, D.L. Steelesmith, G. Brock, J.A. Bridge, C.A. Fontanella, Risk factors for deliberate self-harm and suicide among adolescents and young adults with first-episode psychosis. Schizophr. Bull. **48**(2), 414–424 (2022)
23. F. Nan, R. Ding, R. Nallapati, B. Xiang, Topic modeling with wasserstein autoencoders, in *Proceedings of the 57th Annual Meeting of the Association for Computational Linguistics* (2019), pp. 6345–6381
24. R. Rehurek, P. Sojka et al., Gensim-statistical semantics in python. Retrieved from genism.org (2011)
25. P. Resnik, W. Armstrong, L. Claudino, T. Nguyen, V.A. Nguyen, J. Boyd-Graber, Beyond lda: exploring supervised topic modeling for depression-related language in twitter, in *Proceedings of the 2nd Workshop on Computational Linguistics and Clinical Psychology: From Linguistic Signal to Clinical Reality* (2015), pp. 99–107
26. N. Rezaii, P. Wolff, B.H. Price, Natural language processing in psychiatry: the promises and perils of a transformative approach. Br. J. Psych. 1–3 (2022)
27. M. Röder, A. Both, A. Hinneburg, Exploring the space of topic coherence measures, in *Proceedings of the Eighth ACM International Conference on Web Search and Data Mining* (2015), pp. 399–408
28. H.Y. Shum, X.D. He, D. Li, From eliza to xiaoice: challenges and opportunities with social chatbots. Front. Inf. Technol. Electron. Eng. **19**(1), 10–26 (2018)
29. R. Wang, X. Hu, D. Zhou, Y. He, Y. Xiong, C. Ye, H. Xu, Neural topic modeling with bidirectional adversarial training, in *Proceedings of the 58th Annual Meeting of the Association for Computational Linguistics* (2020), pp. 340–350
30. M.T. ZEMČÍK, A brief history of chatbots. DEStech Trans. Comput. Sci. Eng. **10** (2019)
31. Q.T. Zeng, D. Redd, T. Rindflesch, J. Nebeker, Synonym, topic model and predicate-based query expansion for retrieving clinical documents, in *AMIA Annual Symposium Proceedings*, vol. 2012 (American Medical Informatics Association, 2012), p. 1050

BAUFER: A Baseline-Enabled Facial Expression Recognition Pipeline Trained with Limited Annotations

Charlotte von Numers, Yinan Yu, Aleksandra Petkova, Emmette Hutchison, and Jesper Havsol

Abstract Social science theories suggest that facial expressions serve as a valuable indicator of one's emotions, well-being, and overall functioning. Recent research has found that the facial expressions of participants in clinical trials can be linked to their self-reported quality of life. Since manual facial expression annotation and interpretation is time and cost intensive, automated *facial expression recognition* (FER) tools have the potential to make it quicker and more consistent to study an individual's emotional responses. This paper introduces BAUFER, Baseline-enabled Action Unit identification for Facial Expression Recognition, with the following features: (1) a personalized baseline component to calibrate for the neutral expression of a participant; (2) predictions for anatomically-based facial muscle movement labels (Action Units), which have been reliably linked to emotional experiences in prior research, to enhance interpretability; and (3) a multi-stage training approach with several types of annotations from different datasets to overcome the known challenge of insufficient labeled data. While developed with non-clinical data, an intended future application of BAUFER is in the clinical domain to enhance our understanding of the patient experience.

C. von Numers (✉) · J. Havsol
Data Science and Advanced Analytics, BioPharmaceuticals R&D, AstraZeneca, Gothenburg, Sweden
e-mail: charlotte.vonnumers@astrazeneca.com

J. Havsol
e-mail: jesper.havsol@astrazeneca.com

Y. Yu
Department of Computer Science and Engineering, Chalmers University of Technology, Gothenburg, Sweden
e-mail: yinan@chalmers.se

A. Petkova · E. Hutchison
Human-centered AI & ML, BioPharmaceuticals R&D, AstraZeneca, Gaithersburg, USA
e-mail: aleksandra.petkova@astrazeneca.com

E. Hutchison
e-mail: emmette.hutchison@astrazeneca.com

A. Shaban-Nejad et al. (eds.), *Artificial Intelligence for Personalized Medicine*, Studies in Computational Intelligence 1106,
https://doi.org/10.1007/978-3-031-36938-4_17

Keywords Facial Expression Recognition (FER) · Action Unit (AU) · Limited training data · Deep learning · Baselining

1 Introduction

Facial expressions are among the most powerful signals humans use to convey emotions, stances and intentions [23]. Automated facial expression analysis and emotion detection have gained increased scientific interest during the past few decades, with a growing body of research suggesting a relationship between one's emotions and well-being [12]. Understanding this relationship is especially relevant in a clinical setting given that the ultimate goal of treatment is the improvement of health outcomes.

In recent years, the pharmaceutical and healthcare industries have increasingly leveraged digital tools to better understand the patient experience. Many clinical trials nowadays incorporate digital devices, custom apps, and wearable technology to improve the experience, and thereby the retention, of participants [10]. This transformation gives rise to opportunities for measuring endpoints by less invasive methods and at a higher frequency in the comfort of the participant's own home. For example, *Facial Expression Recognition* (FER) systems were used for monitoring of pain [9, 17] and detection of depression [3]. In addition, FER-based tools can potentially be used with both self-recordings and in real-time (e.g. during virtual care) to gain insight into the underlying emotions and cognitive states of individuals. Therefore, FER tools also have the potential to be incorporated into clinical trial endpoint development, targeting emotional quality and enhancing traditional questionnaire-based reporting tools, such as the common quality of life measurement [25].

This paper describes the development of BAUFER, an automated FER pipeline, designed to address three challenges: (1) baseline calibration of participant facial expressions, (2) the demand for granular facial expression analysis and interpretability, and (3) a lack of training data with fine-grained labels. In the future, this pipeline is intended to be further adapted and validated with clinical data, potentially in novel digital endpoint development or for use in applied clinical settings.

1.1 Baseline Calibration

The concept of a baseline is a key component in automating facial expression recognition as it enables a calibration for each individual's own facial expressions. Without a baseline, i.e., without taking into account a participant's own neutral facial expression, a machine learning model might be limited in its ability to accurately predict change from baseline, which is what indicates the potential experience of emotions of both positive and negative valence. Therefore, BAUFER incorporates a personalized baseline component into the pipeline to calibrate the system to the neutral expressions of each individual.

1.2 Action Units for Granular Analysis and Interpretability

Understanding human emotions through examining facial expressions requires a framework that provides highly granular analytical tools. In the domains of clinical psychology and computer vision alike, there has been a paradigm shift from the analysis of categorical emotions labels, such as happiness, sadness, disgust, etc., to the use of gold-standard *Action Unit* (AU) labels to examine anatomically-based facial behavior that correlates with the experience of positive, negative, and mixed emotions. AUs are a part of the *Facial Action Coding System* (FACS) and denote anatomically-based facial muscle movements (those movements are, in turn, correlated with emotions). FACS itself is the most precise and rigorous system for annotating and measuring noticeable facial movement [6, 7]. AUs are based on the contraction of individual facial muscles that lead to visible changes in the appearance of the face, thus producing rapid signs of emotion expression [2]. BAUFER focuses on detecting individual AUs for enhanced interprerability and precision.

1.3 Scarcity of Training Data

Developing a deep learning model that is capable of performing automated facial expression and emotion analysis requires vast amounts of data. Research and development of these models requires access to datasets that are sufficiently labeled and can be utilized to build a model that generalizes to a diverse set of participants.

However, FACS annotations require trained professionals to label AUs on a second-by-second basis for video data or for each individual static image with image-based data. This process takes a significant amount of time and resources, as a result there are relatively few FACS-coded datasets and, among those with FACS labels, the number of participants is typically low. In addition, not all these datasets are widely available for further research.

In this paper, we mitigate this issue by leveraging the strengths of different types of annotations from three different datasets: (1) identity labels for facial recognition, (2) categorical labels for emotion recognition and (3) FACS-coded AU labels for emotion recognition. These datasets are utilized at different stages in the training process to best leverage their strength. For instance, the training stage has an impact on the problem of *identity bias*, where a model fails to generalize the understanding of facial expressions between different participants. Therefore, we restrict the facial recognition dataset to the first training stage only in order to minimize this risk, while achieving the goal of adapting the model to the domain of human faces.

To choose the most suitable components for the BAUFER pipeline, different deep learning models and pretrained weights were compared. A series of pretraining approaches that are fit-for purpose using FACS labels are proposed. The outcome of this work, BAUFER, is a pipeline that comprises a face localization tool, as well as a *Convolutional Neural Network*-based (CNN) multilabel AU classifier, developed based on an open-source CNN architecture. BAUFER aims to be simple and easy to use for FER use cases and, in the future if validated with clinical data, is intended to serve as a proof-of-concept tool for introducing FER tools in both clinical research and applied settings.

2 Related Work

Deep learning techniques are commonly used in state-of-the-art FER pipelines for feature extraction. We divide the relevant literature into two areas: *facial data annotation* and *deep learning models*.

2.1 Facial Data Annotation

Supervised deep learning training requires annotated datasets. For a dataset containing human faces, there are mainly three types of annotations: *facial recognition for identification*, *categorical emotion annotation* and anatomically-based facial muscle movement *annotations produced with FACS*.

2.2 Facial Recognition for Identification

These annotations are used in building *Facial Recognition* (FR) models to recognize an individual human subject in applications ranging from automatic identification in video-surveillance systems to face tagging on social media platforms [19]. Unlike FACS, the annotation process requires no professional training and can be easily automated in many scenarios. However, in the context of FER, such datasets are often considered irrelevant or even counterproductive since the annotations may introduce identity bias as the network is originally trained to identify an individual and not an expression that might have the same emotional sentiment across many different individuals [15].

2.3 Categorical Emotion Annotation

This type of annotation reflects the categorical theory of emotion which proposes six universal emotions. These are *happiness, anger, sadness, surprise, disgust,* and *fear* [5]. Much like the FR labels, the annotation process does not necessarily require professional training; however, reliability between human annotators should be monitored. While the use of categorical labels have been favoured historically, more granular representations of facial expressions (and, in turn, of emotions), such as FACS AUs, are increasingly used in traditional computer vision [23]. However, datasets with categorical emotion labels and a large number of participants are more common than FACS-coded datasets, potentially due to the low annotation cost compared to FACS annotation.

2.4 Annotations Produced with FACS

Emotions are often communicated by subtle changes in one or a few discrete facial features, which are not sufficiently represented in high-level categorical emotion labels [23]. This motivates the use of granular FACS labels in studying the link between facial expressions and emotions. FACS is anatomically-based and includes annotations for nearly all possible facial movements [21]. It is a human-observer-based system where trained annotators manually code facial appearance changes referred to as AUs. There are a total of 44 AUs. Thirty of those AUs result from contractions of specific facial muscles: 12 correspond to the upper part of the face and 18 correspond to the lower part of the face [6]. A subset of frequently-occurring AUs are displayed in Fig. 1.

While it is not uncommon for FER models to predict high-level categorical emotions such as *happy*, *sad* and *angry*, this approach is possibly not the most relevant in the clinical and social science domains as humans often display subtle and sometimes mixed emotions through their facial expressions. Therefore, modern approaches favor the FACS system due to the higher granularity and the emotion expressions are rarely discrete (e.g. someone might smile while nervous/afraid).

In this paper, the objective is to combine different types of annotated datasets for developing a deep learning model capable of classifying AUs in an efficient way.

2.5 Deep Learning Models

Modern FER models use deep learning architectures to extract relevant features from facial images [15]. The domain is divided into *static* and *dynamic* approaches, where the former deal with static images or videos where each frame is processed independently, whereas the latter treat data as a time series [15]. This paper describes an implementation of a static deep learning-based FER model.

Feature extractors often are complex and data hungry models and thus challenging to train when data is scarce. State-of-the-art pipelines thus often utilize pretrained feature extractors due to the low availability of FER datasets. Pretrained CNN architectures are most commonly used. *Vision Transformers* (VT) have further been applied with good results [13, 18] as they have the advantage of automatically learning key facial areas such as the eyebrows or mouth. However, such models typically require

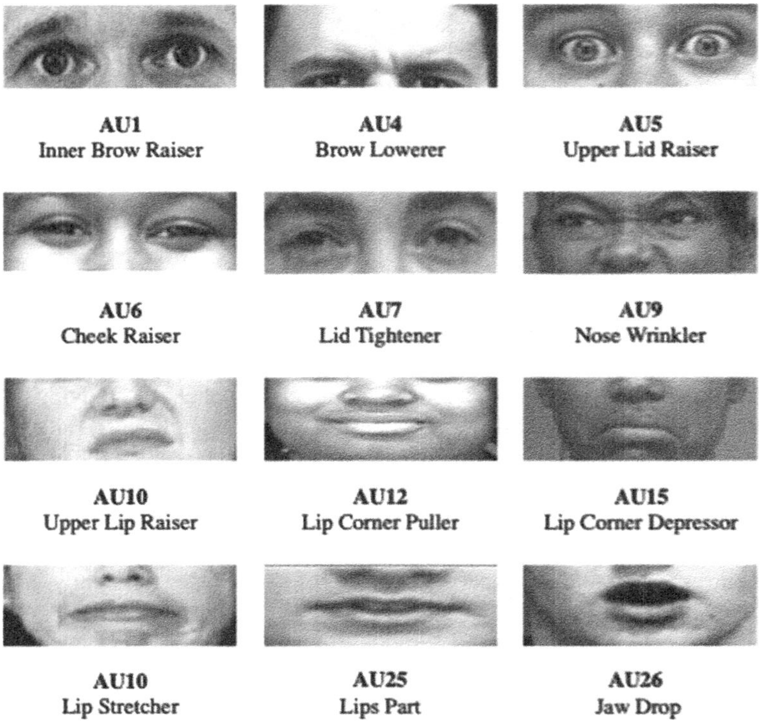

Fig. 1 A subset of frequently occurring FACS Action Units (AUs)

more training data compared to CNNs due to the lack of inductive bias [4], which makes them less relevant in the context of training with limited data.

Ensembles comprising different feature extractors are further commonly utilized given that different FER backbones often have different weaknesses. Supervised and self-supervised learning can be combined in this fashion, like in [8] where a supervised CNN was successfully combined with a self-supervised Variational Autoencoder (VAE). Other ensemble-approaches include the same CNN model with changes made to the filter size, the number of neurons and layers as well as multiple random seeds for network initialization [15].

Multi-stage pretraining approaches such as [11, 14] have been designed to deal with the identity bias challenge which occurs when a FER model is overly attentive to the identity of an individual. VGG Face is utilized as an initialisation in both approaches and the convolutional layers are finetuned as a pre-step to the end-to-end training with fully connected layers. Both approaches highlight the superior performance of VGG Face [1], which is an FR model, compared to ImageNet weights when applied in multi-stage transfer learning. However, neither [11] nor [14] perform classification of AU labels, which is the gap that has motivated the cu0rrent work.

3 Methodology

3.1 *BAUFER Pipeline Overview*

BAUFER takes images that contain a human face as input and outputs one *multi-hot encoded vector*. More specifically, as a deep learning based FER pipeline, BAUFER performs preprocessing, deep feature learning and classification. First, during preprocessing, the impact of factors not related to the facial expression itself, such as background, illumination or pose, is mitigated [15]. The target face is localized in the input frame after which the background is removed by cropping. The input data is assumed to only contain one face per frame. The cropped and altered face is then passed to a deep CNN feature extractor followed by a set of dense layers. Multi-label classification is carried out in the final dense layer. The layer shape corresponds to the multi-hot encoded vector representing the prediction of presence or absence of each AU. The full pipeline is displayed in Fig. 2.

3.2 *AU Classification*

The AU detection problem is posed as multi-label classification, where a multi-hot vector is constructed as the output of the pipeline. For a given input image, the multi-hot vector is then predicted to detect the presence of each AU.

3.3 *Baseline Implementation*

In BAUFER, a simple subtraction-based baselining approach is utilized. The idea is to use the model's own predictions of neutral images of a participant to correct for any false positives. The ideal outcome is for participants without any AUs present that may appear to the model as e.g. sad or angry to be adjusted accordingly. Activation for associated AUs would thus be decreased to ideally only be over a predefined threshold if actually present.

The baseline is defined as the output layer activation vector averaged over a number of neutral subject images. For the prediction of an input, the baseline of this subject is then subtracted from the predicted activation.

3.4 *Multi-stage Training*

The scarcity of FACS-coded datasets calls for creative use of other label types to achieve identity invariant FER models that are capable of AU classification. In

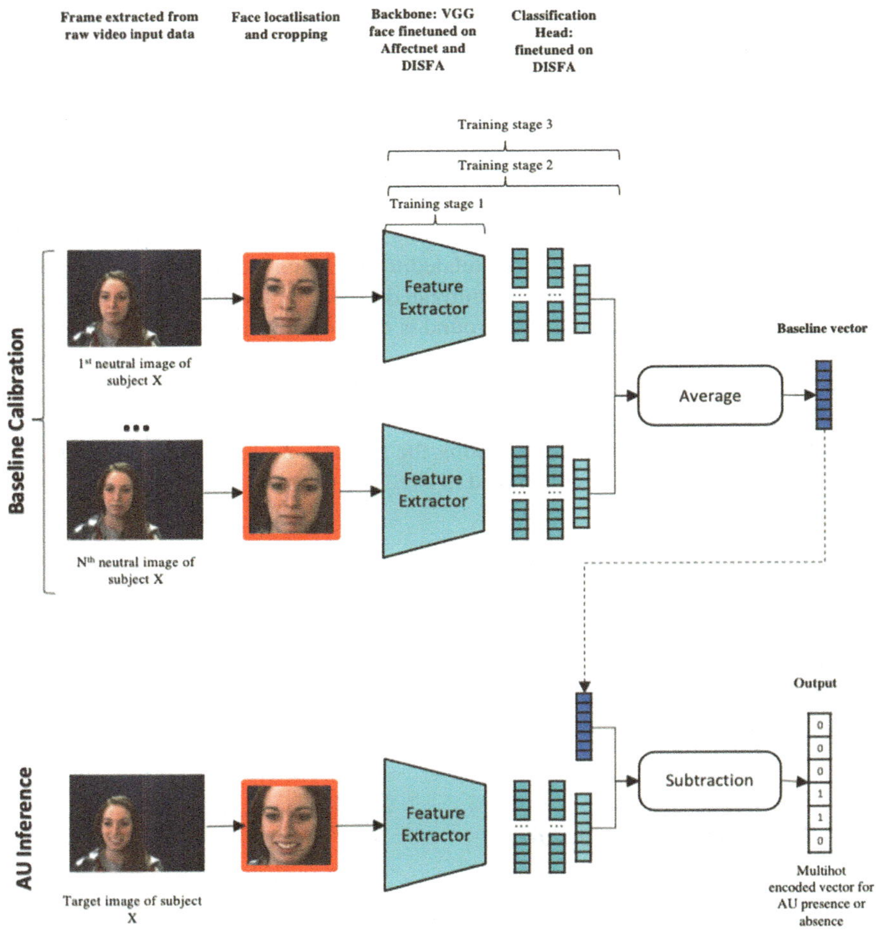

Fig. 2 The full BAUFER pipeline architecture

BAUFER, bootstrapping of features is done by combining Facial Recognition (FR) data and categorically labeled FER data. FR data has the advantage of comprising many different expressions for the same subject with the disadvantage of being focused on the identity while ignoring expressions. On the other hand, despite being more challenging to annotate compared to FR data, categorical FER data has the advantage of representing subject expressions even if the labelling is crude. By leveraging FR features relevant to the human face and continue training the network to pay attention to the expression rather than the identity, one achieves a robust model that is unlikely to collapse into identity bias. Without this intermediate training stage, identity bias is inevitable if the model is only finetuned on a small FACS-coded dataset. In this work, we propose a multi-stage training strategy that combines Facial

Table 1 Multi-stage training and data utilization

Stage	Purpose	Dataset	Annotation	# of subjects	# of images
Stage 1: backbone pretraining	Domain adaptation	VGG Face	Facial recognition for identification	2,622	2.6M
Stage 2: backbone finetuning	Reducing identity bias	AffectNet	Categorical emotion annotation	450,000	1M
Stage 3: classifier finetuning	AU detection	DISFA	FACS AU encodings	27	130788

Recognition (FR) data, categorically labeled FER data and AU annotated data for training the feature extractor in an efficient way.

To best leverage different types of annotations, the feature extractor is trained using the above-mentioned multi-stage training approach. The training dataset used for each stage is described in Table 1.

More specifically, there are three training stages. At the first stage, the FR dataset, VGG Face, is used for pretraining the Visual Geometry Group (VGG) [22] CNN. Despite the fact that this dataset is not tailored to training FER algorithms, it turns out to be beneficial to pretrain the backbone of FER feature extractors using images that focus on human faces instead of generic objects. Moreover, since facial recognition datasets are generally more available compared to FER annotated datasets, and pretrained weights are often readily available for transfer learning purposes, it makes this type of dataset and the corresponding pretrained-backbone a valid choice for adapting to the domain of human faces. In this work, we use the model developed in [1], which is hereafter referred to as VGG Face.

As stage 2, the pretrained convolutional layers are appended with three dense layers for multi-class classification. These layers are pretrained on a categorical dataset with several magnitudes more different subjects than FACS encoded datasets typically have. This pretraining step has the objective of forcing the FR-specific convolutional layers to start ignoring identity but paying attention to the expression of an individual.

After convergence on the categorical dataset, the convolutional feature extractor layers are again appended with new dense layers specific to multilabel classification. This is stage 3, where the network is finetuned on the dataset *Denver Intensity of Spontaneous Facial Action (DISFA)* [20] with FACS-encoded AU annotations for the final task.

4 Experiments and Results

The objective of the experiments is to evaluate the effect of each component of BAUFER in terms of AU detection capability.

4.1 Dataset

The DISFA dataset is used for final finetuning and evaluation. DISFA has 27 subjects recorded during 4 minutes each. Twelve AUs are included in the annotations, of which six are retained for prediction: *AU1: Inner Brow Raiser, AU4: Brow Lowerer, AU5: Upper Lid Raiser, AU6: Cheek Raiser, AU12: Lip Corner Puller* and *AU15: Lip Corner Depressor*. This sub-selection is motivated by research suggesting that those AUs tend to occur more frequently and generally associated with both positive and negative emotions [3, 7, 23]. Since BAUFER in the future may be used in clinical trial research, it was important to focus on AUs that are commonly encountered in emotion expression. To enable automatic detection of AUs, each AU annotation value is cast to 1 if the intensity score is larger than 1 and 0 otherwise.

4.2 Experiments

4.2.1 Frame Selection and Data Balancing

The AU detection problem is posed as multi-label classification. Due to the fact that occurrences of AUs are typically correlated, the training dataset can not be perfectly balanced with standard upsampling approaches. A total of 300 images per subject are gathered for the 26 subjects used for training, giving precedence to rare AUs. The procedure results in a total test set size of 7800 images. Frames with rare AUs are then further upsampled so that the maximum difference in AU frequency is 0.6 between the most rare and most common AUs. Weights are further applied in part of the training procedure to mitigate the imbalance between the positive and negative class of each AU. The weights are based on the relative frequencies within each class, which are displayed in Table 2.

4.2.2 Preprocessing

As the preprocessing step, the Haar Cascade [24] model is used for cropping the human face from a video frame. Unsuccessful crops are removed by an unsupervised clustering approach based on embedding generated by the convolutional layers of the open source FaceNet FR model. Furthermore, during the training process, data

Table 2 Weights for positive and negative classes of each AU

AU#	n_{pos}	n_{neg}	w_{pos}	w_{neg}
AU1	1772	6968	1.595	0.405
AU4	2474	6266	1.434	0.566
AU5	1769	6971	1.595	0.405
AU6	1827	6913	1.582	0.418
AU12	2928	5812	1.33	0.67
AU15	1767	6973	1.596	0.404

augmentation techniques are applied to artificially increase the number of training instances and make the network more robust towards variations in factors such as lighting, camera quality and object orientation. In particular, the following augmentation steps are applied: horizontal flipping; rotation at a maximum of 0.1 radian; empty spaces are filled by reflecting the original image to avoid the sharp angle of the otherwise black frame revealing the degree of rotation; cropping with a factor between 0.8 and 0.98; brightness adjustments with a maximum factor of 0.2; saturation and contrast adjustments with factors between 0.6 and 1.4; hue adjustments with a maximum factor of 0.1.

4.3 Evaluation and Ablation Study

The performance of BAUFER is evaluated on the DISFA dataset. The standard metric *Receiver Operating Characteristic Area Under Curve* (ROC AUC) is chosen for both the ablation study and evaluation of the end-to-end pipeline.

4.4 Multi-stage Training

For the purpose of feature extractor evaluation, a series of 60 consecutive frames per subject are selected for validation and the rest are used for training. The index for the start of the validation data is selected randomly for each subject.

4.4.1 Stage 1: ImageNet Verses VGG Face

First, to study the effect of stage 1 pretraining, backbones pretrained on ImageNet (Benchmark) and VGG Face (VGGF) are compared, where the datasets are designed for object classification and facial recognition tasks, respectively. The key difference between these two datasets is that the former contains generic objects, ranging from

Table 3 Model results for the EfficientNet B0 benchmark model that is only trained on the DISFA subjects, the VGGF model, the EfficientNet model pretrained on Affectnet and the VGGF model pretrained on Affectnet. The Best performance is achieved by VGGF pretrained on affectnet with a training strategy that neither makes use of weighted classes nor a frozen start

Name	Frozen start	Learning rate	ROC AUC
Benchmark	No	$1 \cdot 10^{-6}$	0.820
VGGF	No	$1 \cdot 10^{-6}$	0.909
VGGF-A	Yes	$1 \cdot 10^{-5}$	0.918
	No	$1 \cdot 10^{-6}$	**0.942**
	No	$1 \cdot 10^{-6}$	0.8854
VGGF-R	Yes	$1 \cdot 10^{-5}$	0.906
	No	$1 \cdot 10^{-6}$	**0.931**

volcanoes to ants, whereas the latter is specific to the domain of human faces. The result shows that domain adaptation plays a crucial role in the first stage backbone pretraining, even if the annotation itself (i.e. facial recognition) is not exactly the final task (i.e. AU classification).

4.4.2 Stage 2: AffectNet Verses RAF-DB

To analyze the training strategy for the second stage, the feature extractor is further evaluated using two different categorical datasets; Affectnet and RAF-DB [16]. The labels of the former are notoriously far from ground truth due to the limited annotation resources, while the latter has a fewer number of subjects but with more accurate labels. The purpose of the evaluation with these two datasets is to get an indication of the importance of the *label quality* versus the *data quantity* for stage 2. The result shows that at this stage, it is more beneficial to pretrain the pipeline with a large data quantity instead of cherry picking fewer images with high quality annotations.

4.4.3 Stage 3: Finetuning with DISFA

The final training stage is to finetune BAUFER on the AU annotations, which is a necessary step to enable the final AU detection. During the ablation study, this fine-tuning step is applied to each version of the pipeline presented in Table 3 for comparing the effect of each dataset.

Table 4 ROC AUC for the four most successful models while evaluated on unseen subjects are shown in the table. These are the models pretrained on RAF-DB and Affectnet, with and without a baseline implementation. The results suggest that training with Affectnet is superior to RAF-DB with regards to generalizability

	Model	AU1	AU4	AU5	AU6	AU12	AU15	Average
ROC AUC	VGGF-R	0.757	0.879	0.922	0.893	0.926	0.807	0.864
	VGGF-RB	0.779	0.883	0.920	0.887	0.927	0.824	0.870
	VGGF-A	0.839	0.867	0.949	0.884	0.940	0.796	0.879
	VGGF-AB	0.866	0.884	0.944	0.885	0.928	0.815	**0.887**

4.5 Baseline Versus No Baseline

The purpose of the baseline implementation is to further mitigate the problem of identity bias. Therefore, the effect of the baseline on AU classification performance is evaluated on unseen subjects using *Cross Validation* (CV). The result can be found in Table 4. Specifically, two subjects at a time are excluded from the training data. The CV is thus run for a total of 13 splits on the 26 DISFA training subjects, while subject 11 is retained for testing. The evaluation results of each CV model is aggregated to give an indication of model performance on identity bias. The performance of VGG-Face when pretrained on both RAF-DB and AffectNet is better with a baseline.

4.6 Performance of BAUFER

VGGF-A has a clear advantage over both the Benchmark model and VGGF trained without intermediate steps. The model achieves a ROC AUC of 0.942 when evaluated on familiar subjects. At 0.879 without a baseline and 0.887 with a baseline, the performance remains high on unseen subjects. The Benchmark network trained on ImageNet never reaches a ROC AUC above 0.83 regardless of training strategy making it inferior to its facial image pretrained counterpart. The results imply the importance of domain adaptation: FR-specific weights are superior to transfer learning with few or no faces.

The ROC AUC curves per AU for VGGF-AB are displayed in Fig. 3 to provide a more granular view on BAUFER's capacity for detecting each AU.

5 Discussion

VGG Face is already capable of encoding relevant facial features even if late layers are overly FR-specific due to the original objective of ignoring expression and focusing on identity. While training on AffectNet, the VGG Face architecture unlearns the

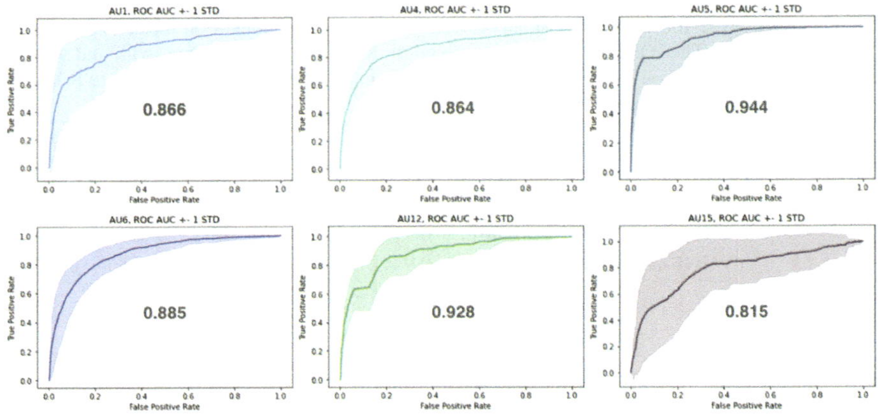

Fig. 3 CV ROC AUC curves and the 95% confidence interval for VGGF-AB on each AU

relevance of identity and learns the relevancy of expressions. The pretrained feature extractor requires limited fine-tuning in order to perform well on DISFA given features now being FER-specific. The network does not collapse into predicting one class per AU and individual due to the low learning rate and proximity to a more generalizable local optima.

The fact that the AffectNet labels are far from ground truth does not seem to impact performance on DISFA. The results indicate that neither the label type of the pretraining dataset nor the validity of its labels have a high impact on performance when pretraining a feature extractor for transfer learning.

The baseline model in BAUFER turns out to be an efficient way to further mitigate identity bias. One should note that the baselining approach is designed to correct for false positives while the opposite problem of false negatives is not addressed, i.e. subjects that indeed do express an emotion while the model perceive them as neutral. Correcting such errors is likely to require a more advanced approach where the the relative intensity of emotion is utilized to normalize over subjects.

Interestingly, we discover that despite being more coarse compared to AUs, datasets annotated with categorical emotions in fact have enhanced the capability of the model for recognizing the more fine-grained AU labels. This provides the opportunity of bootstrapping more resource-intense labels from cheaper and easier ones.

For instance, the model is capable of identifying AU5 (Upper Lid Raiser), a minority class in DISFA, with a high detection rate due to the fact that AU5 distinctively appears in AffectNet as shown in Fig. 4.

Fig. 4 This image illustrates how AU5 (Upper Lid Raiser) appears distinctively in AffectNet within the category *fear*

6 Conclusion

Facial Expression Recognition (FER) can provide valuable information about human emotions. In this paper, we propose a deep learning based FER pipeline, named BAUFER and developed with non-clinical but emotionally expressive data, for detecting the presence of Action Units (AUs), which are anatomically-based facial muscle movements that are reliably linked to emotional experiences. To minimize false positives, BAUFER incorporates a personalized baseline that provides a means to calibrate for each individual's neutral facial expression. Due to the scarcity of AU labeled datasets, BAUFER is developed with a multi-stage training strategy utilizing different types of annotated data. From the experiments, we observe that by first pretraining the backbone model with facial recognition dataset for domain adaptation, followed by a second stage pretraining using a categorical emotion dataset to reduce identity bias, it improves the performance for the final AU detection task. In addition, the baseline model is effective for improving the system performance. The development of the baselining mechanism is an interesting topic for future research. Future work should also examine how FER pipelines can bolster research aimed to understand participants' overall experience and well-being in clinical trials and beyond.

References

1. Q. Cao, L. Shen, W. Xie, O.M. Parkhi, A. Zisserman, Vggface2: a dataset for recognising faces across pose and age, in *2018 13th IEEE International Conference on Automatic Face & Gesture Recognition (FG 2018)* (IEEE, 2018), pp. 67–74
2. J.F. Cohn, P. Ekman, Measuring facial action (2005)
3. J.F. Cohn, T.S. Kruez, I. Matthews, Y. Yang, M.H. Nguyen, M.T. Padilla, F. Zhou, F. De la Torre, Detecting depression from facial actions and vocal prosody, in *2009 3rd International Conference on Affective Computing and Intelligent Interaction and Workshops* (IEEE, 2009), pp. 1–7
4. A. Dosovitskiy, L. Beyer, A. Kolesnikov, D. Weissenborn, X. Zhai, T. Unterthiner, M. Dehghani, M. Minderer, G. Heigold, S. Gelly, et al., An image is worth 16x16 words: transformers for image recognition at scale (2020). arXiv:2010.11929

5. P. Ekman, W.V. Friesen, Constants across cultures in the face and emotion. J. Person. Soc. Psychol. **17**(2), 124 (1971)
6. P. Ekman, W.V. Friesen, Facial action coding system. Environ. Psychol. Nonverbal Behav. (1978)
7. P. Ekman, W.V. Friesen, J.C. Hager, *Facial Action Coding System: Facial Action Coding System: The Manual: on CD-ROM* (Research Nexus, 2002)
8. D. Hamester, P. Barros, S. Wermter, Face expression recognition with a 2-channel convolutional neural network, in *2015 International Joint Conference on Neural Networks (IJCNN)* (IEEE, 2015), pp. 1–8
9. Z. Hammal, J.F. Cohn, Automatic detection of pain intensity, in *Proceedings of the 14th ACM International Conference on Multimodal Interaction* (2012), pp. 47–52
10. Kushal Kadakia, Bakul Patel, Anand Shah, Advancing digital health: Fda innovation during covid-19. Npj Digital Med. **3**(1), 1–3 (2020)
11. H. Kaya, F. Gürpınar, A.A. Salah, Video-based emotion recognition in the wild using deep transfer learning and score fusion. Image Vis. Comput. **65**, 66–75 (2017)
12. Corey LM. Keyes, Jonathan Haidt, *Flourishing: Positive psychology and the life well-lived* (American Psychological Association, Washington, DC, 2003)
13. J-H. Kim, N. Kim, C.S. Won, Facial expression recognition with swin transformer (2022). arXiv:2203.13472
14. B. Knyazev, R. Shvetsov, N. Efremova, A. Kuharenko, Convolutional neural networks pretrained on large face recognition datasets for emotion classification from video (2017). arXiv:1711.04598
15. S. Li, W. Deng, Deep facial expression recognition: a survey. IEEE Trans. Affect. Comput. (2020)
16. S. Li, W. Deng, J. Du, Reliable crowdsourcing and deep locality-preserving learning for expression recognition in the wild, in *2017 IEEE Conference on Computer Vision and Pattern Recognition (CVPR)* (IEEE, 2017), pp. 2584–2593
17. G.C. Littlewort, M.S. Bartlett, K. Lee, Faces of pain: automated measurement of spontaneousallfacial expressions of genuine and posed pain, in *Proceedings of the 9th International Conference on Multimodal Interfaces* (2007), pp. 15–21
18. F. Ma, B. Sun, S. Li, Facial expression recognition with visual transformers and attentional selective fusion. IEEE Trans. Affect. Comput. (2021)
19. I. Masi, Y. Wu, T. Hassner, P. Natarajan, Deep face recognition: a survey, in *2018 31st SIBGRAPI Conference on Graphics, Patterns and Images (SIBGRAPI)* (IEEE, 2018), pp. 471–478
20. S.M. Mavadati, M.H. Mahoor, K. Bartlett, P. Trinh, J.F. Cohn, Disfa: a spontaneous facial action intensity database. IEEE Trans. Affect. Comput. **4**(2), 151–160 (2013)
21. K. Niinuma, I.O. Ertugrul, J.F. Cohn, L.A. Jeni, Systematic evaluation of design choices for deep facial action coding across pose. Front. Comput. Sci. **3**, 636094 (2021)
22. K. Simonyan, A. Zisserman, Very deep convolutional networks for large-scale image recognition (2014). arXiv:1409.1556
23. Y-I. Tian, T. Kanade, J.F. Cohn, Recognizing action units for facial expression analysis. IEEE Trans. Pattern Anal. Mach. Intell. **23**(2), 97–115 (2001)
24. P. Viola, M. Jones, Rapid object detection using a boosted cascade of simple features, in *Proceedings of the 2001 IEEE Computer Society Conference on Computer Vision and Pattern Recognition. CVPR 2001*, vol. 1 (IEEE, 2001), pp. I–I
25. B. Zhang, R. Buendia, N. Iannoti, E. Ramsden, P. O'Regan, J. Swift, S. Lockwood, D.J. Jackson, G. Dennis, L. Hagger, J. Havsol, Home-based digital assessments with applied sentiment & emotion AI capture improved quality-of-life in asthma patients (2021)

Robustness for ECG Classification by Adversarial Training Over Clinical Features

Suparshva Jain, Amit Sangroya, Lovekesh Vig, and C. Anantaram

Abstract Recent work employing deep learning for ECG signal classification has achieved state of the art performance on benchmark datasets, comparable to the accuracy achieved by cardiologists in some cases. However, whether these models discover and attend to the same clinical features that cardiologists utilize for diagnosis remains unexplored. This paper looks at a well known state of the art deep learning model for ECG classification, and observes its performance on high level perturbations. Surprisingly, we find that the model is not always sensitive to these high level perturbations suggesting that they may be relying on medically meaningless correlations to make predictions. We then perform adversarial training on these clinically perturbed ECG signals to enhance model robustness. Additionally, we perform conventional adversarial training against low-level perturbations simultaneously to ensure robustness against adversarial attacks. Experimental results show that the proposed training regimen can improve both model accuracy and the adversarial robustness by a significant margin. We demonstrate that the resulting models are (1) more sensitive to clinical features, (2) robust to adversarial attacks, and (3) yield state of the art performance and (4) clinical perturbations add to the robustness of the model over standard adversarial training.

S. Jain · A. Sangroya (✉) · L. Vig · C. Anantaram
TCS Research, Chennai, India
e-mail: amit.sangroya@tcs.com

S. Jain
e-mail: suparshva.jain@tcs.com

L. Vig
e-mail: lovekesh.vig@tcs.com

C. Anantaram
e-mail: c.anantaram@tcs.com

A. Shaban-Nejad et al. (eds.), *Artificial Intelligence for Personalized Medicine*, Studies in Computational Intelligence 1106,
https://doi.org/10.1007/978-3-031-36938-4_18

1 Introduction

A notable deep learning healthcare application has been the analysis of electrocardiogram (ECG) signals via deep models for detecting arrhythmias. These models are often embedded within wearable devices with recent reports testifying to the technology's life saving capabilities. As with other deep learning applications, training models to detect arrhythmias on patient data requires large volumes of real world patient data. However, a subtle nuance is often overlooked with regard to the generation of training labels. ECG datasets for arrhythmia detection are annotated by teams of cardiologists but unlike common annotation protocols, annotation follows the application of strict well defined rules over high level clinical features [16]. For example, a rule which cardiologists follow to detect atrial fibrillation is the "absence of P-waves in the signal and presence of irregular and narrow QRS complexes". In effect, we have knowledge of the label generation process and ideally would expect the deep model trained on raw ECG signals to discover and exploit the same clinical features. In this paper, we attempt to verify whether this is indeed the case and find that even state of the art deep models for arrhythmia prediction are insensitive to clinical features, and rely on potentially spurious correlations to make a prediction.

This raises the question "why not force the network to extract these clinical features directly?" and simply apply the rules on top of these features. The problem is that these intermediate clinical features are often difficult to identify from noisy time series data and annotating signals for these clinical features at scale is both unreliable and expensive. Further, the resulting clinical feature extractors trained on one dataset may need to be fine tuned for different populations. Thus, forcing the network to predict clinical features via direct supervision may be impractical. In this paper, we explore the possibility of introducing clinical perturbations to the training data to enhance model sensitivity to the desired clinical features. We exploit the consistent patterns present in ECG data for normal, healthy patients to accurately detect and perturb clinical features like P-waves, R-R intervals, and R-R regularity using signal processing. Our experiments demonstrate that by training on these perturbations, the resulting models are significantly more sensitive to clinical features while retaining accuracy.

Another desirable property of a deep model is that its performance be robust to conventional adversarial attacks. Grave security concerns still plague models embedded in healthcare devices that use ECG classification algorithms, which can potentially be tricked into misdiagnosing patients. Unlike traditional non-deep learning models where no gradient information can be exploited, deep learning based ECG classification models are susceptible to both white-box and black box attacks by exploiting their gradient information. We train our ECG classification model against both Fast Gradient Method (FGM) and Projected Gradient Descent (PGD) attacks along with training on the high level clinical perturbations. We find that the training against high level feature perturbations not only improves sensitivity to clinical features but also improves model robustness against gradient based adversarial attacks.

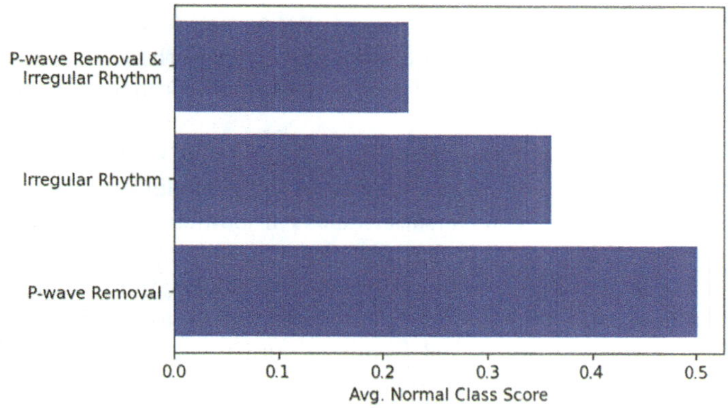

Fig. 1 Sensitivity of SOTA model to high level perturbation

Table 1 Accuracy (using F1-Score) when applying adversarial attacks on selected deep learning models

	SOTA model [[14]]
Original test set	0.80
After FGM attack ($\epsilon = 0.1$)	**0.34**
After PGD attack ($\epsilon = 0.1$)	**0.32**

We conducted an experiment using a well known state of the art (SOTA) deep learning model for ECG classification [14], where we fed 'Normal' ECG Signals with high level clinical perturbations. Surprisingly, we found that the model was not sensitive to these high level perturbations suggesting that it may be relying on medically meaningless correlations (i.e. correlation is not obvious) to make predictions (Fig. 1). Moreover, when this model is subjected to conventional low level perturbations (using FGM and PGD attacks), its performance drops significantly (See Table 1).

To address these problems, we perform adversarial training on these clinically perturbed ECG signals to enhance model robustness. Additionally, we perform conventional adversarial training against low-level perturbations to ensure robustness against adversarial attacks. We compare the accuracy and robustness of the adversarially trained model with current benchmarks. Our results show that the proposed training regimen improve both model accuracy and the adversarial robustness by a significant margin.

Prior work in the area of generating adversarial samples for time series classification has not focused on higher level clinical perturbations for training. To the best of our knowledge, this is the first instance of a deep learning based ECG classification model that is adversarially trained to support sensitivity to high level feature perturbations. The key contributions of this work are:

1. Firstly, we demonstrate that the current SOTA deep learning model for ECG classification is not sensitive to clinical feature perturbations and adversarial training using clinical/high level feature perturbations make the model more sensitive to these features.
2. Further, training against conventional adversarial attacks to prevent malicious misdiagnosis provides additional robustness to the model.
3. Training against low level adversarial attacks in addition to clinical perturbations results in even greater model robustness than training against either individually.
4. Finally, our model that combines conventional adversarial training using low level perturbations with high level (clinical) perturbations, results in models that are (i) more sensitive to clinical feature perturbations; (ii) more robust to low level perturbations; and (iii) yield performance comparable to SOTA.

2 Background and Preliminaries

2.1 Low and High Level Clinical Features

In literature, the terms "high-level" and "low-level" are generally used to refer to the features generated by a deep model like a CNN (convolution neural network) via its intermediate representations [20]. For instance, in image classification, as the CNN learns low-level features (such as edges, corners) through the first hidden layers, mid-level features (squares, circles, etc.) through intermediates hidden layers, and high-level features (faces, text, etc.) through the final hidden layers. The **feature level** is primarily related to the content of the feature maps based on the task. In this paper, we propose perturbations at following two levels for ECG domain:

1. **Low Level Features**: These are the features which are at a finer granular level and normally difficult to interpret. For example: pixels, dots and lines in case of images. Feature maps that correspond to a learned motif of an ECG signal window are "high-level" and small oscillations in time series are "low-level".
2. **High Level Clinical Features**: These are the features which are understandable by end users (medical professionals in our case). For example: P wave, QRS complex, Rhythm etc. We also call them **clinical features**, since they are interpretable by clinicians. These features have been used by classical AI based systems such as **Kardio** where a medical expert system is designed based upon a model of the human heart. The Kardio rule based expert system was designed for the diagnosis of cardiac arrhythmias [3, 16]. One interesting aspect about the Kardio system is that the explanations of arrhythmias generated by the system are based on domain concept features such as *P waves*, and *Rhythm*.

2.2 *Adversarial Attacks*

Malicious manipulation of data/gradients can create a mismatch between training and test data distributions, mislead models, and significantly harm their performance. The community refers to this kind of malicious manipulation as an adversarial attack. Several approaches have been proposed to increase a model's robustness against adversarial attacks. Reference [12] proposed augmenting the training set with adversarial examples. At training time, they minimize the loss for real and adversarial examples, while adversarial examples are chosen to fool the current version of the model.

A popular kind of adversarial attack is performed by altering the data with very minute perturbations (like gaussian noise) that can cause misclassifications but are often imperceptible to humans and therefore may go through undetected. Another possibility is perturbing the hidden layers of a deep network. Researchers have compared the impact of perturbation of the input with perturbation of the hidden layers. In [25], and found that it is usually better to just perturb the original input.

2.2.1 High Level Perturbation Examples

Figure 2 exhibits the ECG signal before and after 'P-wave Removal' perturbation is applied. Similarly, Fig. 3 exhibits the ECG signal before and after 'P-wave removal and Irregular Rhythm' perturbation is applied.

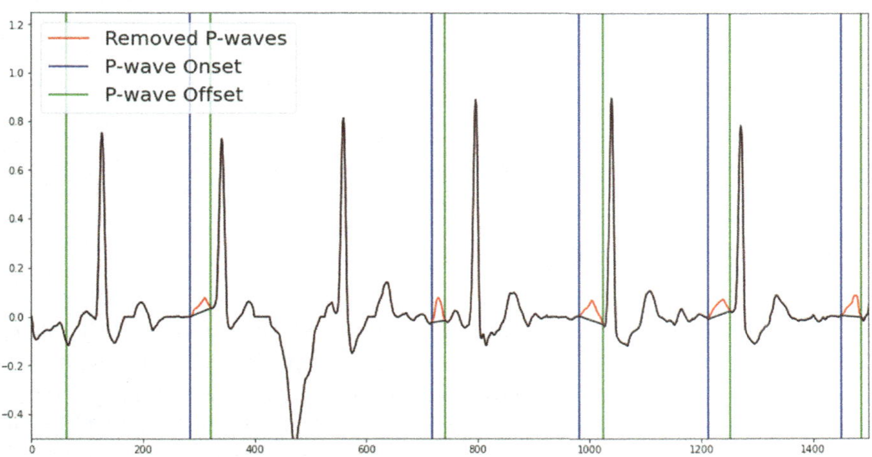

Fig. 2 Example for P-wave removal

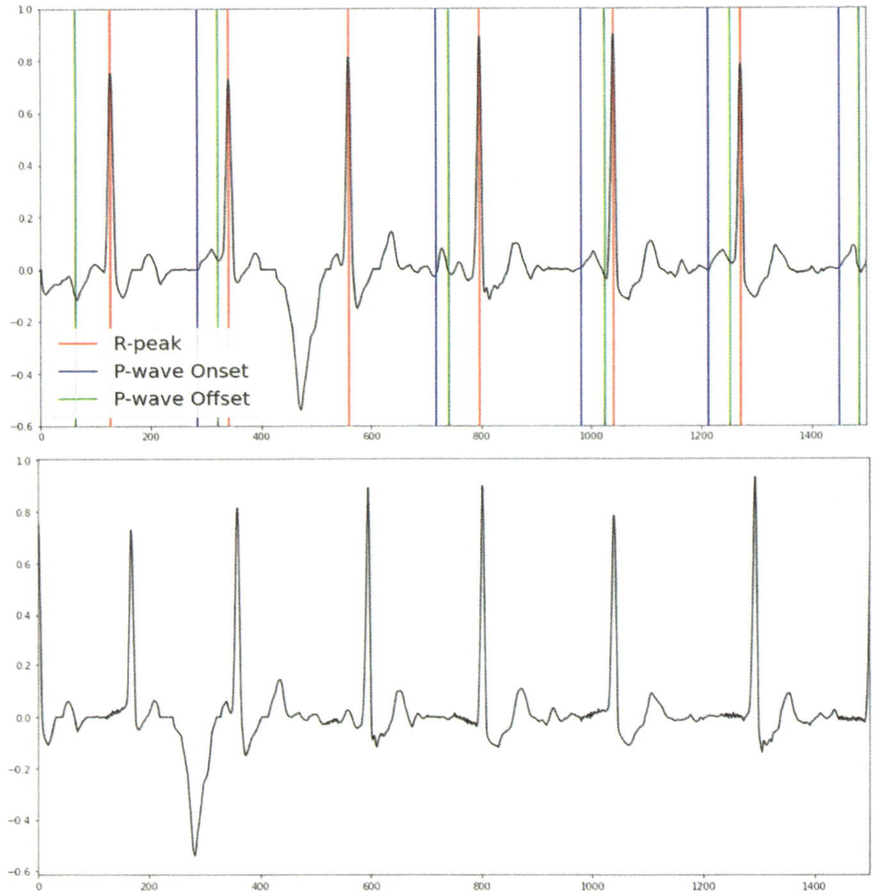

Fig. 3 Example for P-wave removed (Top) and Irregular rhythm (Bottom)

3 Proposed Methodology

Convolution neural network models have achieved SOTA results for ECG signal classification. In this work, firstly, we try to assess the performance of these models against perturbations to low and high level features. Thereafter, we perform adversarial training including conventional training on these clinically perturbed ECG signals to enhance model robustness. The overall architecture is shown in Fig. 4.

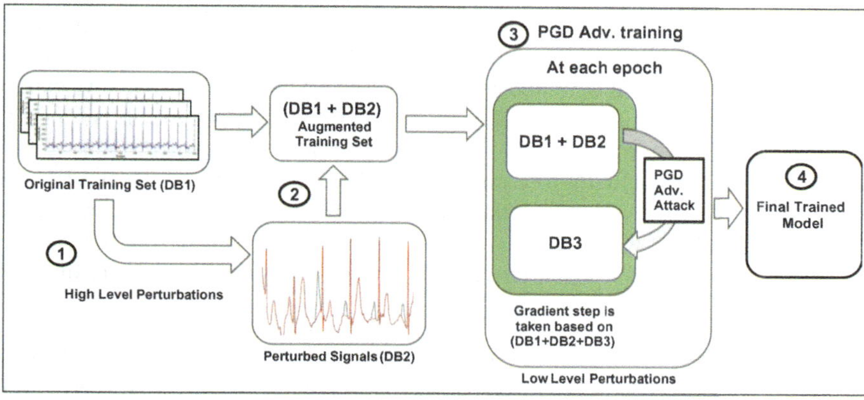

Fig. 4 PGD Adversarial training with High level perturbations: High level perturbations as mentioned in Sect. 3.2 are performed on the training set (DB1) to form an augmented training set (DB1+DB2). PGD Adversarial training is performed on this augmented training set (DB1+DB2). During PGD Adv training at each epoch, PGD attack is use to create signals (DB3) for Adversarial training

3.1 Low Level Perturbations

In this step, very minute perturbations are introduced into the original ECG signals and the performance of the model on these perturbed signals is recorded. Since the introduced changes are quite small, a robust model is expected to exhibit minimal drop in performance. However, if there is a significant drop in the performance of the model, we argue that the model is not robust, as it vulnerable to misclassification even with small changes to the input signal. In order to make these minuscule changes to a signal, gradient based adversarial attacks (namely FGM and PGD) are being used here. The performance of a model against these gradient based attacks is a commonly used practice to examine robustness. Gradient based adversarial attacks attempt to unearth the perturbations that maximise loss on a particular input, while keeping the size of these perturbations smaller than a threshold (ϵ). Hence, these gradient based adversarial attacks are often used to gauge the robustness of the model.

3.2 High Level Perturbations

The other aspect we seek to examine is the sensitivity of the model to perturbations of high level clinical features. This is meant to gauge if the model is actually focusing on the features that a clinician would consider important. Our hypothesis is that if one of the high level features is perturbed, then an important aspect to the signal has been changed. Therefore we expect to see a change in the output of the model, as opposed to traditional adversarial perturbations where we expect the model not to change its predictions. This change in the output should be predictable based on medical theory. For example absence of p-waves and an irregular rhythm are typically signs of *Atrial Fibrillation (AF)* hence, if we make these changes to a *Normal* ECG signal we expect the model's output would observe a considerable drop in '**Normal**' class score. If the model's response to these high level perturbation is as expected, then we can say that the model is actually focusing on these high level features. Then, we want the model to be sensitive to these high level perturbations. In order to make these specific perturbations we only use signals that are labeled 'Normal' in the dataset. We leverage the *Neurokit* signal processing library to perturb one or more of these features.

1. **P-wave removal**: The presence of P-waves are an important factor in determining the presence of Atrial fibrillation in an ECG signal. In case of Atrial Fibrillation (AF), the P-wave is either not consistently present or is absent all together. Therefore, when perturbing a normal signal by removing P-waves we would expect the 'Normal' class probability score to decrease. In order to perform this perturbation, we first identify the onsets and offsets of the P-waves and then we use linear interpolation to connect the P-wave onset and offset points, effectively flattening out the P-waves.
2. **Irregular Rhythms**: Irregular Rhythm here refers to the irregular time intervals between the R-peaks in an ECG signal. Irregular rhythm in an ECG signal can also be a good indicator for the presence of Atrial fibrillation in a signal. Therefore, again after performing this perturbation, we expect the model's 'Normal' class probability score to decrease. In order to make this perturbation, we first identified the R-peaks within the signal. Thereafter, the interval between two R-peaks is stretched or squeezed in accordance with a normal distribution.
3. **P-wave removal and Irregular Rhythms**: Absence of P-waves in conjunction with irregular rhythm is a clear indicator that Atrial fibrillation is present in an ECG signal. Hence, not only should the 'Normal' class probability score decrease, but also the model should largely classify these signals as "Atrial fibrillation". In order to achieve this perturbation, we first remove the P-waves within the signal (as described above) and then proceed to make the rhythm irregular (as described above). A sample signal before and after these perturbations is shown in Figs. 2 and 3.

3.3 Adversarial Training Against Perturbations

We found that the original model did not perform well against these low and high level perturbations. Therefore, in order to make the model more robust against low level perturbations while making the model more sensitive to high level perturbations we employed adversarial training. In our experiments, we try two types of adversarial training as explained in the following two subsections.

1. **Adversarial Training using Low level perturbations**: Adversarial training against FGM and PGD attacks is one of the most commonly used methods for adversarial training to ensure model robustness against these low level attacks. We use our base model and train it for 10 additional epochs. During this additional training, at each epoch, new adversarial samples are generated based on the samples in the training set and then these adversarial samples are used to train the model.
2. **Adversarial Training using High and Low level perturbations**: In order to train the model against both high level and low level perturbations, first we generate signals using 'Normal' signals in the training set, where the P-waves are removed and the rhythm is made irregular. These newly generated signals are now labeled as AF (Atrial fibrillation), as absence of P-waves and irregular rhythm is characteristic of an ECG signal with Atrial Fibrillation. Further, these generated signals are then added to the training set which is used for adversarially training our base model against PGD attack for an additional 10 epochs.
   In order to ensure that the adversarial samples that we are adding to the training data are clinically meaningful, we performed a human validation with the help of a senior cardiologist. In this experiment, we validated a subset of perturbed signals via a cardiologist who confirmed that for 85% of the generated data samples resemble an abnormal ECG signal.

4 Experiments

We use Physionet 2017 dataset, which includes 8528 ECG samples. Each data sample is labeled as one of the four possible classes: Normal, Atrial Fibrillation, Other and Noise. For our experiments we are ignoring the 'Noise' class since it contained only a handful of signals. Finally, we used 5154 Normal signals, 771 AF, 2557 Other rhythm and 46 Noisy signals. For training and test purposes, we have divided this data into training, validation and test sets in 80:10:10 ratio respectively.

We have used the Stanford cardiologist-level ECG classifier [14] as a base model for our experiments. This model was shown to perform better than an average cardiologist in various ECG classifications tasks. This model is a ResNet based classification model with 34 layers capable of classifying ECG of arbitrary length. Since a pre-trained version of this model was not available, we have trained this model from scratch on Physionet 2017 dataset.

4.1 Adversarial Training Against Low Level Perturbations

Since our base model is susceptible to gradient based adversarial attacks, we employed FGM adversarial training as well as PGD adversarial training which are two of the most commonly used defence against gradient based attacks. We trained the model for an additional 10 epochs for both of these methods. We set the maximum perturbation limit as 0.1 according to l_∞ norm, and for the PGD Adversarial training we set the maximum iterations that the PGD attack can make as 15. Once the model was trained, we evaluate it against FGM & PGD attacks with the same hyper-parameters.

Table 2 suggests that the PGD adversarial trained model performed much better against both PGD and FGM attacks. Therefore, we can conclude that the PGD Adversarial trained model is more robust against these adversarial attacks. Next, we also evaluate the performance of the PGD trained adversarial model against high level clinical feature perturbations. This is performed in a similar manner as we had done for the our base model against three types of high level perturbations: (1) P-wave removed, (2) Irregular Rhythm, and (3) P-wave removed and Irregular rhythm.

From Fig. 6, it is observed that the PGD adversarial trained model is more sensitive to these high level clinical features as well. Beginning, with *P-wave removal* perturbation, we observe that there is a further reduction in the '**Normal**' class score to 0.462 compared to 0.501 for the base model. This suggests that the model has become more sensitive to this particular perturbation. Next, for *Irregular Rhythm* perturbation again there is a slight decrease in the 'Normal' class score to 0.32 compared to 0.361 for the base model, indicating slight increase in the sensitivity to this type of perturbation as well. However, for *P-wave removed and Irregular Rhythm* perturbation, we observe that there is slight increase in the '**Normal**' class score, 0.27 for the current model as opposed to 0.224 for the base model. This suggests there is no improvement in the sensitivity to this particular perturbation.

4.2 Adversarial Training Against High and Low Level Perturbations

We would like to make the model more sensitive towards high level clinical features. To achieve this, we expand our training set to include these high level feature perturbations, similar to adversarial training. [22] have documented that adding generated data along with adversarial training leads to a more robust model. In our case, we have used the 'Normal' samples form our training set to generate new samples by performing the p-wave removal and irregular rhythm perturbation and labeling them as 'AF', (described in Sect. 3). Once the model is trained we report its performance against the test set as well as PGD and FGM attacks. We also compare the performance of the newly trained model with other models, as seen in Table 2.

From Table 2, we observe that the newly trained model (PGD Adv. Trained and High level trained model) is slightly more robust against FGM & PGD attacks when compared against only PGD Adversarial Trained model. While maintaining similar F1-score against original test set. Therefore, this experiment shows that the newly trained model is more robust than the other models, without sacrificing the accuracy on the original test dataset.

Table 2 Comparing robustness of models using F1-score

	Original testset	FGM attack ($\epsilon = 0.1$)	PGD attack ($\epsilon = 0.1$)
Base model	0.809	0.343	0.321
FGM adv. trained	**0.868**	0.649	0.707
PGD adv. trained	0.844	0.751	0.802
PGD Adv. and high-level trained	0.855	**0.785**	**0.803**

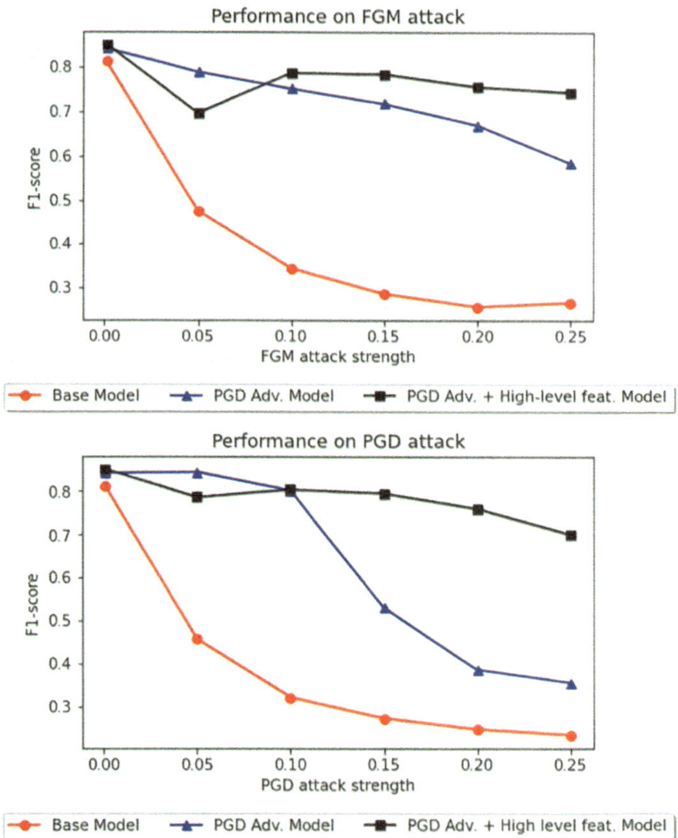

Fig. 5 Performance of the models against adversarial attacks

In order to confirm if the model trained with against high level clinical feature perturbations as well as PGD adversarial attack is more robust than the other two models, we subjected each of these models to PGD and FGM attacks of varying strength ranging from $\epsilon = 0.001$ to $\epsilon = 0.25$, the F1-score obtained on by each model against these attacks is plotted in Fig. 5. Here, its observed that as the attack strength increases the F1-score of our base model falls sharply for both PGD and FGM attacks. The F1-score of the PGD trained model also shows a significant drop, however the performance of this model is much better than our base model's performance. It is observed that the model trained against both high level perturbations and PGD attack performances significantly better against these attacks, there is only a slight drop in the F1-scores. Therefore, we can say that the adversarial training against high level feature perturbations along with PGD adversarial training has yielded a more robust model.

Next, we also examine the sensitivity of the newly trained model to the high level feature perturbations. This is done in the same manner that we had done for the other two models. In order to compare the sensitivity of the three different models to high level feature perturbations, the average 'Normal' class score for each of the models against these high level feature perturbations are plotted in Fig. 6. We can observe that the newly trained model is even more sensitive to these high level feature perturbations. We see that for p-wave removal perturbation the average 'Normal' class score has decreased even more when compared to the PGD adv. trained model.

Similarly, the average 'Normal' class score for irregular rhythm perturbation also decreases further when compared to PGD adv. trained model. And, finally we also see that for p-wave removed and irregular rhythm perturbation, the average 'Normal' class score also decreases to the lowest value of 0.039. Therefore, from these set of experiments we can conclude that the new trained model which used high level perturbed data as well as PGD adversarial training is not only more robust but also more sensitive to high level perturbations.

4.3 Evaluating Robustness to Low Level Perturbations

We use FGM and PGD adversarial attacks to add small perturbations to assess the robustness of our base model. A robust model should be able to achieve high accuracy even when these adversarial attacks are employed. In this experiment, we employed FGM and PGD attacks and we report the F1- score for (AF, Normal and Other class) for each attack as well as on the original test set. The overall F1 scores were calculated by taking the average of the F1 score for each of the three classes. We have used the same test set for each of the attacks.

From Table 2, we observe that the model was performing very well on the original test set and was able to achieve an F1-score of 0.80. But, as soon as we employ gradient based adversarial attacks there is a significant drop in the performance of the model. F1-score after FGM attack reaches 0.34. Similarly, the F1-score drops

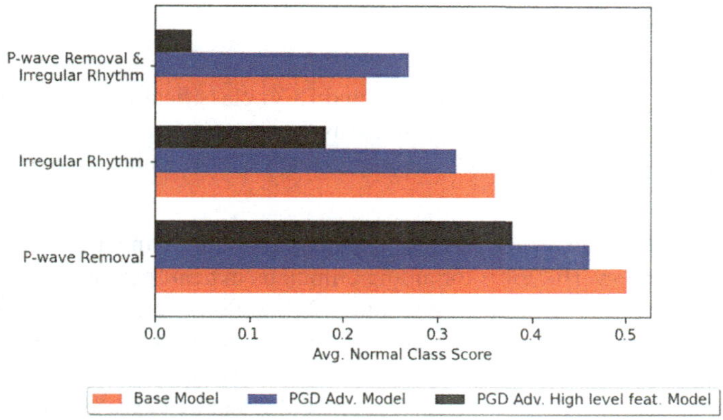

Fig. 6 Sensitivity of models to high level perturbations

to 0.32 after PGD attack. This again highlights that the model is not robust against these low level perturbations.

4.4 Evaluating Sensitivity to High Level Perturbations

In order to assess the sensitivity of our base model to high level clinical features, we generated 250 signals for each type of perturbation, from 'Normal' signals and then we report the probability score for each class averaged over these 250 samples. We do this for both the original set of signals as well as the perturbed signals. We examine how the model classification changes as each feature is perturbed (See Fig. 6).

From this experiment, we can observe that our base model is not very sensitive to the high level feature perturbations. This is evident in the p-wave removal perturbation. We can clearly see that even after removing all the p-waves in the signals, a lot of the signals are still being classified as normal. Further, the model is slightly more sensitive to the irregular rhythm perturbations, but the classification the 'Normal' class score is still quite high. Finally for P-waves removed and irregular rhythm perturbation, it is observed that the 'Normal' class score is 0.224. This suggests that the model is more sensitive to this perturbation, when compared to the other perturbations. This experiment shows that the base model is actually not very sensitive to these high level clinical feature perturbations.

5 Related Work

While a host of prior techniques exist around ECG classification, very few of these have looked at adversarial robustness and sensitivity to clinical features. This section highlights past contributions to ECG classification, Model Robustness and Explainability.

1. **ECG classification**: Machine learning algorithms, including deep learning, have proved to be powerful tools for aiding clinicians in heart patient screening using ECG data [5, 15, 19]. In a recent study [27], authors generate clean and noisy versions of an ECG dataset before applying various systematic image transformations to the signal. A convolutional neural network is used to classify these image transforms. They highlight that physiological ECG noise impacts classification using deep learning methods and careful consideration should be given to the inclusion of noisy ECG signals in the training data when developing supervised networks for ECG classification. Our work can address some of these issues by giving a better estimate of robustness under noisy conditions.
2. **Robustness in Deep Learning**: Reference [25] first propose the concept of adversarial examples that can mislead deep neural networks with small malicious perturbations. Besides generating adversarial examples to attack models [11], existing studies also concern improving the adversarial robustness and generalization ability of neural networks via adversarial training [7, 23, 26]. Adversarial training is widely adopted in both the computer vision field [18, 29] and the natural language processing field [17, 30]. However there have been only a handful of papers related to adversarial attacks and adversarial training for ECG classification [13, 24].
   Deep learning networks have achieved SOTA performance on ECG classification. These models employ various architectures which include CNN based [1, 14], LSTM based architectures [8, 28] and combinations thereof [4]. Some approaches have even used hand crafted features along with deep learning models [10]. A CNN based ECG classification model developed by [14] was even shown to out perform cardiologists for detection of a wide range of heart arrhythmias from single-lead ECG records. However, the susceptibility of the deep learning models to various adversarial attacks have raised a lot of concern over the development and deployment of such deep learning models in the medical field [2, 9]. Further, CNN based ECG classification models have also been shown to be vulnerable against adversarial attacks [13]. They have formulated a type of adversarial attack which constructs smoothed adversarial examples that are invisible to a human expert. However, to the best of our knowledge, prior work has not examined the performance of ECG Classification model against high level feature perturbations, which is what we seek to explore in this paper.
3. **Robustness & Explainability using Clinical features**:
   Despite some work on visualizing high level features by using the weight filters in a CNN [6, 21], most researchers use deep learning approaches as a black box without the possibility to explain results or without the ability to apply mod-

ifications in case of misclassification. In general, automatic feature extraction from noisy ECG signals requires large scale clinical feature annotations which can be used for training a machine learning network. Obtaining consistent large scale clinical annotations for the purpose of training and explainability is highly impractical and expensive for each use case. We first leverage traditional feature extraction techniques for obtaining higher level features (conceptual/clinical features) from normal ECG signals and then exploit the for improving model robustness and explainability. While abnormal ECG signals can exhibit significant variation in the feature shapes and motifs, for normal, healthy patients the ECG signals follow a very predictable pattern. Thus, we focus on ECG signals from normal patients to extract high level clinical features.

6 Conclusions and Future Work

In this work, we looked at the sensitivity of a SOTA deep learning model to high level clinical feature perturbations. Our experiments indicate that current deep learning models are susceptible to low level adversarial attacks and are not sensitive to high level clinical features. To the best of our knowledge, this is the first instance of work where a deep model is adversarially trained, to ensure sensitivity to clinical features. From an explainability perspective, it is important to understand, if the model is actually focusing on the correct clinical features. We intend to further explore this as part of our future work.

References

1. U.R. Acharya, S.L. Oh, Y. Hagiwara, J.H. Tan, M. Adam, A. Gertych, R.S. Tan, A deep convolutional neural network model to classify heartbeats. Comput. Biol. Med. **89**, 389–396 (2017)
2. G. Bortsova, C. González-Gonzalo, S.C. Wetstein, F. Dubost, I. Katramados, L. Hogeweg, B. Liefers, B. van Ginneken, J.P. Pluim, M. Veta, C.I. Sánchez, M. de Bruijne, Adversarial attack vulnerability of medical image analysis systems: unexplored factors. Med. Image Anal. **73**, 102141 (2021). https://www.sciencedirect.com/science/article/pii/S1361841521001870
3. I. Bratko, I. Mozetič, N. Lavrač, *KARDIO: A Study in Deep and Qualitative Knowledge for Expert Systems* (MIT Press, Cambridge, MA, USA, 1990)
4. C. Chen, Z. Hua, R. Zhang, G. Liu, W. Wen, Automated arrhythmia classification based on a combination network of CNN and LSTM. Biomed. Signal Process. Control **57**, 101819 (2020)
5. F.M. Dias, H.L. Monteiro, T.W. Cabral, R. Naji, M. Kuehni, E.J.D.S. Luz, Arrhythmia classification from single-lead ECG signals using the inter-patient paradigm. Comput. Methods Programs Biomed.**202**, 105948 (2021). https://doi.org/10.1016/j.cmpb.2021.105948, https://www.sciencedirect.com/science/article/pii/S0169260721000225
6. D. Erhan, Y. Bengio, A. Courville, P. Vincent, Visualizing higher-layer features of a deep network. Technical Report, Univeristé de Montréal (2009)
7. C. Etmann, S. Lunz, P. Maass, C. Schoenlieb, On the connection between adversarial robustness and saliency map interpretability, in *Proceedings of the 36th International Conference*

on Machine Learning, Proceedings of Machine Learning Research ed. by K. Chaudhuri, R. Salakhutdinov, vol. 97 (PMLR, 2019), pp. 1823–1832. https://proceedings.mlr.press/v97/etmann19a.html
8. O. Faust, A. Shenfield, M. Kareem, T.R. San, H. Fujita, U.R. Acharya, Automated detection of atrial fibrillation using long short-term memory network with RR interval signals. Comput. Biol. Med. **102**, 327–335 (2018)
9. S.G. Finlayson, J.D. Bowers, J. Ito, J.L. Zittrain, A.L. Beam, I.S. Kohane, Adversarial attacks on medical machine learning. Science **363**(6433), 1287–1289 (2019) https://doi.org/10.1126/science.aaw4399, https://www.science.org/doi/abs/10.1126/science.aaw4399
10. Z. Golrizkhatami, A. Acan, ECG classification using three-level fusion of different feature descriptors. Expert Syst. Appl. **114**, 54–64 (2018). https://doi.org/10.1016/j.eswa.2018.07.030, https://www.sciencedirect.com/science/article/pii/S0957417418304469
11. I.J. Goodfellow, J. Shlens, C. Szegedy, Explaining and harnessing adversarial examples (2014). https://doi.org/10.48550/ARXIV.1412.6572, arXiv:1412.6572
12. I.J. Goodfellow, J. Shlens, C. Szegedy, Explaining and harnessing adversarial examples. CoRR (2015). arXiv:1412.6572
13. X. Han, Y. Hu, L. Foschini, L. Chinitz, L. Jankelson, R. Ranganath, Deep learning models for electrocardiograms are susceptible to adversarial attack. Nat. Med. **26**(3), 360–363 (2020) https://doi.org/10.1038/s41591-020-0791-x
14. A.Y. Hannun, P. Rajpurkar, M. Haghpanahi, G.H. Tison, C. Bourn, M.P. Turakhia, A.Y. Ng, Cardiologist-level arrhythmia detection and classification in ambulatory electrocardiograms using a deep neural network. Nat. Med. **25**(1), 65–69 (2019) https://doi.org/10.1038/s41591-018-0268-3
15. S.H. Jambukia, V.K. Dabhi, H.B. Prajapati, Classification of ECG signals using machine learning techniques: a survey, in *2015 International Conference on Advances in Computer Engineering and Applications*, pp. 714–721 (2015). https://doi.org/10.1109/ICACEA.2015.7164783
16. N. Lavrac, I. Bratko, I. Mozetič, B. Čerček, A. Grad, M. Horvat, KAEDIO-E-an expert system for electrocardiographic diagnosis of cardiac arrhythmias. Expert. Syst. **2**, 46–55 (2007). https://doi.org/10.1111/j.1468-0394.1985.tb00449.x
17. X. Liu, H. Cheng, P. He, W. Chen, Y. Wang, H. Poon, J. Gao, Adversarial training for large neural language models (2020). https://doi.org/10.48550/ARXIV.2004.08994, arXiv:2004.08994
18. A. Madry, A. Makelov, L. Schmidt, D. Tsipras, A. Vladu, Towards deep learning models resistant to adversarial attacks (2017). https://doi.org/10.48550/ARXIV.1706.06083, arXiv:1706.06083
19. A. Mincholé, J. Camps, A. Lyon, B. Rodríguez, Machine learning in the electrocardiogram. J. Electrocardiol. **57**, S61–S64 (2019) https://doi.org/10.1016/j.jelectrocard.2019.08.008, https://www.sciencedirect.com/science/article/pii/S0022073619304571
20. T. Nisia, S. Rajesh, Extraction of high-level and low-level feature for classification of image using Ridgelet and CNN based image classification. J. Phys.: Conf. Ser. **1911**, 012019 (2021). https://doi.org/10.1088/1742-6596/1911/1/012019
21. Z. Qin, F. Yu, C. Liu, X. Chen, How convolutional neural network see the world—a survey of convolutional neural network visualization methods. Math. Found. Comput. **1**, 149–180 (2018)
22. S.A. Rebuffi, S. Gowal, D.A. Calian, F. Stimberg, O. Wiles, T. Mann, Fixing data augmentation to improve adversarial robustness (2021). https://doi.org/10.48550/ARXIV.2103.01946, arXiv:2103.01946
23. A.S. Ros, F. Doshi-Velez, Improving the adversarial robustness and interpretability of deep neural networks by regularizing their input gradients, in *Proceedings of the Thirty-Second AAAI Conference on Artificial Intelligence and Thirtieth Innovative Applications of Artificial Intelligence Conference and Eighth AAAI Symposium on Educational Advances in Artificial Intelligence, AAAI'18/IAAI'18/EAAI'18* (AAAI Press, 2018)
24. J. Shao, S., Geng, Z. Fu, W. Xu, T. Liu, S. Hong, Defending against adversarial attack in ecg classification with adversarial distillation training (2022). https://doi.org/10.48550/ARXIV.2203.09487, arXiv:2203.09487

25. C. Szegedy, W. Zaremba, I. Sutskever, J. Bruna, D. Erhan, I. Goodfellow, R. Fergus, Intriguing properties of neural networks, in *International Conference on Learning Representations* (2014). arXiv:1312.6199
26. D. Tsipras, S. Santurkar, L. Engstrom, A. Turner, A. Madry, Robustness may be at odds with accuracy (2018). https://doi.org/10.48550/ARXIV.1805.12152. arXiv:1805.12152
27. J. Venton, P.M. Harris, A. Sundar, N.A. Smith, P.J. Aston, Robustness of convolutional neural networks to physiological electrocardiogram noise. Phil. Trans. R. Soc. A **379**(2212), 20200262 (2021)
28. Ö. Yildirim, A novel wavelet sequence based on deep bidirectional LSTM network model for ECG signal classification. Comput. Biol. Med. **96**, 189–202 (2018) https://doi.org/10.1016/j.compbiomed.2018.03.016. https://www.sciencedirect.com/science/article/pii/S0010482518300738
29. H. Zhang, Y. Yu, J. Jiao, E. Xing, L.E. Ghaoui, M. Jordan, Theoretically principled trade-off between robustness and accuracy, in *Proceedings of the 36th International Conference on Machine Learning, Proceedings of Machine Learning Research*, ed. by K. Chaudhuri, R. Salakhutdinov, vol. 97 (PMLR, 2019), pp. 7472–7482. https://proceedings.mlr.press/v97/zhang19p.html
30. W.E. Zhang, Q.Z. Sheng, A. Alhazmi, C. Li, Adversarial attacks on deep-learning models in natural language processing: a survey. ACM Trans. Intell. Syst. Technol. **11**(3) (2020). https://doi.org/10.1145/3374217

A Transformer-Based Deep Learning Algorithm to Auto-Record Undocumented Clinical One-Lung Ventilation Events

Zhihua Li, Alexander Nagrebetsky, Sylvia Ranjeva, Nan Bi, Dianbo Liu, Marcos F. Vidal Melo, Timothy Houle, Lijun Yin, and Hao Deng

Abstract As a team studying the predictors of complications after lung surgery, we have encountered high missingness of data on one-lung ventilation (OLV) start and end times due to high clinical workload and cognitive overload during surgery. Such missing data limit the precision and clinical applicability of our findings. We hypothesized that available intraoperative mechanical ventilation and physiological time-series data combined with other clinical events could be used to accurately predict missing start and end times of OLV. Such a predictive model can recover existing miss-documented records and relieves the documentation burden by deploying it in clinical settings. To this end, we develop a deep learning model to predict the

---

Z. Li (✉) · N. Bi · L. Yin
Binghamton University, Binghamton, NY, USA
e-mail: zli191@binghamton.edu

N. Bi
e-mail: nbi1@binghamton.edu

L. Yin
e-mail: lijun@cs.binghamton.edu

A. Nagrebetsky · S. Ranjeva · T. Houle · H. Deng
Massachusetts General Hospital, Boston, MA, USA
e-mail: anagrebetsky@mgh.harvard.edu

S. Ranjeva
e-mail: sranjeva@mgh.harvard.edu

T. Houle
e-mail: thoule1@mgh.harvard.edu

H. Deng
e-mail: hdeng1@mgh.harvard.edu

D. Liu
Mila AI Institute, Quebec, Canada
e-mail: dianbo.liu@mila.quebec

M. F. Vidal Melo
Columbia University Irving Medical Center, New York City, NY, USA
e-mail: mv2869@cumc.columbia.edu

A. Shaban-Nejad et al. (eds.), *Artificial Intelligence for Personalized Medicine*,
Studies in Computational Intelligence 1106,
https://doi.org/10.1007/978-3-031-36938-4_19

occurrence and timing of OLV based on routinely collected intraoperative data. Our approach combines the variables' spatial and frequency domain features, using Transformer encoders to model the temporal evolution and convolutional neural network to abstract frequency-of-interest from wavelet spectrum images. The performance of the proposed method is evaluated on a benchmark dataset curated from Massachusetts General Hospital (MGH) and Brigham and Women's Hospital (BWH). Experiments show our approach outperforms baseline methods significantly and produces a satisfactory accuracy for clinical use.

Keywords One-lung-ventilation · Transformer · Medical records · Deep learning

1 Introduction

Among two million people diagnosed with lung cancer each year [4], approximately one-third need lung resection surgery [21]. An operation on the lung requires one-lung ventilation (OLV) to deflate and immobilize the operative lung for surgical visualization. OLV, in turn, presents unique challenges for mechanical ventilation and for prevention of postoperative pulmonary complications [17]. Transition to OLV during thoracic surgery is a distinct risk factor for post-operative acute lung injury, ranging in severity from mild atelectasis to severe acute respiratory distress syndrome (ARDS) [15]. Strategies for lung-protective management during two-lung ventilation have evolved from studies of ARDS in ICU populations [1, 19]. These protective ventilation strategies aim to provide sufficient oxygenation while minimizing ventilator-induced alveolar trauma, inflammation, and cyclic collapse [23]. While lung-protective strategies for two-lung ventilation are well-described, patients receiving OLV during lung resection, a cohort that is inherently vulnerable to pulmonary complications, suffer from a paucity of clinically meaningful evidence.

The lack of reliable data on the time of transition from two-lung ventilation to OLV is a limiting factor that complicates research into the pathophysiology and prevention of pulmonary complications after lung surgery. The transition from two- to one-lung ventilation is a time of heightened risk for respiratory decompensation, and is thus a time of high cognitive and procedural burden for the anesthesia provider. Therefore, the manually entered documentation of this transition may not be timed correctly and is often missing (as shown in Fig. 1). At the same time, multiple streams of physiological data that are recorded automatically during the start and end of OLV, make it possible to accurately impute the occurrence and timing of OLV. For example, airway pressures, volumes of delivered breaths, and respiratory rate change within seconds after the start and end of OLV. Other physiological metrics such as heart rate, blood pressure, exhaled CO_2, and hemoglobin oxygen saturation (SPO_2) may also change in response to the start and end of OLV. The missing or incorrectly timed documentation of OLV illustrates a common clinical scenario where the need for event documentation in the procedural or emergency care settings competes for clinicians' attention with patient care tasks. Furthermore, the need to recall and document

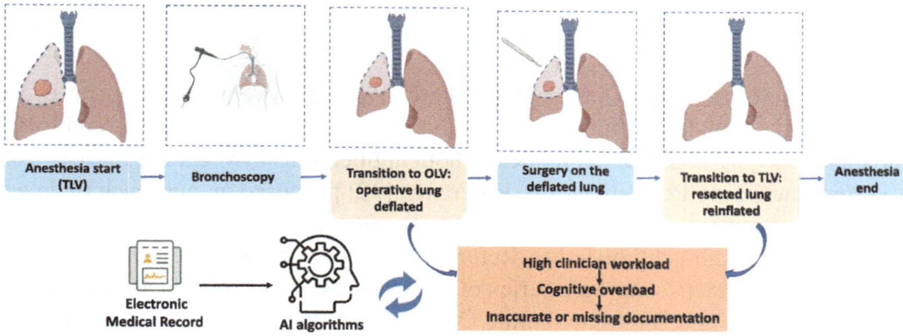

Fig. 1 Intraoperative sequence of events during lung resection with one-lung ventilation. TLV: two-lung ventilation; OLV: one-lung ventilation

clinical events contributes to the cognitive overload of clinicians and can result in burnout [10]. The cognitive burden of clinical documentation may be alleviated by a machine learning algorithm that automatically detects the occurrence and timing of clinical events of interest. If used in real-time or near-real time, such an algorithm can redirect a clinician's attention toward patient care. In retrospective analysis of data, it can aid in clinical quality control and imputing missing data for research.

Given large-scale documented historic medical records, we hypothesized that a data-driven machine learning model could estimate the occurrence and timing of clinical events using routinely available physiological settings and measurement data. We aimed to test this hypothesis by developing a deep learning model and testing its ability to detect the occurrence and timing of OLV start and end in a dataset of patients undergoing lung resection surgery. Specifically, the OLV timestamp prediction can be formulated as a time-series detection task, which takes multiple sequences of 1-D time-series signals as the inputs and outputs the target timestamp of the occurring event. Recognizing a specific segment of the waveform is the key to locating the desired timestamps.

This paper proposes a Transformer-based deep learning framework for OLV timestamp detection. Apart from spatial features, the 1-D signals are transformed into wavelet spectrum images and fed into customized convolutional neural networks to extract the discriminative frequency information. Furthermore, an innovative label temporal smoothing technique is proposed to optimize the procedure to locate the maximum scores from the prediction curves. Experimental results show that our proposed method significantly outperforms the basic time-series change-point detection methods and the recurrent-based deep learning methods.

2 Related Works

The objective of this work is to detect if there is an OLV event occurring for every minute of the signals. In the area of time-series analysis, Twitter [25] employed statistical learning to detect anomalies in both applications (Tweets Per Sec) as well as system metrics (CPU utilization). Reference [22] proposed an approach to detect outliers in streaming univariate time series based on Extreme Value Theory that did not require hand-set thresholds. Reference [7] developed a time-series anomaly detection toolkit by packaging a series of statistic-based methods such as CUSUM (cumulative sum) and Bayesian Online Change Point Detection. However, they are not suitable for the OLV detection tasks because there are no observable change points near the OLV actions for most of the variables. Additionally, some signals have many change points that are not related to the OLV procedure; thus, many false positives would be generated.

Traditional hand-crafted features are expensive to create and require expert knowledge of the field. The performances of traditional statistical models are not satisfactory in real applications. Recently, deep learning approaches have shown superior power in big data analysis with successful applications to computer vision, pattern recognition, and natural language processing [14]. Researchers are investigating data-driven models to improve anomaly detection accuracy [9].

Opprentice [13] used existing detectors to extract anomaly features and fed them to a random forest classifier to automatically select the appropriate detector-parameter combinations and the thresholds. Reference [28] proposed Donut, an unsupervised anomaly detection algorithm based on Variational Auto-Encoder (VAE), making it the first generative-based anomaly detection algorithm. The reconstruction probability was used as an anomaly indicator. Reference [5] employed a contrastive learning strategy for change point detection by learning an embedded representation through self-supervision. Reference [27] proposed an autoencoder-based deep learning network to learn physiological features and use multivariate Gaussian distribution anomaly detection method to detect anomaly data. Similarly, [20] proposed a LSTM-based encoder-decoder network to construct a predicted multivariate "normal" time series and used the reconstruction error for prediction.

The existing methods are mostly unsupervised and based on statistical stationarity assumptions. Therefore, they failed to handle the more complicated scenarios for OLV timestamp estimation and capture the correlations between different variables addressed in this paper. Another barrier using existing methods is that they are incapable of using multiple input signals in a complementary way. The widely used ensemble method that combines predictions of each signal is ineffective since some variables themselves are not discriminative enough for prediction, and they can only provide complementary information for other dominant variables. To handle the above challenges, we design an innovative Transformer-based model that absorbs multivariate time-series signals and predicts the timestamps of an OLV event under supervision. The proposed method enables direct communications from signal to signal; it also provides direct temporal communications in a self-attention manner by introducing Transformer encoders.

3 Data Management

Study Design: Our study followed a retrospective cohort design. The Mass General Brigham (MGB) Institutional review board (IRB) committee had reviewed the research protocol and exempted the requirement of individual informed consent due to consideration of feasibility and minimal risks to study human subjects (Protocol ID: #2021P002173).

Inclusion and exclusion criteria: Inclusion criteria: (1) patients who were 18 years or older at the time of surgery; (2) lung resection with one lung ventilation; (3) admitted to study site on or after June 15th 2016 and discharged prior to or on June 15th 2021; Exclusion criteria: (1) age less than 18 years old; (2) pregnancy; (3) intra-operative death. (4) multiple hospital encounters: (5) multiple OLV-included surgical procedures performed within encounter: (6) multiple OLV episodes in surgery: (7) no OLV timestamp data.

Data Source: Our study team retrospectively extracted Electronic Medical Records (EMR) information from the MGB central data repository named Enterprise Data Warehouse (EDW) utilizing structured SQL queries. We then constructed and maintained a clinical and observational database of all adult patients who received open thoracic surgeries at both Brigham and Women's Hospital (BWH) and Massachusetts General Hospital (MGH) from 2016 to 2021 for this research project. There were no a priori power analyses performed to determine the required sample size due to the innovative model structure. We made all extracted EMR records available for analyses, model development, and validation. The curated dataset consists of 4245 patient admission records.

Predictors and Features: OLV can lead to immediate changes of value in certain physiological measures such as gas exchange, ventilation mechanics, and hemodynamics. Therefore, modeling the patterns of their changes can be utilized to predict the OLV event. Specifically, we consider variables/physiological measurements that could show an indication of OLV status from the signal shape. We divide the variables into two groups setting values and physiological measurements from biomedical sensors. The former includes (1) VT (tidal volume) set: target volume for each breath, (2) respiration set. The latter contains (1) peak airway pressure: maximum pressure during a breath, (2) measured respiration rate, (3) SPO_2, and (4) VT exhaled. Detailed variable descriptions are shown in Table 1.

Missing Data: Our study collected multi-subject multivariate time-series data, and missingness could occur at any feature throughout the entire observational period of surgical procedures. We assumed that the missing mechanism was under a common assumption of Missing At Random (MAR), and imputed the within-sequence missing values for each feature using the classic linear interpolation method. We used the nearest value padding method to fill in the missing head and tail values of different feature sequences to achieve the same fixed length of time series data for followed modeling steps (Fig. 2).

Table 1 Variable descriptions. The variables are categorized into ventilator setting variables and physiology measurement variables

Ventilator setting variables	Tidal volume (VTset)	A mechanical ventilator setting which determines the volume goal of each ventilator-delivered breath
	Respiratory rate (RRset)	A mechanical ventilator setting which determines the number of ventilator-delivered breaths per minute
	Positive end-expiratory pressure (PEEP)	a mechanical ventilator setting which determines the lowest airway pressure after each ventilator-delivered breath
Respiratory physiology variables	Exhaled tidal volume (exhaledVT)	The measured gas volume returning from the patient's lungs to the mechanical ventilator after each breath
	Respiratory rate (RR)	Measured number of ventilator-delivered breaths per minute, based on the cyclical variations in composition of gas mixture returning to mechanical ventilator
	Peak inspiratory pressure (PIP)	Maximum pressure in the airway during each breath
	Hemoglobin oxygen saturation (SPO_2)	The measured % of hemoglobin that is saturated with oxygen which indicated how well oxygen is delivered to blood from the lungs

4 Method

At a high level, our deep learning model consists of two spatial and frequency branches. The former is primary, and the latter acts as a supplementary. For the spatial feature extraction, our method first transforms the variable intensity of each minute to high dimensional feature embeddings, which improves the representation capability. Furthermore, several transformer encoder blocks are stacked to enable spatial-temporal fusion, where the inter-signal relationships are also exploited. In terms of the frequency branch, we apply the wavelet transform to each signal and obtain 2D spectrum images. In addition, a Gaussian smoothing operation is performed on the ground-truths in order to produce more continuous predictions.

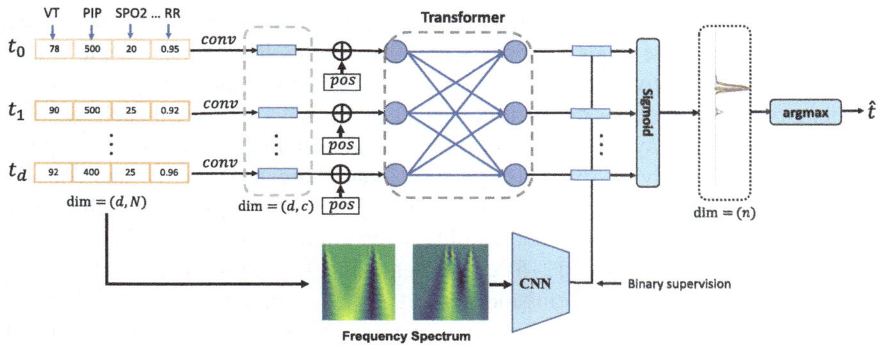

Fig. 2 Framework overview. The upper branch leverages signal morphological characteristics, and the lower branch exploits the frequency spectrum from wavelet transform. First, the multivariate signals are transformed into feature embeddings and fed into the transformer block. We use the combined features from spatial and frequency for the last layer detection. A label smoothing technique is deployed to generate the supervision signals. The timestamp corresponding to the highest OLV occurrence probability is selected as the predicted timestamp. Detailed training and testing steps are shown in Algorithm 1

4.1 Problem Definition

The goal of the OLV documentation system is to automatically predict the start and end timestamp of OLV during lung surgery. Formally, the dataset P with M number of patients is represented by $P = \{p_1, p_2, \ldots, p_M\}$. Each patient p_m consists of N variables, and $p_m = \{s_0, s_1, s_2, \ldots, s_{N-1}\}$, where $s_i \in \mathbb{R}^d$, and d is the sequence length (in minute) after pre-processing. This paper proposes a deep learning method that attempts to learn a function $\phi(\cdot)$ that absorbs p and predicts the OLV start and end timestamp y_s and y_e, respectively, which can be described as $\widehat{y} = \phi(p)$, where $\widehat{y}$ is an integer and $0 \leq y \leq d$.

4.2 Variable Embeddings

In practice, the operation time (from surgery start time to surgery end time) varies significantly from patient to patient, causing the patient-wise signal length to unfixed. Feeding an arbitrary length of data to machine learning models needs additional padding steps and brings about more complexity in model building. Therefore, we propose to segment the entire sequence into fixed-length windows of length l_{ws}, namely a sliding window method with a certain step size l_{step}. Specifically, for each training iteration, we randomly sample a start timestamp t where $t < l - l_{ws}$, then select the values from time t to $t + l_{ws}$ as the input to the deep learning model. Since there exist multiple synchronized signal recordings for each patient, we concatenate all the N signals into a channel dimension, resulting in an input $\mathbf{x} \in \mathbb{R}^{N \times l_{ws}}$.

In general, among the N recorded physiological signals, some signals (e.g., tidal volume) provide more clues to address the OLV events than others, while it also happens that informative signals are noisy and the other signals could provide complementary information for prediction. Hence, we decide to represent each signal in a multidimensional way in order to improve the representation capability and capture more complex underlying relationships and signal characteristics. To this end, we employ three linear layers to transform the raw signal recordings $\mathbf{x}$ into high dimensional embeddings, and the embedding dimension gradually increases from N to 64, and then to 512, and another 512–512. After encoding the low-dimension raw signal intensities to high-dimension embeddings $\mathbf{f} \in \mathbb{R}^{512 \times l_{ws}}$, the densely distributed representation can better represent the signal patterns. Next, the feature embeddings $\mathbf{f}$ are ready for sequential information extraction.

4.3 *Transformer Blocks for Temporal Learning*

To capture the temporal patterns existing around the OLV events, introducing a sequential machine learning model is demanded. Although the recurrent neural networks (RNNs) such as long short-term memory (LSTM) [12] could handle the sequential input, the gating and recurrent steps have shortcomings in modeling long-distance dependencies and could get trapped into gradient vanishing problems. Instead, we deploy a self-attention-based transformer encoder framework. It enables direct interactions among each value in a sequence, thus overcoming the limitations of RNNs and their variants.

The Transformer encoder [6, 26] is composed of multiple identical sub-blocks, and each block consists of a multi-head self-attention module and a fully connected feed-forward layer. Each sublayer is also succeeded by a normalization layer and residual layer. Since the transformer blocks are not aware of the input order, we add a learnable positional embedding to the input similar to BERT [6]. And the last layer output from the network can be described as:

$$\mathbf{f} = \mathrm{MLP}\left(\mathrm{Transformer}\left(\mathbf{f} + \mathbf{x}_{pos}\right)\right) \quad (1)$$

where $\mathbf{x}_{pos} \in \mathbb{R}^{l_{ws} \times 512}$ is the positional embedding. The final features are obtained by applying a fully connected layer to the output of the Transformer.

4.4 *Frequency Domain Features*

In the previous sections, the methods we present are mainly for spatial feature extraction, where we focus on morphological patterns. On the other hand, the change in the spatial domain, namely the frequency domain feature, is a more subject-invariant cue for time-series event detection. For example, the power spectrum is a solid fea-

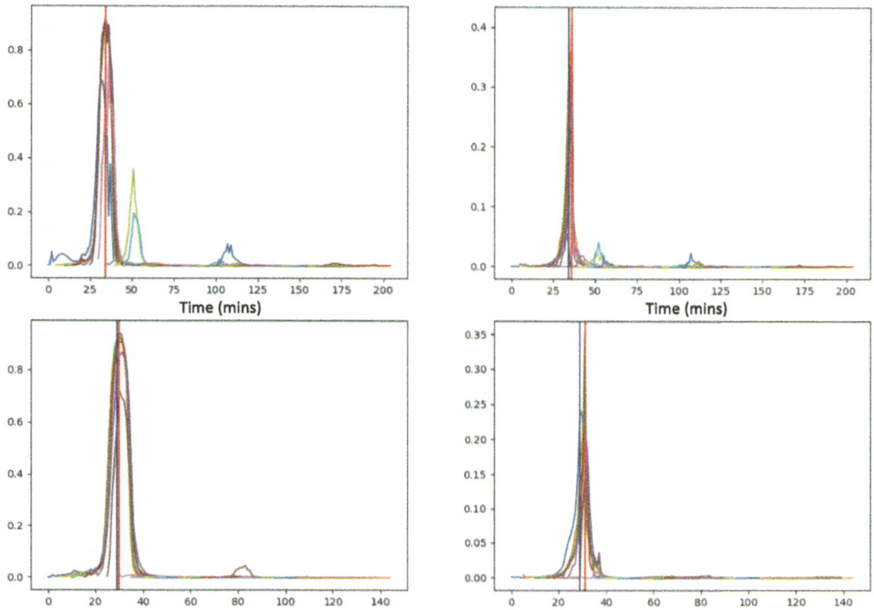

Fig. 3 Scores predicted by the model with (left column) and without (right column) label temporal smoothing. Different color of the curves denotes a different sliding window. Vertical blue line: ground truth timestamps; Vertical red lines: predicted timestamps by finding the maximum

ture representation for electrocardiogram (ECG) and electroencephalography (EEG) [2, 3] for classification. Since the temporal resolution is crucial for timestamp prediction, we adopt the wavelet spectrum [18] 2D image of the input variables as the initial frequency data. By concatenating the N wavelet spectrum images, which are calculated from N input signals, we obtain multi-channel 2D feature maps of size $\mathbb{R}^{100\times 100}$. Then, it's straightforward to utilize convolutional neural network (CNN) to extract prediction-related information from the 2D spectrum image. Specifically, we deploy three layers of 2D convolutions together with BatchNorm and non-linear ReLU operations. The abstracted frequency features $\mathbf{f}_{freq} \in \mathbb{R}^{64}$ are further concatenated with the spatial features f in Eq. 1 to detect the presence of an OLV event. We describe the calculation as:

$$\hat{y} = \sigma\left(\text{CLS}\left(\mathbf{f};\mathbf{f}_{freq}\right)\right) \tag{2}$$

where σ denotes the Sigmoid operation, and ; is feature concatenation. The output logits $\hat{y} \in \mathbb{R}^{l_{ws}}$ are obtained by applying a fully connected classification layer (CLS).

Algorithm 1 Pseudo code for the proposed model

Training stage:

1: data_loader samples batches of training records
2: **for** variables in batch **do**
3: $t \leftarrow random()$
4: $\mathbf{x}_{signal} \leftarrow$ variables$[t : t + l_{ws}]$
5: $\mathbf{x}_{freq} \leftarrow$ wavelet$(\mathbf{x}_{signal})$
6: $\mathbf{f} \leftarrow$ Transformer$(\mathbf{x}_{signal})$
7: $\mathbf{f}_{freq} \leftarrow$ CNN$(\mathbf{x}_{freq})$
8: $\hat{y} \leftarrow$ MLP$(f;\, f_{greq})$
9: init_label = zeros[0 : ws]
10: **if** $t < t_{OLV} < t + l_{ws}$ **then**
11: init_label$[t_{OLV} - t] \leftarrow 1$
12: init_label $\leftarrow$ Gaussian(init_label)
13: **end if**
14: Update the model by cross_entropy(init_label, $\hat{y}$)
15: **end for**

Testing stage:

1: data_loader samples batches of testing records
2: **for** variables in batch **do**
3: **for** $i = 1$ to n_{seg} **do** ▷ calculated by Equation 4
4: $\mathbf{x}_{signal} \leftarrow$ variables$[i : i \times ws]$
5: Repeat step 4 $\sim$ 8 in the training stage.
6: Obtain $\hat{y}^{i}_{seg}$
7: **end for**
8: $\hat{y} = \text{argmax}([\hat{y}^{0}_{seg};\, \hat{y}^{1}_{seg}, ..., \hat{y}^{n_{seg}-1}_{seg}])$
9: **end for**

4.5 Label Smoothing

Recall that after the entire sequence is split into segments, we need to prepare the ground-truths specifically for each segment. First, if the OLV start timestamp y_{start} occurs inside the sampled t to $t + l_{ws}$ segment, we set the value at $y_{start} - t$ to 1, and set all the other timestamps to 0. Second, if there is no OLV event performed in the sampled segment, we keep the zero-filled label list as the corresponding ground truths.

However, the above approach to preparing for the labels is sub-optimal. The first reason is that, due to the burden of documenting clinical procedures, it happens that clinicians documented the OLV event a few minute later or earlier than the exact timestamp of the OLV procedural. In other words, the previous approach to generating the labels is vulnerable to label noise. Second, optimizing the model using a label list that has a sudden high probability value at the exact timestamp and zeros nearby causes the model to generate discrete predictions. Then, it becomes more challenging to locate the OLV timestamps from the full-length signals. To tackle the weaknesses, we propose to utilize the Gaussian distribution function to smooth the ground truths. Specifically, we use a Gaussian distribution centered at the ground truth timestamp with the standard deviation σ, to construct the label distribution y as shown in Eq. 3.

$$y_i = \begin{cases} \frac{4}{\sqrt{2\pi}\sigma}\exp\left(-\frac{(t_i - t_{OLV})^2}{2\sigma^2}\right), \text{if } -3 \leq t_i - t_{OLV} \leq 3 \\ 0, \quad \text{otherwise} \end{cases} \tag{3}$$

Note that the smoothing step only applies to the ±3 min near the ground-truth timestamp.

As for the inference phase, similar to training, a sliding window with size l_{ws}, and a step size l_{step} are used to split the signals into smaller equal-length segments. The total number of segments is calculated by:

$$n_{seg} = \text{floor}\left(\frac{d - l_{ws}}{l_{step}}\right) \tag{4}$$

where d is the original sequence length. To obtain the final predictions for each patient, we find the maximum response among all the n_{neg} outputs and the corresponding timestamp. Formally,

$$\hat{y} = argmax\left(\hat{y}_{seg}^{0}; \hat{y}_{seg}^{1}; \ldots; \hat{y}_{seg}^{n_{seg}-1}\right). \tag{5}$$

4.6 *Loss Function*

The model predicts the probability of an occurring OLV event for each timestamp. The binary cross entropy loss is used for optimization. In addition, we apply an auxiliary cross-entropy classification loss to the frequency feature to classify the binary occurrence of an OLV event.

5 Results

In this section, we first describe our performance metrics and implementation details and then provide the comparisons with baseline methods. Following that, we analyze the importance of each variable through leaving-one-variable-out training and model interpretation. Lastly, we show the cross-institution generalization ability by training the model on Site A and testing on Site B, and vice versa.

5.1 Model Performance

Metrics. Model performance is evaluated in the testing set. We use mean absolute error (MAE) and accuracy with error margin as the OLV timestamp estimation criteria. The justification of the event prediction accuracy is based on a pre-defined margin value, which is an allowable temporal range before or after the ground-truth timestamp. If the predicted timestamp is within this range, the prediction of the OLV event is a true positive. We use $acc3$ to represent accuracy with a clinically significant margin of 3 min. The MAE is calculated from the distance between the ground-truth timestamps and the predicted timestamps.

Implementation details. Our framework is trained using PyTorch. The mini-batch size is 24, and the learning rate is set to 0.0005. We employed an Adam [11] optimizer with 0.9 Nesterov momentum, where the weight decay rate is set to 0.0001. The sliding window size ws is 40 min. We split the data into five folds randomly, and each fold includes 845 records. Five-fold cross-validation is applied for all experiments. The reported results are the average of every fold. Two identity models are trained separately to predict the respective start and end timestamps. All experiments are done in a GeForce RTX 2080Ti GPU. Detailed training and testing steps are shown in Algorithm 1.

Overall performance and ablation study. Table 2 shows the performance using the full set of variables for training. Figure 4 describes the histogram of the prediction errors for five folds of the testing set. The results show that most of the errors are centered around 0. A majority (81.7%) of the predictions are within the 3 min error margin. The highest portion is at 0 min error, where the model has successfully predicted the timestamp at the exact minute of the ground truth, and as the distance increases, the error percentage decreases significantly. We claim that most of the predictions are within a 3 min error margin.

As compared to the baseline models, it can be seen from Table 2 that our proposed consistently outperforms all the baselines by a considerable margin. We implemented

Table 2 Model performances and ablation studies. **Change-point**: locating the maximum change point from the most OLV-correlated variable; **Base-NN**: baseline model using basic fully connected layers; **Base-LSTM**: baseline model by replacing transformer with LSTM cells. acc(n): the accuracy under a n minutes margin, as discussed in the metrics section. The left and right side of "|" denotes OLV start and end, respectively

Method	MAE↓	acc5↑	acc4↑	acc3↑	acc2↑	acc1↑
Change-point	15.3 | 20.3	71.1 | 73.1	68.5 | 71.4	65.1 | 67.6	55.8 | 55.7	34.5 | 31.5
Base-NN	19.0 | 27.5	67.0 | 62.8	63.9 | 61.3	59.7 | 58.0	53.3 | 51.1	41.0 | 36.8
Base-LSTM	6.0 | 9.2	80.0 | 84.3	77.1 | 82.7	72.9 | 80.4	66.7 | 74.7	52.2 | 57.8
w/o smoothing	5.8 | 5.5	80.7 | 89.1	78.1 | 87.5	73.4 | 85.4	67.1 | 80.8	52.7 | 67.1
w/o spectrum	4.5 | 5.1	82.7 | 89.9	79.6 | 88.5	75.0 | 86.1	68.6 | 80.9	53.3 | 64.5
Full model	**4.4 | 4.1**	**83.6 | 91.3**	**80.2 | 90.0**	**75.8 | 87.6**	**69.4 | 82.3**	**55.7 | 67.2**

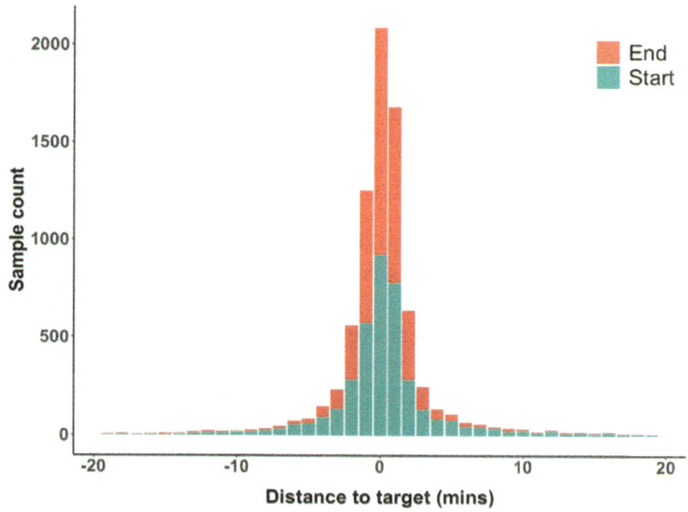

Fig. 4 Histogram of prediction errors for OLV start and end timestamps

a heuristic method that locates the top 2 change points from the most OLV-correlated variable (i.g., VT_set). The results show that the hand-crafted feature based method produces inferior MAE and accuracy than data-driven learning-based methods. It's worth noting that the basic neural network with fully connected layers is not comparable to the sequential models such as LSTM (Long short-term memory) and Transformers. This indicates that incorporating temporal information is crucial for OLV detection. Although the LSTMs can learn temporal evolution using recurrent gates, they are sub-optimal compared to Transformers. Transformers enable direct connections between each node of timestamps, and the direct attention mechanism allows longer sequential modeling.

In terms of the ablations as shown in Table 2, we observe an obvious decreasing in accuracy (from 83.6|91.3 to 80.7|84.3) and increasing of MAE (from 4.4|4.1 to 5.8|5.5) after removing the label temporal smoothing training. It demonstrates that smoothing the target timestamps during training produces score curves that have smoother transitions from OLV occur to not occur, consequently making it easier to locate the maximum response from the prediction curves. As shown in Fig. 3, after smoothing the supervision signals with the Gaussian function, the predicted scores at each timestamp are also curved and noise reduced. On the contrary, without smoothing, the output scores present noticeable noise that interferes with locating the maximum score, thus yielding a biased predicted timestamp.

In order to see how much the frequency domain features affect the performance, we train our model again without the spectrum input and compare it to our original method. Utilizing the frequency domain features decreases the MAE from 4.5|5.1 to

4.4|4.1, and acc5 increases from 82.7|89.9 to 83.6|91.3. It shows the effectiveness of frequency features that can provide solid extra information for time-series OLV detection.

5.1.1 Cross-Site Generalizability

Recall that our dataset is curated from two institutions. Regarding the within-site results, we observe that site B yields better performance than site A. This makes intuitive sense because Site B consists of more patients' records than Site A. To investigate the model generalization ability across two sites, we use the data from one site for training and the data from the other for testing. Regarding cross-site validation, the MAE increases to 5.8|5.5 and 8.7|6.5. While underlying reasons remain unclear for further investigation, we attribute this to the distribution shift between the data from the two sites introduced by differences in patient demographics and clinical characteristics. We leave it as our future work to incorporate domain adaptation modules to fine-tune our models to improve the model generalizability [8] (Table 3).

5.2 Feature Importance Decomposition

To explore how important each variable contributes to the prediction model training and which variable has more information to predict the OLV start and end timestamp. We conduct our experiments in two aspects: (1) training without one of the variables and (2) computing the integral of the gradients of the output prediction for the predicted label with respect to the input variables [24].

As can be seen from Table 4, after removing either of the VT_set or the $VT\_exhaled$, the MAE drops more significantly than the others. In contrast, the rest variables have a minor contribution to accuracy. Some variables contain easier-recognized signal patterns driven by OLV operation than others, which is consistent with our observations. By feeding all the variables to the model, the best performance is achieved. We can see that although some variables are much less informative than VT_set or $VT\_exhaled$, they still provide supplementary information for

Table 3 Cross-institution validation results. The left and right side of "|" denotes OLV start and end, respectively

Train	Test	MAE↓	acc5↑	acc4↑	acc3↑	acc2↑	acc1↑
Site A	Site A	7.5 | 5.9	67.4 | 88.2	60.7 | 86.5	52.5 | 82.9	39.7 | 77.7	26.9 | 65.0
Site B	Site B	3.6 | 3.9	87.8 | 91.7	85.3 | 90.4	81.9 | 88.8	76.3 | 84.5	61.0 | 71.5
Site A	Site B	5.8 | 5.5	83.4 | 89.5	80.0 | 88.0	72.9 | 84.9	59.7 | 75.8	38.4 | 52.1
Site B	Site A	8.7 | 6.5	67.0 | 86.4	60.8 | 83.6	52.4 | 78.0	42.7 | 66.2	29.3 | 42.4

Table 4 Training by leaving one variable out. "s" and "m" means setting variables and measurement variables, respectively. Detailed variable descriptions are shown in Table 1. The left and right side of "|" denotes OLV start and end, respectively

Metrics	MAE↓	acc5↑	acc4↑	acc3↑	acc2↑	acc1↑
w/ all	4.4 | 4.1	83.6 | 91.3	80.2 | 90.0	75.8 | 87.6	69.4 | 82.3	55.7 | 67.2
w/o VT(m)	6.5 | 5.8	78.2 | 88.0	75.1 | 86.5	70.6 | 84.2	64.8 | 78.3	50.4 | 62.4
w/o VT(s)	5.6 | 5.1	82.0 | 90.0	78.5 | 88.4	74.2 | 86.1	66.6 | 80.5	49.8 | 65.2
w/o PIP (m)	4.6 | 5.2	82.2 | 89.6	79.4 | 88.2	74.7 | 85.6	68.2 | 79.1	53.3 | 63.1
w/o PEEP (m)	4.5 | 4.5	83.0 | 90.7	79.8 | 89.2	75.3 | 87.0	69.0 | 81.7	53.6 | 66.6
w/o RR (m)	4.3 | 4.7	83.4 | 90.4	80.4 | 89.1	75.6 | 86.6	69.1 | 81.5	53.4 | 65.8
w/o RR (s)	4.6 | 4.7	83.2 | 90.6	80.3 | 89.3	76.0 | 86.7	69.0 | 80.8	53.3 | 65.8
w/o SPO_2 (m)	4.4 | 4.3	83.3 | 90.8	80.3 | 89.3	75.8 | 87.0	69.8 | 81.9	54.4 | 66.5

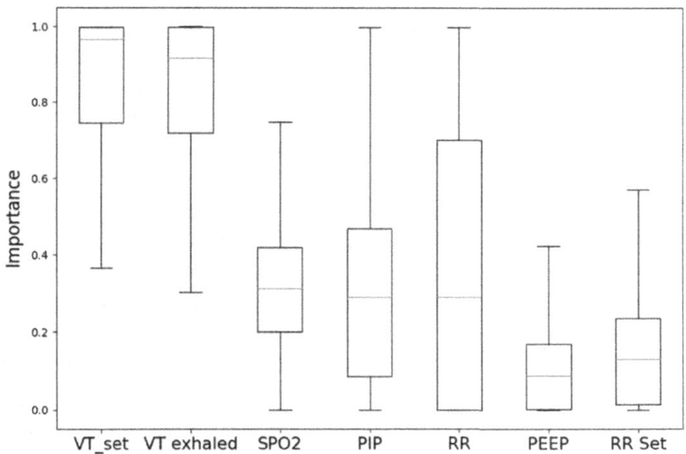

Fig. 5 Normalized attribution scores of each variable

OLV prediction and contribute to higher accuracy. Figure 5 shows the average of the attribution score of each variable of the patients using the integral of the gradients. Consistently, we observe higher attribution scores for VT_Set and $VT\_exhaled$. Following them, PIP and RR present a high importance score less frequently. Moreover, the scores are widely scattered between 0 to 1, which indicates that there exists a small portion of cases when PIP and RR contribute the most and act as dominant variables. The deep learning model adaptively attends to the variables that are correlated most to OLV events, and it lowers the reliance on the variables that hardly show OLV-correlated patterns. In contrast, the anomaly detection methods using hand-craft statistical patterns are incapable of weighing the reliance on the variables except by using pre-defined weights and thresholds.

6 Discussion and Limitation

In this paper, we developed and validated an innovative Transformer-based deep learning model for predictions of start/end timestamps of OLV procedures utilizing objective physiological monitoring data. We obtained a satisfactory predictive performance for this DL algorithm, and this framework can be potentially extended to applications of clinical auto-documentation of other OR clinical events or procedures in other medical settings such as Intensive Care Units (ICU) or Post Anesthesia Care Units (PACU). Additionally, our extended experiments allow us to track down the contributing factors of observed between-institution variance, providing insights for interventions to improve the quality of collected data.

Our study has limitations. First, this work utilizes a retrospective observational clinical dataset, which is inevitably exposed to selection bias when sampled from the target patient population. As the first pilot study of this program of research, we decided to first focus on clinical cases containing only one OLV event due to practical considerations of data quality control and conditioning, and the convenience of modeling. This choice was based on assumptions that the majority of thoracic cases would undergo only one OLV procedure, and that OLV timestamps of one-OLV cases would be documented more reliably compared to multiple-OLV cases. However, such choices might have ruled out cases with higher clinical complexity and more/shorter OLV procedures hence introducing selection biases into our study sample. It partially explains the discrepancy of predictive performance in cross-site generalizability analysis due to different clinical characteristics by selection. We plan to further include and adjust clinical confounding factors to overcome this issue in our next research. Second, deep learning models are known to be difficult to interpret and prone to overfit the observed data. We will address this limitation by introducing the game-theory-based SHAP [16] values for explanations and minimizing overfitting by utilizing internal cross-validation techniques and performing cross-sample (Site A vs. Site B) validations to assure model generalizability across samples from different institutions. We plan to refine our algorithm in followed studies to predict both non-event and multiple-events by incorporating external validation data sources from other institutions such as Columbia University Hospital systems.

References

1. M.G. Allison, M.C. Scott, K.M. Hu, M.D. Witting, M.E. Winters, High initial tidal volumes in emergency department patients at risk for acute respiratory distress syndrome. J. Crit. Care **30**(2), 341–343 (2015)
2. H. Alquran, A.M. Alqudah, I. Abu-Qasmieh, A. Al-Badarneh, S. Almashaqbeh, ECG classification using higher order spectral estimation and deep learning techniques. Neural Netw. World **29**(4), 207–219 (2019)
3. S. Banerjee, M. Mitra, Application of cross wavelet transform for ECG pattern analysis and classification. IEEE Trans. Instrum. Meas. **63**(2), 326–333 (2013)

4. F. Bray, J. Ferlay, I. Soerjomataram, R.L. Siegel, L.A. Torre, A. Jemal, Global cancer statistics 2018: GLOBOCAN estimates of incidence and mortality worldwide for 36 cancers in 185 countries. CA: Cancer J. Clin. **68**(6), 394–424 (2018)
5. S. Deldari, D.V. Smith, H. Xue, F.D. Salim, Time series change point detection with self-supervised contrastive predictive coding, in *Proceedings of the Web Conference 2021* (2021), pp. 3124–3135
6. J. Devlin, M.W. Chang, K. Lee, K. Toutanova, Bert: pre-training of deep bidirectional transformers for language understanding (2018). arXiv:1810.04805
7. Facebook (2022) Kats. https://facebookresearch.github.io/Kats/
8. A. Farahani, A. Voghoei, K. Rasheed, H.R. Arabnia, A brief review of domain adaptation, in *Advances in Data Science and Information Engineering* (2021), pp. 877–894
9. J.C.B. Gamboa, Deep learning for time-series analysis (2017). arXiv:1701.01887
10. M. Iskander, Burnout, cognitive overload, and metacognition in medicine. Med. Sci. Educ. **29**(1), 325–328 (2019)
11. D.P. Kingma, J. Ba, Adam: a method for stochastic optimization (2014). arXiv:1412.6980
12. Z.C. Lipton, J. Berkowitz, C. Elkan, A critical review of recurrent neural networks for sequence learning (2015). arXiv:1506.00019
13. D. Liu, Y. Zhao, H. Xu, Y. Sun, D. Pei, J. Luo, X. Jing, M. Feng, Opprentice: towards practical and automatic anomaly detection through machine learning, in *Proceedings of the 2015 Internet Measurement Conference* (2015), pp. 211–224
14. W. Liu, Z. Wang, X. Liu, N. Zeng, Y. Liu, F.E. Alsaadi, A survey of deep neural network architectures and their applications. Neurocomputing **234**, 11–26 (2017)
15. J. Lohser, Evidence-based management of one-lung ventilation. Anesthesiol. Clin. **26**(2), 241–272 (2008)
16. S.M. Lundberg, S.I. Lee, A unified approach to interpreting model predictions, in *Advances in Neural Information Processing Systems 30, Curran Associates, Inc.*, ed. by I. Guyon, U.V. Luxburg, S. Bengio, H. Wallach, R. Fergus, S. Vishwanathan, R. Garnett (2017), pp. 4765–4774. http://papers.nips.cc/paper/7062-a-unified-approach-to-interpreting-model-predictions.pdf
17. K. Marseu, P. Slinger, Peri-operative pulmonary dysfunction and protection. Anaesthesia **71**, 46–50 (2016)
18. Y. Meyer, *Wavelets and Operators: Volume 1*, vol. 37 (Cambridge University Press, Cambridge, 1992)
19. A.S. Neto, S.N. Hemmes, C.S. Barbas, M. Beiderlinden, A. Fernandez-Bustamante, E. Futier, O. Gajic, M.R. El-Tahan, A.A. Al Ghamdi, E. Günay et al., Association between driving pressure and development of postoperative pulmonary complications in patients undergoing mechanical ventilation for general anaesthesia: a meta-analysis of individual patient data. Lancet Respir. Med. **4**(4), 272–280 (2016)
20. P. Malhotra, G. Anand, L. Vig, P. Agarwal, G. Shroff, A. Ramakrishnan, LSTM-based encoder-decoder for multi-sensor anomaly detection (2016). arXiv:1607.00148
21. S.K. Perera, S. Jacob, R. Sullivan, M. Barton, Evidence-based benchmarks for use of cancer surgery in high-income countries: a population-based analysis. Lancet Oncol. **22**(2), 173–181 (2021)
22. A. Siffer, P.A. Fouque, A. Termier, C. Largouet, Anomaly detection in streams with extreme value theory, in *Proceedings of the 23rd ACM SIGKDD International Conference on Knowledge Discovery and Data Mining* (2017), pp. 1067–1075
23. A.S. Slutsky, V.M. Ranieri, Ventilator-induced lung injury. N. Engl. J. Med. **369**(22), 2126–2136 (2013)
24. M. Sundararajan, A. Taly, Q. Yan, Axiomatic attribution for deep networks, in *International Conference on Machine Learning* (PMLR, 2017), pp. 3319–3328
25. O. Vallis, J. Hochenbaum, A. Kejariwal, A novel technique for {Long-Term} anomaly detection in the cloud, in *6th USENIX Workshop on Hot Topics in Cloud Computing (HotCloud 14)* (2014)
26. A. Vaswani, N. Shazeer, N. Parmar, J. Uszkoreit, L. Jones, A.N. Gomez, Ł. Kaiser, I. Polosukhin, Attention is all you need, in *Advances in Neural Information Processing Systems*, vol. 30 (2017)

27. K. Wang, Y. Zhao, Q. Xiong, M. Fan, G. Sun, L. Ma, T. Liu, Research on healthy anomaly detection model based on deep learning from multiple time-series physiological signals. Sci. Program. (2016)
28. H. Xu, W. Chen, N. Zhao, Z. Li, J. Bu, Z. Li, Y. Liu, Y. Zhao, D. Pei, Y. Feng, et al., Unsupervised anomaly detection via variational auto-encoder for seasonal KPIs in web applications, in *Proceedings of the 2018 World Wide Web Conference* (2018), pp. 187–196

Analyzing the Trends of Responses to COVID-19 Related Tweets from News Stations: An Analysis of Three Countries

Andrew Fisher, Rajesh Sharma, and Vijay Mago

Abstract During the COVID-19 pandemic, news stations have used social media platforms such as Twitter to deliver information to the general public. To understand the trends as well as impact of these posts, we analyze 500 k tweets and responses across 15 news outlets from USA, Canada, and UK, through three research questions. The first question is related to topic popularity where a zero-shot classification algorithm is used to determine what type of COVID-19 related tweets users are mostly interacting with. Then, the second question looks to determine how the audiences differ in their responses between news stations within each country by using a sentiment, emotion, and stance analysis algorithm as well as statistical hypothesis test. Lastly, the third question uses the previous analyzes' results along with the political leanings of each news station to see if there is a correlation in differences. As a result, we discover that the topic of *vaccine* is the most popular, audiences in the USA and UK have a considerable amount of differences in their responses, and that the differences in political leanings strongly match with differences in audience response.

A. Fisher (✉)
Department of Mathematics and Computing Science, Saint Mary's University, Halifax, Canada
e-mail: andrew.fisher@smu.ca

R. Sharma
Institute of Computer Science, University of Tartu, Estonia and Department of Computer Science, Lakehead University, Thunder Bay, Canada
e-mail: rajesh.sharma@ut.ee

V. Mago
Department of Computer Science, Lakehead University, Thunder Bay, Canada
e-mail: vmago@lakeheadu.ca

A. Shaban-Nejad et al. (eds.), *Artificial Intelligence for Personalized Medicine*,
Studies in Computational Intelligence 1106,
https://doi.org/10.1007/978-3-031-36938-4_20

1 Introduction

Since the emergence of COVID-19 at the beginning of 2020, there has been a mixed reaction from the public in regards to health related topics such as vaccines, public mandates, and political responses [1, 2]. This can be observed on social media platforms such as Twitter where users have the ability to respond to news stories posted by various outlets. However, due to the differing biases and opinions of these channels, divisions in the responses can lead to harmful outcomes such as vaccine hesitancy, protests, and distrust in public health officials [3–5]. To better understand these trends and the impact of such posts, it would be beneficial to have a framework that can compare how the audiences of different news stations react to COVID-19 tweets. Therefore, this work proposes a solution and applies it to three different countries to analyze the results.

To begin, a dataset of nearly 500 k tweets is created via the Twitter Academic API by collecting 15 news stations' tweets and the responses to them from the first two years (January 1, 2020–January 1, 2022) of the pandemic in the USA, Canada, and UK. Then, the following three research questions are defined to understand the trends and impact of these posts using a framework with pre-trained language models:

- **RQ1**: Which COVID-19 related topics from different news stations were the most popular overall?
- **RQ2**: Is there a difference in the responses to COVID-19 related tweets from different news stations?
- **RQ3**: Do differences in the responses to COVID-19 related tweets occur between news stations with different political stances?

To answer these questions, a zero-shot model is ran to classify the tweets into a set of COVID-19 related topics such that the first research question can be answered by looking at overall engagement. Next, models that perform sentiment, emotion, and stance analysis are ran to determine the respective information for each of the responses. From this, the second research question can be answered by performing a statistical hypothesis test to observe whether or not the means of the results differ between news stations. Lastly, based on the observations made and political stance information obtained from reputable websites, the third research question can be answered to see if there is a correlation.

As a result, this work provides a detailed overview of the trends in responses to COVID-19 related tweets, as well as the impact different news stations have on their audiences. Our overall goal is to demonstrate that tweets discussing similar stories can be conveyed in different ways by analyzing the reactions from the general public. This is important to consider as strong biases or stereotyping for example may elicit a more negative response whereas another news station discussing the same story in a more positive manner may not receive the same attention [6].

2 Related Works

2.1 COVID-19 Pandemic

The use of deep learning for COVID-19 problems has become a very popular application area in recent times. Due to the amount of engagement on social media platforms such as Twitter, analyzes can be ran to better understand how the general population is responding to certain topics. For example, Durazzi et al. [7] looked to analyze $\sim$350 m English tweets from $\sim$26 m users between January 13, 2020 through June 7, 2020 to identify COVID-19 related clusters and see what similarities they have. This was accomplished by using a community detection algorithm to see how users are connected to one another in terms of (1) classification (e.g., science or media), (2) presumed physical location, and (3) engagement (i.e., retweets and responses). As a result, they found that there were four main "super-communities" (science-health, national elite, political, and other), several country-specific communities related to politics, and that the science-health community received significantly more attention at the beginning of the pandemic.

In a more focused approach, Han et al. [8] gathered 34,352 COVID-19 related tweets from UK news stations and citizens between April 2, 2020 through April 8, 2020 to see if similar topics were discussed. They used unsupervised text analysis methods via structural topic modelling to find that there was a difference between them: news stations focused on safety advice, death news, and international COVID-19 news while citizens mainly posted about COVID-19 discussions, feelings towards the pandemic, and activities they were doing. Since previous works had focused on English tweets only, Garcia et al. [9] looked to propose a multilingual framework to also deal with Portuguese. This was accomplished by using BERT [10] to perform emotion analysis and an algorithmic approach for topic modelling to compare $\sim$7 m COVID-19 related tweets between the USA and Brazil. To be able to input the Portuguese tweets, a dataset from Kaggle[1] was used to pre-train a separate instance of the model. From this, they found that the majority of emotions were negative (anger, fear, and sadness) and that 70% of the main topics discussed were equivalent between the two countries- the most popular in USA was "economic impacts" while Brazil's was "proliferation care".

2.2 Political Campaigns

Although the application of natural language processing on pandemic datasets has become common in recent times, another related area is analyzing the tweets during political campaigns. For example, Mueller et al. [11] performed sentiment analysis on 97,909 tweets from 342 US politicians during the 2018 midterm elections to see

[1] https://www.kaggle.com/datasets/augustop/portuguese-tweets-for-sentiment-analysis.

if more negativity results in more interaction with the posts. They developed a novel statistical measurement to determine *negativity incentive* and used statistical models including negative binomial, ordinary least squares, as well as Naive Bayes to analyze the tweets. After performing the analysis, their hypothesis was demonstrated to be true as the amount of interaction a negative tweet received was substantially higher than others. Similarly, Sahly et al. [12] analyzed how Trump and Clinton differed in the development of their posts on Twitter and Facebook during the 2016 presidential election to see if it affected audience engagement. They used two annotators (kappa score of 0.91) that were trained with frame definitions to label their data so they could determine if the candidates differed in the frequency of conflict, morality, and attribution of responsibility frames. The results showed that, on Twitter, the politicians' use of tweets framed in a negative way were statistically different with more negativity resulting in more interaction. To understand the data preparation steps and deep learning methods used in our work, the next section will discuss the methodology.

3 Methodology

3.1 Data Preparation

Using pre-existing algorithms that will be described in the next subsection, our analysis focuses on the following three countries: USA, Canada, and UK. Following Fig. 1, step 1 begins with selecting five news stations, for each country, to gather tweets from between January 1, 2020 through January 1, 2022 (step 2) that contain the keywords *pandemic*, *vaccine*, *virus*, *outbreak*, *coronavirus*, or *covid*: USA- ABC, CNN, Fox News, MSNBC, NBC, Canada- CBC, CTV, Globe and Mail, National Post, Global News, and UK- 5 News, Channel 4 News, ITV News, Sky News, BBC. The keywords were developed after manually extracting relevant high-level words from a subset of the tweets, and news stations were selected based on high follower counts. In step 3, for each tweet, all replies are collected to allow for analysis on the audience engagement across different COVID-19 related topics. At this stage in the process, there are just over 10 million tweets and replies in total before any additional steps.

Using the raw dataset, sentiment analysis is performed to determine how many common tweets there are across each news station within each country. This is done in step 4 by selecting the news station that had the most tweets with responses: USA- NBC, Canada- Global News, and UK- BBC. Then in step 5, for each of their tweets, the contents are compared to each of the other news stations' tweets within their respective country using a semantic similarity algorithm [13] to extract the most similar results. After setting a minimum threshold of 60–70% in step 6, the remaining number of tweets and replies are as follows: Canada- 1,290 tweets and 81,533 replies, US- 1,180 tweets and 324,246 replies, and UK- 725 tweets and 85,226 replies. Using this information, the three research questions can then be answered by analyzing the topics of the tweets and contents of the replies [14].

3.2 *Prerequisites*

3.2.1 Semantic Similarity

In order to perform the analyses, the tweets from each news station were first grouped together such that a set of common stories can be extracted. To do this, an algorithm referred to as *Sentence-Bert* (SBERT) [13] is used which is a modified version of *BERT* [10] and performs semantic similarity on a pair of text. Their framework consists of a siamese (i.e., dual) and triplet network structure that aims to produce better sentence embeddings than previous approaches so it can be compared using consine similarity. The former network works to produce these results while the latter network applies the following loss function when training:

$$max(||s_a - s_p|| - ||s_a - s_n|| + \epsilon, 0) \tag{1}$$

where s_a is an anchor sentence, s_p is a positive sentence, s_n is a negative sentence, and ϵ is a Euclidean distance set to 1. In this work, a pre-trained model from the authors is used that learned from a combination of two datasets which contain a total of ~1 million sentence pairs [13]. Consider the following example from our dataset where the similarity between the two tweets are ~88.18%:

```
NBC NEWS: The FDA has told Johnson & Johnson to discard about
60,000,000 doses of its Covid-19 vaccine that were produced at a
troubled plant in Baltimore, according to 2 people familiar with the
matter.

FOX NEWS: FDA reportedly tells Johnson & Johnson to toss 60M COVID-19
vaccine doses over contamination concerns LINK
```

For quality assurance, the following two conditions must be met in order for an extracted story across all 5 news stations within a given country to be considered for further analysis: (1) the most similar tweets must meet a minimum similarity threshold (Fig. 1, Step 6) and (2) the most similar tweets must be within a week range of each other at most.

3.2.2 Topic Identification

Once the tweets have been grouped into a set of common stories, the next step is to determine the different topics that are being discussed. For the purpose of this work, consider the following possibilities: vaccine, restriction, infection, death,

Fig. 1 An overview of the data preparation process

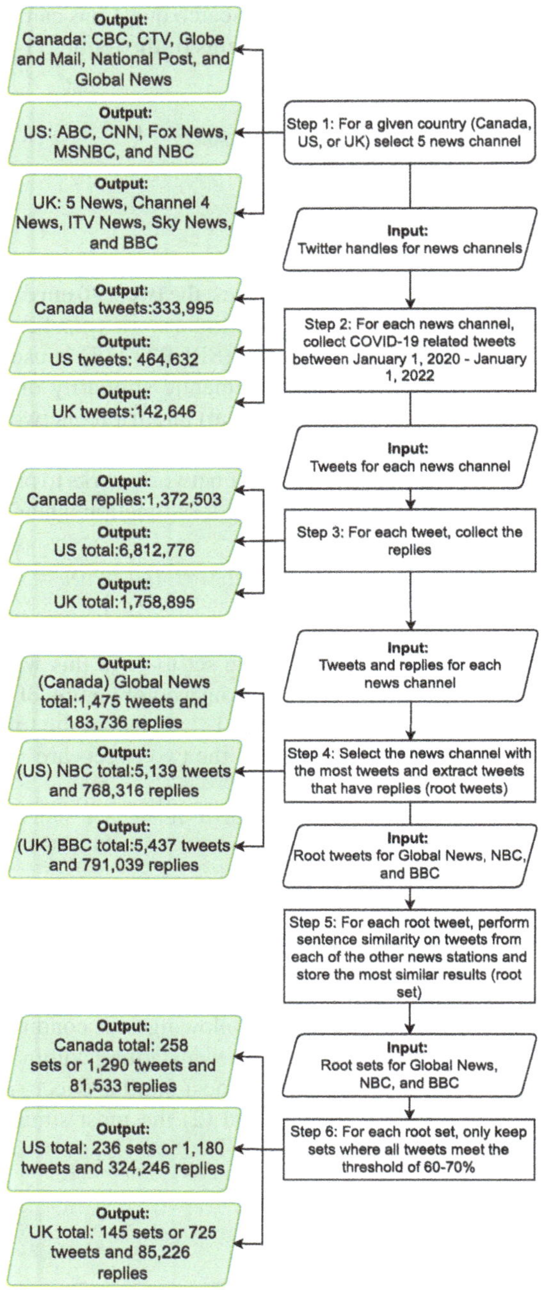

and other. In order to perform this classification, an algorithm referred to as *XLM-RoBERTa* (XLM-R) [15] is used which is based on the XLM approach [16] and *RoBERTa* architecture [17]. The authors looked to improve over existing models by utilizing a large dataset consisting of 100 languages while ensuring that overall performance would not be impacted. This was achieved by using transformers trained with a multilingual masked language model (MLM) objective [10, 16], and not using language embeddings to better deal with *code-switching* between inputs. As a result, a *CommonCrawl Corpus* was developed which contains significantly more data in comparison to previous approaches which typically used English data from Wikipedia articles. Although the tweets in this work do not utilize the multilingual aspect of this model, it is a more desirable approach since it was able to learn from a large variety of data. For instance, in the previous example shown for semantic similarity between two tweets discussing the Johnson & Johnson vaccine, the zero-shot classifier placed the story under the *vaccine* category.

3.2.3 Sentiment Analysis

To determine the sentiments of responses to new stations' tweets, an approach referred to as *Time Language Models* (TimeLMs) [18] is used which implements the *RoBERTa* algorithm [17]. The novelty of this framework is that the models are continuously updated with labelled Twitter data every 3 months to ensure that they are up-to-date with current trends and improve overall accuracy. When the article was initially published in 2019, they had used a dataset consisting of 90 million tweets for pre-training to see how adding additional data over time would affect performance. As a result, they have been able to demonstrate that the additional training does improve accuracy on pre-exisiting data, with the version used in this paper being trained on a total of 124 million tweets. To demonstrate this algorithm, consider the following example from our dataset that was labelled *positive*:

```
I think he is doing a great job and anyone ask how is doing cause he is
human like us.
```

3.2.4 Emotion Analysis

Similar to the previously discussed TimeLMs [18], our emotion analysis uses the *RoBERTa* algorithm [17] which was identified by the *TweetEval* framework [19] as being a top performer for the problem. This was done by comparing the performance of the model on a dataset consisting of ~11 k labelled tweets [18, 19]. After analyzing the results of five other models, we chose to use their pre-trained version of this algorithm for our work and demonstrate it on the following tweet from our dataset that was labelled *anger*:

```
Vaccine bogus. I wonder how many MORE UNDERLING HEALTH ISSUES It does
not protect.
```

3.2.5 Stance Analysis

To determine the stance that replies are taking, we chose to use the *Knowledge Enhance Masked Language Modeling* (KE-MLM) method [20] which fine-tunes a *BERT* model to incorporate contextual knowledge for a given topic. This was accomplished by utilizing a weighted log-odds-ratio technique with informed Dirichlet priors, which helps extract important words for determining the classification (i.e., against, in favor, neutral) [20]. In traditional approaches, the authors noted that TF-IDF was typically used for this aspect which they felt was a limitation in achieving better performance. As a result, after training on a dataset related to the 2020 US Presidential election, their approach was demonstrated to outperform previous methods which is why we use their pre-trained model for this analysis [20]. As an example, consider the following tweet from our dataset that was labelled *in favor*:

```
Let's focus on safety first. Measures need to put in place to prevent
the spread of the virus. Contain this virus and no racism. We can do
both.
```

4 Results

4.1 *RQ1: Which COVID-19 Related Topics from Different News Stations Were the Most Popular Overall?*

After extracting sets of similar tweets between news stations, the zero-shot algorithm [15] is ran to classify them into one of the following five topics: vaccines, infections, restrictions, deaths, and other. As this step is being entirely labeled by an algorithm, we chose to use a concise list of high-level topics to help reduce the loss in overall accuracy. Figure 2 shows how the overall responses are distributed for each news station within each country where it can observed that *vaccines* are the most engaged (see folder *Topics*). Furthermore, the new stations with the highest distribution on this topic are CNN (USA), Global News (Canada), and Channel 4 News (UK) meaning that their audiences were the most responsive to it. Another observation that can be made is that in Canada, the topic of *vaccines* are significantly more popular than others whereas in the UK, it is relatively close to the topic of *infections*.

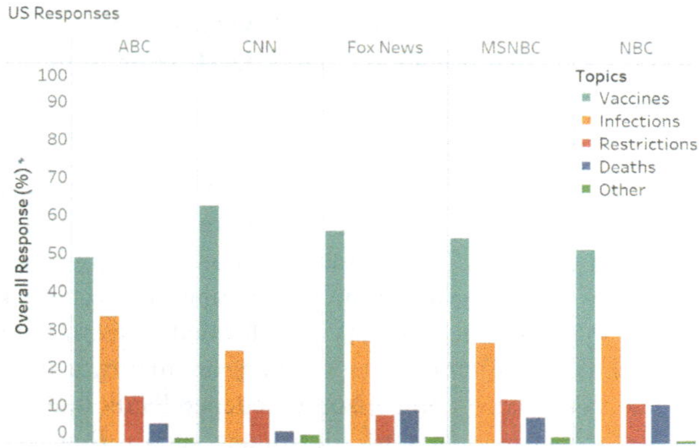

(a) A summary of the response distribution for USA

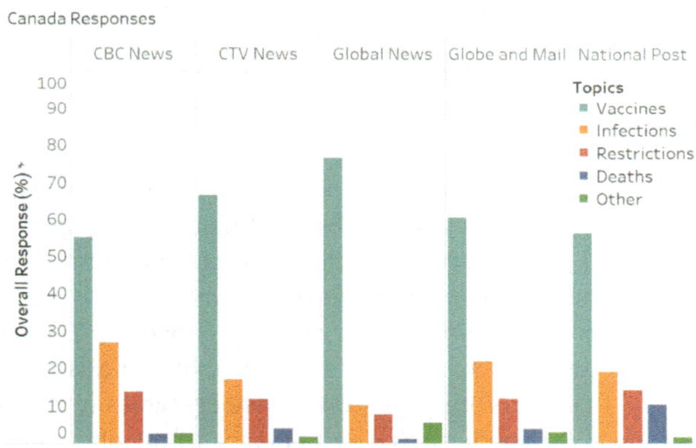

(b) A summary of the response distribution for Canada

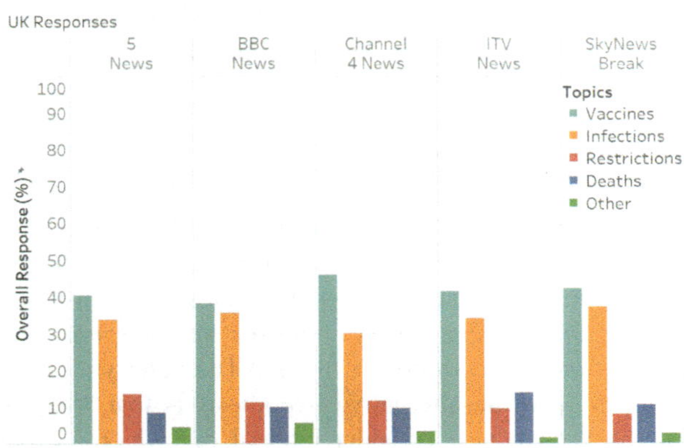

(c) A summary of the response distribution for UK

Fig. 2 Summaries of the response distribution for each country

4.2 RQ2: Is There a Difference in the Responses to COVID-19 Related Tweets from Different News Stations?

To determine whether or not there is a difference in the responses to COVID-19 related tweets, three analyzes are performed to determine (1) sentiments (positive, neutral, negative), (2) emotions (anger, sadness, optimism, joy), and (3) stances (against, neutral, in favor). Then, using the total counts for each respective classification, a statistical hypothesis test (t-test) is performed to see if the news stations have statistically significant differences between one another by using a threshold of 0.05 for the p-value. Since each comparison is independently done (i.e., there is a separate hypothesis for each comparison), adjustments such as *Bonferroni correction* are not made [21]. To visualize the label distribution of these analyses, a repository is available at the URL: https://github.com/andrfish/COVID-Tweet-Response-Trends.

4.2.1 Sentiment Analysis

The classification distribution for sentiment analysis is shown in the *Sentiments* folder of the repository where it can be observed that, on average, *negative* is the most prominent overall. When performing the statistical hypothesis test, we consider the following hypothesis: *Is the mean of {negative, neutral, positive} responses to news station A's tweets different than news station B's tweets?* The null hypothesis is that "they have the same mean" and the alternative hypothesis is that "they have different means". After performing the test, consider Tables 1, 2, 3 which shows the rejections of the null hypothesis (i.e., acceptance of the alternative hypothesis) for USA, Canada, and UK respectively. It can be observed that a large amount of rejections between news stations happen in the USA while in the UK, a majority of the rejections occur with one news station (Sky News Break).

4.2.2 Emotion Analysis

The classification distribution for emotion analysis is shown in the *Emotions* folder of the repository where it can be observed that, on average, *anger* is the most prominent overall. When performing the statistical hypothesis test, we consider the following hypothesis: *Is the mean of {angry, joyful, optimistic, sad} responses to news station A's tweets different than news station B's tweets?* The null hypothesis is that "they have the same mean" and the alternative hypothesis is that "they have different means". After performing the test, consider Tables 4, 5, 6 which shows the rejections of the null hypothesis (i.e., acceptance of the alternative hypothesis) for USA, Canada, and UK respectively. Similar to the sentiment analysis, it can be observed that a large amount of rejections between news stations happen in the USA while in the UK, a majority of the rejections occur with one news station (Sky News Break).

Table 1 Rejections of the sentiment analysis null hypothesis for responses in USA

News station A	News station B	Label	P-value
ABC	CNN	Negative	0.033
ABC	Fox news	Negative	0.001
Fox news	MSNBC	Negative	0.023
Fox news	NBC	Negative	0.002
ABC	CNN	Neutral	0.032
ABC	Fox news	Neutral	0.001
Fox news	MSNBC	Neutral	0.003
Fox news	NBC	Neutral	0.0
ABC	Fox news	Positive	0.007
Fox news	MSNBC	Positive	0.013
Fox news	NBC	Positive	0.003

Table 2 Rejections of the sentiment analysis null hypothesis for responses in Canada

News station A	News station B	Label	P-value
CBC news	Globe and mail	Negative	0.027
CBC news	Globe and mail	Neutral	0.044

Table 3 Rejections of the sentiment analysis null hypothesis for responses in UK

News station A	News station B	Label	P-value
5 news	SkyNews break	Negative	0.044
5 news	BBC news	Negative	0.04
Channel 4 news	SkyNews break	Negative	0.047
ITV news	SkyNews break	Negative	0.044
5 news	SkyNews break	Neutral	0.036
5 news	BBC news	Neutral	0.037
Channel 4 news	SkyNews break	Neutral	0.044
ITV news	SkyNews break	Neutral	0.039
5 news	SkyNews break	Positive	0.028
Channel 4 news	SkyNews break	Positive	0.033
ITV news	SkyNews break	Positive	0.027

Table 4 Rejections of the emotion analysis null hypothesis for responses in USA

News station A	News station B	Label	P-value
ABC	CNN	Anger	0.018
ABC	Fox news	Anger	0.001
Fox news	MSNBC	Anger	0.028
Fox news	NBC	Anger	0.002
ABC	Fox news	Joy	0.002
CNN	MSNBC	Joy	0.034
Fox news	MSNBC	Joy	0.002
Fox news	NBC	Joy	0.001
ABC	CNN	Optimism	0.021
ABC	Fox news	Optimism	0.002
Fox news	NBC	Optimism	0.001
ABC	Fox news	Sadness	0.003
Fox news	MSNBC	Sadness	0.0
Fox news	NBC	Sadness	0.001

Table 5 Rejections of the emotion analysis null hypothesis for responses in Canada

News station A	News station B	Label	P-value
CBC news	Globe and mail	Anger	0.028
CBC news	Globe and mail	Optimism	0.047

4.2.3 Stance Analysis

The classification distribution for stance analysis is shown in the *Stances* folder of the repository where it can be observed that, overwhelmingly, *none* is the most prominent overall. When performing the statistical hypothesis test, we consider the following hypothesis: *Is the mean of responses {against, in favor, neutral} to news station A's tweets different than news station B's tweets?* The null hypothesis is that "they have the same mean" and the alternative hypothesis is that "they have different means". After performing the test, consider Tables 7, 8, 9 which shows the rejections of the null hypothesis (i.e., acceptance of the alternative hypothesis) for USA, Canada, and UK respectively. Similar to the previous two analyzes, it can be observed that a large amount of rejections between news stations happen in the USA while in the UK, a majority of the rejections occur with one news station (Sky News Break).

Table 6 Rejections of the emotion analysis null hypothesis for responses in UK

News station A	News station B	Label	P-value
5 news	SkyNews break	Anger	0.043
5 news	BBC news	Anger	0.039
Channel 4 news	SkyNews break	Anger	0.047
ITV news	SkyNews break	Anger	0.044
5 news	SkyNews break	Joy	0.032
5 news	BBC news	Joy	0.046
Channel 4 news	SkyNews break	Joy	0.04
ITV news	SkyNews break	Joy	0.032
5 news	SkyNews break	Optimism	0.045
5 news	BBC news	Optimism	0.039
Channel 4 news	SkyNews break	Optimism	0.047
ITV news	SkyNews break	Optimism	0.048
ITV news	BBC news	Optimism	0.047
5 news	SkyNews break	Sadness	0.03
5 news	BBC news	Sadness	0.038
Channel 4 news	SkyNews break	Sadness	0.038
ITV news	SkyNews break	Sadness	0.032

Table 7 Rejections of the stance analysis null hypothesis for responses in USA

News station A	News station B	Label	P-value
ABC	CNN	Against	0.027
ABC	Fox news	Against	0.008
CNN	MSNBC	Against	0.004
CNN	NBC	Against	0.0
Fox news	MSNBC	Against	0.003
Fox news	NBC	Against	0.0
ABC	MSNBC	Favor	0.038
ABC	CNN	None	0.035
ABC	Fox news	None	0.001
Fox news	MSNBC	None	0.012
Fox news	NBC	None	0.001

Table 8 Rejections of the stance analysis null hypothesis for responses in Canada

News station A	News station B	Label	P-value
CBC news	Globe and mail	None	0.035

Table 9 Rejections of the stance analysis null hypothesis for responses in UK

News station A	News station B	Label	P-value
5 news	SkyNews break	Against	0.04
5 news	BBC news	Against	0.046
Channel 4 news	SkyNews break	Against	0.045
ITV news	SkyNews break	Against	0.04
ITV news	BBC news	Against	0.043
5 news	SkyNews break	None	0.039
5 news	BBC news	None	0.039
Channel 4 news	SkyNews break	None	0.045
ITV news	SkyNews break	None	0.041

4.3 RQ3: Do Differences in the Responses to COVID-19 Related Tweets Occur Between News Stations with Different Political Stances?

To begin, each news stations' political leaning is first determined. After examining the methodology of prior works [22, 23], three websites are used[2,3,4] to compile the following list:

- **Left**: CNN, MSNBC
- **Lean Left**: ABC, NBC, CBC News, Channel 4 News, 5 News
- **Center**: Global News, Globe and Mail, CTV News, BBC News, SkyNews Break
- **Lean Right**: National Post, ITV News
- **Right**: Fox News

When reviewing the results shown in Tables 1, 2, 3, 4, 5, 6, 7, 8, 9, it can be observed that every null hypothesis rejection (i.e., difference in responses) has different political leanings except for CNN and MSNBC when they differed in the emotion analysis for *joy* as well as the stance analysis for *against*. To summarize, in the USA, 34/36 ($\sim$94.4%) rejections of the null hypothesis had different political leanings, 5/5 in Canada, and 37/37 in the UK. This demonstrates a strong correlation between differences in responses to differences in news stations' political leanings to positively answer the third research question.

[2] https://www.allsides.com/media-bias.

[3] https://adfontesmedia.com/interactive-media-bias-chart.

[4] https://mediabiasfactcheck.com.

5 Discussion and Future Work

After developing a dataset from the first two years of the pandemic, RQ1 looked to determine what the most popular topic was overall. To do this, for each country, a semantic similarity algorithm was ran to group the news stations' tweets such that sets of common stories could be extracted. Then, a zero-shot classification algorithm was ran to group these stories into COVID-19 related topics so the amount of user interaction could be observed. From this, it was discovered that *vaccines* was the most popular across all news stations and countries. Because of this, it may be possible for future work to find that social media plays a role in influencing people about whether or not to get vaccinated. As for RQ2, the goal was to determine if the audiences of each news station react differently to COVID-19 related tweets by performing sentiment, emotion, and stance analyzes. Overall, it was observed that the USA had many differences, Canada had few differences, and the UK primarily differed with one news station. Lastly, based on RQ2's results, the purpose of RQ3 was to observe if the differences if responses matched with differences in political stances between the news stations. This was discovered to be true in all instances except for the differences within USA between CNN and MSNBC. Therefore, future work may find that audiences are influenced by the different stances news stations have on COVID-19 topics. However, to provide more evidence for this statement, another analysis should be performed to determine if there is division *within* political leanings, or to cover a wider variety of topics by expanding the keyword list. Furthermore, another extension of this work could be to compare country to country or to check for sarcasm in the replies to ensure that the labels extracted are accurate.

6 Conclusion

In this article, three research questions were answered to better understand how Twitter users responded to COVID-19 related tweets from news stations in the USA, Canada, and UK. As a result, an analysis on the popular topics and differences in user engagement were presented by using natural language processing techniques as well as statistical hypothesis tests. From these results we hope to demonstrate that news stations play a role in influencing their audiences' response to the COVID-19 pandemic, and provide future work with ideas on how the analysis can be expanded.

Acknowledgements This work is supported by NSERC Discovery Grant RGPIN-2017-05377 held by Dr. Vijay Mago.

References

1. Z. Hou, et al., Cross-country comparison of public awareness, rumors, and behavioral responses to the COVID-19 epidemic: infodemiology study. J. Med. Internet Res. **22**(8), e21143 (2020)
2. A. Singhal, M.K. Baxi, V. Mago, et al., Synergy between public and private health care organizations during COVID-19 on twitter: sentiment and engagement analysis using forecasting models. JMIR Med. Inform. **10**(8), e37829 (2022)
3. M.K. Baxi, J. Philip, V. Mago, Resilience of political leaders and healthcare organizations during COVID-19. Peer J. Comput. Sci. **8**, e1121 (2022)
4. R. Goel, R. Sharma, Studying leaders & their concerns using online social media during the times of crisis-A COVID case study. Soc. Netw. Anal. Min. **11**(1), 1–12 (2021)
5. M.M.A. Qudar, V. Mago, Tweetbert: a pretrained language representation model for twitter text analysis (2020). arXiv:2010.11091
6. M. Sandhu, et al., From associations to sarcasm: mining the shift of opinions regarding the supreme court on twitter. Online Soc. Netw. Media **14**, 100054 (2019)
7. F. Durazzi, et al., Clusters of science and health related twitter users become more isolated during the COVID-19 pandemic. Sci. Rep. **11**(1), 1–11 (2021)
8. C. Han, M. Yang, A. Piterou, Do news media and citizens have the same agenda on COVID-19? an empirical comparison of twitter posts. Technol. Forecast. Soc. Chang. **169**, 120849 (2021)
9. K. Garcia, L. Berton, Topic detection and sentiment analysis in twitter content related to COVID-19 from Brazil and the USA. Appl. Soft Comput. **101**, 107057 (2021)
10. J. Devlin, et al., Bert: pre-training of deep bidirectional transformers for language understanding (2018). arXiv:1810.04805
11. S.D. Mueller, M. Saeltzer, Twitter made me do it! twitter's tonal platform incentive and its effect on online campaigning. Inf., Commun. Soc. **25**(9), 1247–1272 (2022)
12. A. Sahly, C. Shao, K.H. Kwon, Social media for political campaigns: an examination of Trump's and Clinton's frame building and its effect on audience engagement. Soc. Media+ Soc. **5**(2), 2056305119855141 (2019)
13. N. Reimers, I. Gurevych, Sentence-BERT: sentence embeddings using Siamese BERT-networks (2019). arXiv:1908.10084
14. C.H. Mendhe, et al., A scalable platform to collect, store, visualize, and analyze big data in real time. IEEE Trans. Comput. Soc. Syst. **8**(1), 260–269 (2020)
15. A. Conneau, et al., Unsupervised cross-lingual representation learning at scale (2019). arXiv:1911.02116
16. G. Lample, A. Conneau, Cross-lingual language model pre-training (2019). arXiv:1901.07291
17. Y. Liu, et al., Roberta: a robustly optimized bert pretraining approach (2019). arXiv:1907.11692
18. D. Loureiro, et al., Timelms: diachronic language models from twitter (2022). arXiv:2202.03829
19. F. Barbieri, et al., Tweeteval: unified benchmark and comparative evaluation for tweet classification (2020). arXiv:2010.12421
20. S. Ghosh, et al., Stance detection in web and social media: a comparative study, in *International Conference of the Cross-Language Evaluation Forum for European Languages* (Springer, Berlin, 2019), pp. 75–87
21. R.A. Armstsrong, When to use the Bonferroni correction. Ophthalmic Physiol. Opt. **34**(5), 502–508 (2014)
22. T. Spinde, et al., Automated identification of bias inducing words in news articles using linguistic and context-oriented features. Inf. Process. Manag. **58**(3), 102505 (2021)
23. P. Stefanov, et al., Predicting the topical stance and political leaning of media using tweets, in *Proceedings of the 58th Annual Meeting of the Association for Computational Linguistics* (2020), pp. 527–537

Understanding the Role of Questions in Mental Health Support-Seeking Forums

Aylin Gunal, Ian Stewart, Rada Mihalcea, and Verónica Pérez-Rosas

Abstract People who seek mental health help online often receive supportive comments from other users, but their intentions may not be clear, as when someone asks a question that does not require a response. In this work, we explore the role of questions asked in response to support-seeking posts during online interactions centered around mental health support. We introduce a new dataset consisting of 1,089 mental health related post-response pairs from Reddit containing response questions annotated as rhetorical, information-seeking or not applicable. Through several experiments, we find that we can effectively distinguish between rhetorical and information seeking questions using linguistic features. Our findings highlight the importance of surrounding context and functional features (e.g., auxiliary verbs) as opposed to semantic (e.g., words related to mental processes) being significant predictors of question type.

1 Introduction

Online mental health communities are growing as more people seek support or offer it to others. Several works have explored the nature of interactions within these communities, including expressions of empathy and members' motivations for participation [6, 25, 28]. However, few studies have examined the role and characteristics

A. Gunal (✉) · R. Mihalcea · V. Pérez-Rosas
University of Michigan, Ann Arbor, MI, USA
e-mail: gunala@umich.edu

R. Mihalcea
e-mail: mihalcea@umich.edu

V. Pérez-Rosas
e-mail: vrncapr@umich.edu

I. Stewart
Pacific Northwest National Laboratory, Richland, WA, USA
e-mail: ian.stewart@pnnl.gov

A. Shaban-Nejad et al. (eds.), *Artificial Intelligence for Personalized Medicine*,
Studies in Computational Intelligence 1106,
https://doi.org/10.1007/978-3-031-36938-4_21

of *questions* posed by members of these communities [8, 27], which can reflect how people request information and make arguments about mental health.

In this work, we study the linguistic cues associated with different types of questions posed in mental health forums on Reddit. We focus on interactions where supporters ask questions in response to support-seekers, since understanding the nature of these questions can help counselors and non-professionals alike understand how to connect meaningfully with patients. While online help-seeking interactions might differ from their counseling counterparts they also share important characteristics such as information probing through questioning and expressing empathy. We believe that our study can contribute to understanding what types of questions are asked by support givers around mental health topics in both peer and counseling settings. Furthermore, since questions play a key role in counseling techniques such as Motivational Interviewing [16], our work presents a first step into understanding the intent and nature of questions in psychotherapy.

We introduce Mental heAlTh quEstions (MATE), a new dataset of questions from mental health subreddits that are labeled as information-seeking or rhetorical.[1] We evaluate the use of linguistic features in the task of classifying rhetorical versus information-seeking questions. Our findings show that functional features (e.g., auxiliary verbs) are significant predictors of question type, and that question classification models benefit from the addition of context-related features. This study provides insights into differences in the way questions are asked in the mental health domain. In addition, our work contributes an important resource for question generation studies, where training with low-quality data can lead to harmful generated text [9].

2 Related Work

Proposed taxonomies of questions often focus on sentence structure (e.g., wh-word appearance; Kearsley [11]), and on the intent of the speaker (comparison versus explanation; Nielsen et al. [17]). Two particularly common question categories in online discussions are information-seeking, where the inquirer looks for new information, and rhetorical, where the inquirer uses the implied answer to the question as a rhetorical device [23]. Prior work on the classification of rhetorical versus information-seeking questions has demonstrated that a bag-of-words representation provides a strong baseline [2, 29], but has done less to explore the relative importance of different linguistic features. Our work provides an analysis of different semantic, functional, and contextual features in question classification, specifically within mental health discussions where determining question intent can prove critical to a successful conversation.

[1] We will release the code and annotated dataset upon completion of screening the dataset for personal identifiable information, in line with previous similar work [9].

3 Data Collection

We use the dataset collected by [13], consisting of posts and responses discussing mental health issues on Reddit. This dataset both aligns with the domain as well as the dialogue structure we are interested in observing, i.e. questions asked in response to common mental health concerns. From the full set, we focus on posts with responses that include follow-up questions. We identify the post-response pairs by filtering responses that contain question marks–a simple approach that previous works have successfully applied [1, 29] to identify questions on social media.

We further filter the resulting set to remove posts written by known bots, questions duplicating content from the original post, and URLs that were incorrectly identified as questions due to their use of question marks. Because some comments include multiple questions, we create multiple post-comment pairs by extracting each individual question from the comment. We obtain 76,476 post-response pairs from 48,784 unique responses from which we collect their post content and corresponding author "flairtags" i.e., a label indicating the support provider expertise (e.g., "Mental Health Practitioner"). Table 1 shows an example of a post-comment pair from the resulting set along with its author flairtag.

3.1 Annotation

To enable our experiments, we manually annotate a randomly sampled subset of 1,100 questions with three main types of questions as described below.
Information-Seeking (IS) Questions that explicitly ask for information from the author of the original post.

Rhetorical (R) Questions that address the author of the original post but do not explicitly seek information.

Table 1 Example of a post and a responding comment that contains a question

Post	Comment	Post-author flairtag	Comment-author flairtag
How should I ask my therapist for my diagnosis?	**Are you paying with insurance?** Because many therapists actually are very hesitant to formally diagnose clients. However, when using insurance it is required in order to get paid.	No expertise	High expertise

Table 2 Example questions with context from the dataset

Questions with context	Label
Would therapy help? I'd say there's a good shot.	R
Some people have difficulties with more intimate relationships. **Do you believe you could be on the spectrum?**	IS

Table 3 Dataset statistics

Statistics	IS	R
Avg. sentence length (tokens)	10.9	10.7
Questions with preceding context	325 (67.0%)	224 (75.6%)
Questions with following context	224 (46.1%)	261 (88.2%)

Not Applicable (NA) Questions that were neither IS or R due to one of the following conditions: (1) being unrelated to mental health support (e.g., seeking career advice); (2) not being directed towards the author of the original post (e.g., community post); (3) including harmful content (e.g., trolling, bigotry).

Two annotators independently label each question. The annotators are undergraduate students who read the instructions carefully and practiced the guidelines before annotating. During the annotation task, they are shown the question as well as the sentences preceding and following the question, whenever available. Annotators discard 11 post-responses that do not fit into any of the IS, R, or NA categories (e.g. emojis). Disagreements are resolved by a third annotator. The overall inter-annotator agreement is 79% and Cohen's Kappa was 0.68.

The final dataset includes 1,089 questions, annotated as 485 IS, 296 R, and 308 NA respectively and distributed across 38 different mental health subreddits. The most popular subreddits in the final set include r/ADHD (394), r/mentalhealth (257), r/relationship_advice (166), r/askatherapist (56), and r/offmychest (27). Examples of questions in the dataset including preceding or following context are shown in Table 2, and the dataset statistics are shown in Table 3.

4 Features

We explore the effectiveness of different linguistic features in the classification between information-seeking and rhetorical questions, focusing on lexical, functional, semantic, and contextual features.

Lexical Features We extract unigrams using TF-IDF counts and word embeddings with Word2Vec by obtaining the mean of embeddings for individual words in the question.

Functional Features A question-asker may indicate their intention by changing how their question is framed. We therefore extract words that serve a syntactic function, rather than content, within a question.[2] (1) *Head Nouns.* We extract head nouns i.e., nouns located within the main noun phrase of a sentence, from questions using the heuristic designed by [15]. (2) *Auxiliary Verbs.* We extract the leftmost auxiliary verb i.e., verbs that add context to the main verb, in a target sentence. (3) *Wh-keywords.* We extract the leftmost wh-question word (e.g., "where").

Semantic Features We obtain features that capture semantic aspects of the question's content. (1) *Lexical Diversity metrics.* Lexical diversity measures the diversity of words that a speaker uses to convey an idea, which may indicate a more information-heavy sentence and therefore information-seeking. We use the following metrics: Measure of Textual Lexical Diversity, Hypergeometric Distribution, Maas [14], Type-Token Ratio, Mean Word Frequency, Yules K, and Yules I [18]. (2) *LIWC features.* Prior work in mental health conversations has found that the Linguistic Inquiry Word Count (LIWC) lexicons prove useful in identifying different mental processes [22], which we hypothesize may extend to expressing rhetorical intent. We derive semantic features using the LIWC lexicon [21] for the 80 semantic classes in the lexicon. (3) *Concreteness.* Words that are highly concrete and refer to well-formed concepts (e.g. "hair") may correlate with the intent to seek new information. Concreteness scores are computed per-word using the dataset from [3], and we use the mean over all words in the question as the feature. (4) *Polarity and NPIs.* We calculate the polarity of questions using TextBlob.[3] Following previous findings [24] that consecutive sentences with opposite polarities may indicate rhetorical intent, we calculate negative-polarity indicators (NPIs) by parsing the dependencies of the question and its corresponding context sentences, and note whether the question or either context sentence includes a negative dependency.

Contextual Features We extract both linguistic and social contextual features of a question. (1) *Context.* We extract the unigrams TF-IDF embedding for the sentence immediately preceding the question and for the sentence immediately following the question within the full response. (2) *Context-Question Similarity.* Under the same intuition from the NPI index above, we compute cosine similarity and Word Mover's Distance [12] between the word embedding of question and corresponding context sentences. (3) *User Expertise.* We embed information about the level of user expertise using their Reddit flairtags. We label users as *high expertise* if their tags indicate subject matter expertise (e.g., 'Psychiatry PhD'), *low expertise* for tags that indicate knowledge but not necessarily authority in mental health (e.g., 'depression') and *no expertise* for tags that do not reference mental health experience at all.

---

[2] For extraction of features that require a parser (e.g., auxiliary verbs), we use SpaCy (https://spacy.io).

[3] https://textblob.readthedocs.io/en/dev/.

Table 4 Baseline performances for mental health question classification using SVM

Features	# Feat.	Acc.	F1-IS	F1-R
Word2Vec	100	0.611	0.718	0.365
Uni	1467	**0.743**	**0.796**	**0.644**

5 Experiments and Results

We conduct a series of experiments to distinguish between rhetorical and information seeking questions using the extracted features separately and jointly. Due to the imbalanced class distribution in our dataset, we upsample using SMOTE [5] to synthetically generate minority class labels. We use a linear SVM as our main classifier[4] and evaluate using five-fold cross-validation, with accuracy, precision, recall, and F-score as performance metrics. We use the machine learning algorithm implementations available in the Scikit-learn[5] library with their default parameters.

Baseline Models We establish a baseline performance for the classification task using lexical features only. Classification results when using word embeddings (Word2Vec) and unigrams (Uni) are shown in Table 4. We did not observe a performance gain when combining these features. Similar to prior work (0.69 F1 in [29], 0.76 F1 in [19]), we show that a simple classifier using only the question text achieves reasonable F1 scores, which supports the task's validity.

Feature Groups We conduct experiments using each of the feature groups defined in Section Features separately, and in combination with question unigrams. Results are shown in Table 5. The Functional feature set does the best overall, and auxiliary verbs are often the key features in the set: IS questions typically begin with primary auxiliary verbs ("Do you have a plan to be safe?") whereas some R questions begin with modal auxiliary verbs ("Would I recommend that?"). Overall, the Contextual feature group that includes the question embedding provides the highest performance for IS prediction.

Feature Ablation We conduct a set of ablation experiments in order to find the most important predictive features to classify question type. We start with a model including all features and then we remove one feature set at the time. Results are shown in Table 6. The best performing model includes all features except for the question-context similarity metrics, with an F1-IS of 0.84 and F1-R of 0.71.

Performance falls significantly when context sentences are dropped from the feature set. Interestingly, dropping auxiliary verbs reduces model performance more than dropping either or both context sentences. Disregarding context, the most dramatic drops in performance for both IS and R questions occur when functional features and concreteness are dropped as features.

---

[4] We experimented with other classifiers and found the best performance using the SVM model.

[5] https://scikit-learn.org/stable/modules/generated/sklearn.svm.SVC.html.

Table 5 Grouped feature results. Underlined values are highest in groups including unigrams (uni), and bold values are highest in groups excluding unigrams

Feature set	# Feat.	Acc.	F1-IS	F1-R
Functional	194	0.695	**0.759**	**0.581**
Functional + uni	1702	0.748	0.800	*0.657*
Semantic	107	0.629	0.692	0.530
Semantic + uni	1615	0.748	0.801	0.655
Contextual	3012	0.683	0.743	0.576
Contextual + uni	4520	0.764	*0.821*	0.653

Table 6 Ablation results. Bold numbers indicate greatest drop in performance within the feature group

Feature set	Acc.	F1-IS	F1-R
All features	0.784	0.832	0.696
Functional features			
–Auxiliary verbs	0.766	**0.820**	**0.663**
–Head nouns	0.771	0.821	0.679
–Wh-keywords	0.775	0.826	0.680
Semantic features			
–Concreteness	0.772	**0.823**	**0.680**
–LIWC	0.780	0.830	0.685
–Polarity	0.782	0.831	0.694
–NPIs	0.785	0.833	0.696
–Lexical diversity	0.785	0.832	0.696
Contextual features			
–Expertise	0.777	0.827	0.688
–Cosine + WMD	0.793	0.838	0.710
–Prec.	0.775	0.823	0.688
–Foll.	0.767	0.819	**0.672**
–Prec. + Foll.	0.764	**0.814**	0.677

Question Context Since we observe that dropping context sentences substantially reduces performance, we train and test models with context sentences and question unigrams. Results are shown in Table 7.

Following context predicts both IS questions and R questions more accurately than preceding context, and following context also predicts R questions more accurately than the combination of both preceding and following context. This performance may reflect the fact that rhetorical questions are more likely to have following context than information seeking questions (as shown in Table 3).

Table 7 Effects of adding context sentences. All sentences are represented with the same unigram features as before

Feature set	# Feat.	Acc.	F1-IS	F1-R
Q + Prec.	2975	0.750	0.807	0.647
Q + Foll.	3048	0.766	0.823	**0.667**
Q + Prec. + Foll.	4515	**0.768**	**0.827**	0.643

Adding both preceding and following sentences to the question yields the highest performance for IS question classification. A possible explanation is that IS questions may often appear in groups, not in isolation. For example, our data includes the sequences of IS questions "Are you really depressed? Could it possibly be a personality disorder?" A mental health counselor may therefore choose to ask questions in tandem, rather than in separate turns, to make it clear that they are actually seeking new information.

Additionally, we manually went through the questions that our model fails to predict correctly. We group the main failures into the following categories:

- **Context dependence**. For many questions, the model predicted IS when it should have predicted R given the context. For instance, the R question "Does it sound likely that it would or is happening?" is followed by the context "Not at all to me" but the model predicts IS, possibly because the question's well-formed grammar outweighs the role of context in prediction.
- **Challenges**. Some R questions reflect a *challenge* to the original poster rather than an intention to engage them in good faith: the R question "And really, do the labels—ADHD, personality, or character flaw—really matter?" was predicted as IS by the model. This kind of R question may require more subtle lexical features than those utilized here, such as discourse markers like "really" [20].
- **Non-standard question form**. The model sometimes predicted R instead of IS when the question had a non-standard form, as in "I am curious about what the flare up looks like when it starts to turn for you?" In such cases, more advanced text preprocessing may be needed to identify the structure of the underlying question, i.e. removing embedding phrases like "I am curious about."

6 Conclusion and Future Work

In this work, we developed MATE, a dataset of mental health-focused questions annotated by type, and we evaluated the role of linguistic features in question type classification.

Through extensive feature ablation, we found that functional features–particularly auxiliary verbs–provide more predictive power than semantic features, which reinforces prior findings about the limits of content-based features in question classification [10]. We also found that the combination of preceding and following context best predicts information-seeking questions. Overall, the study shows that subtle choices in wording and sentence order may be more important than content when it comes to predicting supporters' question-asking intentions.

Future work may include further exploration of the relationship between domain experts and the kinds of questions they produce; in the realm of mental health support, it seems logical to assume that professionals in the field are more likely to try to achieve a better understanding of a patient's state rather than make a point, which is a common function of rhetorical questions [1]. Similarly, we note that there are several repeat posters across various subreddits, and it may be interesting to study how similar users respond to different types of posts.

Additionally, future work may consider how user behavior varies between different online communities. The distribution of subreddits in MATE is highly uneven, and a significant portion of the most popular subreddits are more geared towards a general audience (e.g. r/mentalhealth). These subreddits accommodate a variety of users with different degrees of severity of mental health ailments. It may be insightful to analyze how such different users ask or answer questions, because different conditions lend themselves to different information-seeking needs.

7 Limitations

Because our dataset includes online interactions, the language used includes informal phrases, slang, or abbreviations, and the questions may be phrased in non-traditional ways ("You OK?"). The heuristics we used to extract features such as head nouns and auxiliary verbs require a correct parse of the sentence structure, and proper sentence structure is not always to be expected from social media interaction data [7]. A possible way to mitigate this issue as well as reduce bias associated with interactions from Reddit would be to expand our data sources to include formal online counseling platforms such as CrisisTextLine.

Additionally, the validity of our data labels relies on the quality of annotation. We acknowledge the difficulty in achieving high inter-annotator agreement in such a subjective task–even for annotators collaborating closely throughout the process. Future work may be centered around mitigating disagreement by adding the "Ambiguous" label in annotation.

8 Ethics

We acknowledge that mental health is a sensitive area, where the cost of incorrect model predictions can be disproportionately high compared to the general population [4]. We do not intend for any of the models trained in this work to be deployed without more careful testing of possible biases and performance shortcomings (e.g., systematic misclassification of abusive content). The models used in this work should not be used to replace a supporter judgment of what makes a question information-seeking versus rhetorical, but instead to augment their judgment [26] by showing which features may contribute to different perceptions of their questions.

Acknowledgements This material is based in part upon work supported by the John Templeton Foundation (#62256) and by the University of Michigan Precision Health Initiative. Any opinions, findings, and conclusions or recommendations expressed in this material are those of the authors and do not necessarily reflect the views of the John Templeton Foundation or the Precision Health Initiative. We thank members of the LIT Lab, including Ashkan Kazemi, Naihao Deng, and Shinka Mori as well as Ava Luna Pardo-Keegan for their help in developing and carrying out the annotation task, and for providing feedback on early results.

References

1. S. Bagga, A. Piper, D. Ruths, "Are you kidding me?": detecting unpalatable questions on Reddit, in *Proceedings of the 16th Conference of the European Chapter of the Association for Computational Linguistics: Main Volume, Association for Computational Linguistics* (2021), pp. 2083–2099, https://doi.org/10.18653/v1/2021.eacl-main.179, https://aclanthology.org/2021.eacl-main.179
2. S. Bhattasali, J. Cytryn, E. Feldman, J. Park, Automatic identification of rhetorical questions, in *Proceedings of the 53rd Annual Meeting of the Association for Computational Linguistics and the 7th International Joint Conference on Natural Language Processing (Volume 2: Short Papers), Association for Computational Linguistics* (Beijing, China, 2015), pp. 743–749, https://doi.org/10.3115/v1/P15-2122, https://aclanthology.org/P15-2122
3. M. Brysbaert, A.B. Warriner, V. Kuperman, Concreteness ratings for 40 thousand generally known English word lemmas. Behav. Res. Methods **46**(3), 904–911 (2014)
4. S. Chancellor, M.L. Birnbaum, E.D. Caine, V.M. Silenzio, M.D. Choudhury, A taxonomy of ethical tensions in inferring mental health states from social media, in *Proceedings of the Conference on Fairness, Accountability, and Transparency* (2019), pp. 79–88
5. N.V. Chawla, K.W. Bowyer, L.O. Hall, W.P. Kegelmeyer, SMOTE: synthetic minority oversampling technique. J. Artif. Intell. Res. **16**, 321–357 (2002). https://doi.org/10.1613/jair.953
6. Y. Chen, Y. Xu, Exploring the effect of social support and empathy on user engagement in online mental health communities. Int. J. Environ. Res. Public Health **18**(13), 6855 (2021)
7. J. Eisenstein, What to do about bad language on the internet, in *Proceedings of the 2013 Conference of the North American Chapter of the Association for Computational Linguistics: Human Language Technologies* (2013), pp. 359–369
8. J. Frank, You call that a rhetorical question?: forms and functions of rhetorical questions in conversation. J. Pragmat. **14**(5), 723–738 (1990)
9. S. Gupta, A. Agarwal, M. Gaur, K. Roy, V. Narayanan, P. Kumaraguru, A. Sheth, Learning to automate follow-up question generation using process knowledge for depression triage on Reddit posts, in *Proceedings of the Eighth Workshop on Computational Linguistics and Clinical Psychology, Association for Computational Linguistics* (Seattle, USA, 2022), pp. 137–147. https://doi.org/10.18653/v1/2022.clpsych-1.12, https://aclanthology.org/2022.clpsych-1.12
10. A.L. Kalouli, R. Kehlbeck, R. Sevastjanova, O. Deussen, D. Keim, M. Butt, Is that really a question? going beyond factoid questions in NLP, in *Proceedings of the 14th International Conference on Computational Semantics (IWCS), Association for Computational Linguistics, Groningen* (The Netherlands, 2021), pp. 132–143, https://aclanthology.org/2021.iwcs-1.13
11. G.P. Kearsley, Questions and question asking in verbal discourse: a cross-disciplinary review. J. Psycholinguist. Res. **5**(4), 355–375 (1976)

12. M. Kusner, Y. Sun, N. Kolkin, K. Weinberger, From word embeddings to document distances, in *International Conference on Machine Learning* (PMLR, 2015), pp. 957–966
13. A. Lahnala, Y. Zhao, C. Welch, J.K. Kummerfeld, L.C. An, K. Resnicow, R. Mihalcea, V. Pérez-Rosas, Exploring self-identified counseling expertise in online support forums. Find. Assoc. Comput. Linguist.: ACL-IJCNLP **2021**, 4467–4480 (2021)
14. H.D. Mass, Über den zusammenhang zwischen wortschatzumfang und länge eines textes. Zeitschrift für Literaturwissenschaft und Linguistik **2**(8), 73 (1972)
15. D. Metzler, W.B. Croft, Analysis of statistical question classification for fact-based questions. Inf. Retr. **8**(3), 481–504 (2005)
16. W.R. Miller, S. Rollnick, *Motivational Interviewing: Helping People Change*. (Guilford Press, 2012)
17. R.D. Nielsen, J. Buckingham, G. Knoll, B. Marsh, L. Palen, A taxonomy of questions for question generation, in *Proceedings of the Workshop on the Question Generation Shared Task and Evaluation Challenge* (2008)
18. M.P. Oakes, *Statistics for Corpus Linguistics* (Edinburgh University Press, 1998). http://www.jstor.org/stable/10.3366/j.ctvxcrdkd
19. S. Oraby, V. Harrison, A. Misra, E. Riloff, M. Walker, Are you serious?: rhetorical questions and sarcasm in social media dialog, in *Proceedings of the 18th Annual SIGdial Meeting on Discourse and Dialogue, Association for Computational Linguistics* (Saarbrücken, Germany, 2017), pp. 310–319. https://doi.org/10.18653/v1/W17-5537, https://aclanthology.org/W17-5537
20. U. Pavalanathan, J. Fitzpatrick, S.F. Kiesling, J. Eisenstein, A multidimensional lexicon for interpersonal stancetaking, in *Proceedings of the 55th Annual Meeting of the Association for Computational Linguistics (Volume 1: Long Papers)* (2017), pp. 884–895
21. J.W. Pennebaker, M.E. Francis, R.J. Booth, Linguistic inquiry and word count: LiWC 2001. Mahway: Lawrence Erlbaum Assoc. **71**(2001), 2001 (2001)
22. V. Pérez-Rosas, X. Sun, C. Li, Y. Wang, K. Resnicow, R. Mihalcea, Analyzing the quality of counseling conversations: the tell-tale signs of high-quality counseling, in *Proceedings of the Eleventh International Conference on Language Resources and Evaluation (LREC 2018)* (2018)
23. S. Ranganath, X. Hu, J. Tang, S. Wang, H. Liu, Understanding and identifying rhetorical questions in social media. ACM Trans. Intell. Syst. Technol. (TIST) **9**(2), 1–22 (2018)
24. J.M. Sadock, Queclaratives, in *Seventh regional meeting of the Chicago Linguistic Society*, vol. 7 (1971), pp. 223–232
25. A. Sharma, A. Miner, D. Atkins, T. Althoff, A computational approach to understanding empathy expressed in text-based mental health support, in *Proceedings of the 2020 Conference on Empirical Methods in Natural Language Processing (EMNLP)* (2020), pp. 5263–5276
26. A. Thieme, D. Belgrave, G. Doherty, Machine learning in mental health: a systematic review of the HCI literature to support the development of effective and implementable ML systems. ACM Trans. Intell. Syst. Technol. (TIST) **27**(5), 1–53 (2020)
27. T. Zhang, J.H.D. Cho, C. Zhai, Understanding user intents in online health forums, in *Proceedings of the 5th ACM Conference on Bioinformatics, Computational Biology, and Health Informatics, Association for Computing Machinery* (New York, NY, USA, BCB '14, 2014), pp. 220–229. https://doi.org/10.1145/2649387.2649445
28. X. Zhang, S. Liu, Z. Deng, X. Chen, Knowledge sharing motivations in online health communities: a comparative study of health professionals and normal users. Comput. Hum. Behav. **75**, 797–810 (2017)
29. Y. Zhuang, E. Riloff, Exploring the role of context to distinguish rhetorical and information-seeking questions, in *Proceedings of the 58th Annual Meeting of the Association for Computational Linguistics: Student Research Workshop* (2020), pp. 306–312

The manufacturer's authorised representative in the EU is Springer Nature Customer Service Centre GmbH, Europaplatz 3, 69115 Heidelberg, Germany. If you have any concerns regarding our products, please contact ProductSafety@springernature.com

Printed and bound by CPI Group (UK) Ltd, Croydon, CR0 4YY
28/09/2026
02224075-0012